World Soils Book Series

Series Editor

Alfred E. Hartemink
Department of Soil Science, FD Hole Soils Laboratory
University of Wisconsin–Madison
Madison, WI
USA

The World Soils Book Series publishes books containing details on soils of a particular country. They include sections on soil research history, climate, geology, geomorphology, major soil types, soil maps, soil properties, soil classification, soil fertility, land use and vegetation, soil management, soils and humans, soils and industry, future soil issues. The books summarize what is known about the soils in a particular country in a concise and highly reader-friendly way. The series contains both single and multi-authored books as well as edited volumes. There is additional scope for regional studies within the series, particularly when covering large land masses (for example, The Soils of Texas, The Soils of California), however, these will be assessed on an individual basis.

More information about this series at http://www.springer.com/series/8915

Lia Matchavariani
Editor

The Soils of Georgia

Springer

Editor
Lia Matchavariani
Faculty of Exact & Natural Sciences,
Department of Soil Geography
Ivane Javakhishvili Tbilisi State University
Tbilisi, Georgia

ISSN 2211-1255 ISSN 2211-1263 (electronic)
World Soils Book Series
ISBN 978-3-030-18508-4 ISBN 978-3-030-18509-1 (eBook)
https://doi.org/10.1007/978-3-030-18509-1

This Springer imprint is published by the registered company Springer Nature Switzerland AG
The registered company address is: Gewerbestrasse 11, 6330 Cham, Switzerland

This book is dedicated to the 100th anniversary (1918–2018) of Ivane Javakhishvili Tbilisi State University (TSU) establishment.

Preface

Georgia is a country that extends along the Caucasus Mountains in the South and from the Black Sea in the East. Despite the small territory, Georgia is characterized by very different soils because of a diversity of soil-forming factors—various geological, climatic, orographic, and other features are found there.

The main goal of this book is to provide an explanatory review of the soils of Georgia and to summarize the knowledge accumulated over many decades concerning their formation and contemporary management.

The Soils of Georgia gives a comprehensive overview of the country's soil cover being highly complex and distinctive. The book discloses wide diversity and spatial distribution of soils of the various areas of Georgia. The chapters discuss the history of Georgian soils, the role of soil-forming processes and environmental conditions responsible for soil diversity, morpho-chemical properties and various aspects of soils; presents also contemporary diagnosing and scientific interpretation of soil processes based on the micropedological studies, etc. The last chapters focus on soil degradation and land use management in the country. The appendix contains a lot of graphs.

This book is a basic documentation of Georgia's soil properties, written by soil experts and scientists with much experience in different fields of soil researchers. The book contains 145 titles of figures (maps, pictures, graphs, schemes, etc.) and 50 tables. All included topics are relevant and useful for soil scientists, as well as for geographers, ecologists, agronomists, biologists, foresters, territory planners, food producers, etc. The bibliography is included in each chapter.

Acknowledgements The authors would like to thank everybody whose assistance was valuable contributions to this book. Especially, we are grateful to the young scientist Dr. David Svanadze—expert of GIS for assistance with preparation of map series.

We also gladly acknowledge the Vakhushti Bagrationi Institute of Geography at TSU for providing us with five maps from the "National Atlas of Georgia", Stuttgart: Steiner-Verlag, 2018 (Editors: N. Bolashvili, A. Dittmann, L., King, V. Neidze); among them three maps are compiled by the authors of this monograph.

Tbilisi, Georgia　　　　Lia Matchavariani

Contents

1 Introduction

Lia Matchavariani

Abstract

Georgia is located in the mountainous Caucasus Region and washed in the west by the Black Sea. Its geographical position and versatile natural conditions determine diverse soil landscape and wide range of soil spectrum. Despite a relatively small territory, there are different types of soils (from the marshy lands to the semidesert soils), as well as their altitudinal distribution is very diverse (from the sea level to the high mountains). Based on the National classifications, soil cover includes such main types as mountain-meadow, mountain-forest-meadow, mountain chernozems, brown-forest, raw-humus-calcareous, cinnamonic and meadow cinnamonic, gray-cinnamonic and meadow gray cinnamonic, vertisols, salt, solonetz, red, yellow, subtropical podzolic and gley-podzolic, bog, alluvial soils, etc.

Keywords

Pedosphere • Mirror of landscape • Memory of landscape • Soil types • Soil diversity • Georgia

The soil covering the Earth's surface is formed, exists, and develops through the interaction of geospheres existing near the surface of the Earth—the lithosphere, the atmosphere, the hydrosphere, and the biosphere. All these Geospheres are materially represented in the soil: the lithosphere in soil minerals, atmosphere in soil air, hydrosphere in soil water, and the biosphere in biota. Therefore, the soil with its composition and structural diversity is the "copy" of the geosphere. The dynamics of the soil "life" reflects the dynamics of the entire geosphere. Consequently, investigation of soil will give the answer to all questions concerning the complex and extensive geosystems. The soil is rightly called the "mirror of the landscape" because it reflects the conditions of the environment, where it actually exists and where it was formed and developed. This is the main postulate of Soil Science.

However, soil is the mirror not only for the current landscapes but also for those existed in the past. Consequently, the soil is not just a "mirror reflection" of the landscape in the literal sense. The soil properties formed in the past never disappear; rather they remain for more or less long period. Soil profiles are not always adequate for modern conditions. They retained the properties obtained in the past, "remembering" all the events of the landscape life. As a result, the soil is not only "the mirror" but also "the memory" of the landscape retaining the paleogeographical and relict characteristics. Thus, the soil varies not only in space; it is rather unsustainable in time. This is the main difference between the soil and the other components of the landscape. Therefore, the study of paleopedological properties of soils is one of the important tasks of genetic Soil Science.

Georgia, the country in the Mountainous Caucasus, borders with Russia, Azerbaijan, Armenia, and Turkey; from the west it is washed by the Black Sea (Fig. 1.1). Its area is 69 700 km^2. Country divides into two autonomous republics (Abkhazeti and Adjara), nine regions (Samegrelo-Zemo Svaneti, Guria, Imereti, Racha-Lechkhumi and Kvemo Svaneti, Samtskhe–Javakheti, Shida Kartli, Kvemo Kartli, Mtskheta-Mtianeti, Kakheti), and one city—capital Tbilisi (Fig. 1.1 and Table 1.1).

Georgia occupies only 0.02% of the total land area of the world. Due to its small area, it does not outstand either with great areas of forests, agricultural plots of field (38%), protected areas, or great number of flora or fauna species; however, on the other hand, the country is diversified, specific, and even unique in many respects. The major reason for the outstanding nature and uniqueness of Georgia is its geographical location and versatile natural conditions. Its

L. Matchavariani (✉)
Faculty of Exact & Natural Sciences, Department of Soil Geography, Ivane Javakhishvili Tbilisi State University, Tbilisi, Georgia
e-mail: lia.matchavariani@tsu.ge

L. Matchavariani (ed.), *The Soils of Georgia*, World Soils Book Series,
https://doi.org/10.1007/978-3-030-18509-1_1

Fig. 1.1 Location and administrative division. This map is created by D. Svanadze

Table 1.1 Administrative division

Region	Center	Area (km$^{2)}$
Abkhazeti AR	Sokhumi	8 660
Adjara AR	Batumi	2 880
Guria	Ozurgeti	2 033
Imereti	Kutaisi	6 475
Samegrelo-Zemo Svaneti	Zugdidi	7 440
Racha-Lechkhumi and Kvemo Svaneti	Ambrolauri	4 990
Shida Kartli	Gori	5 729
Kvemo Kartli	Rustavi	6 072
Mtskheta-Mtianeti	Mtskheta	6 786
Samtskhe–Javakheti	Akhaltsikhe	6 413
Kakheti	Telavi	11 311
Tbilisi	Tbilisi	720

soil landscape is extremely diverse and characterized by a wide range of soil spectrum. Georgia is distinguished with its interesting soil cover and includes almost all major genetic types existing on the Earth (except for a hot tropical) that can be explained by complex and versatile combinations of the soil-forming factors. On a relatively small territory of the country, different types of soils can be found beginning from the marshy lands of the western subtropics to the eastern semiarid and semidesert soils with the altitudinal distribution beginning from the sea level to the eternal snow zones of high mountains.

According to national soil classification, Georgia occupies the following main soils types (Fig. 1.2): Mountain-Meadow, Mountain-Forest-Meadow, Mountainous-Meadow-Chernozem-like, Yellow-Brown-Forest, Mountain Chernozems, Raw-Humus-Calcareous, Brown-Forest, Cinnamonic and Meadow Cinnamonic, Gray-Cinnamonic and Meadow Gray Cinnamonic, Black soils, Salt soils, Solonetz, Red, Yellow, Subtropical Podzolic and Gley-Podzolic, Marsh and Alluvial soils, etc.

What is the reason for such diversity? Diversity of soil-forming factors determines a quite diverse soil cover of Georgia. The complex geological structure and diversity of underlying parent materials, the different relief, contrast climate, specific vegetation, and the biodiversity are the main reasons for the specific geographical distribution of soils in Georgia. This fact inspired V. Dokuchaev to say that Georgia is a “Natural Museum of Soils under the Open Air”. More importantly, in geographic science, the soil zonality and generally, geographical zonality was based on these studies in the Caucasus Mountains on the example of Georgian soils.

The diversity of the soil cover with respective subregions, zones, and areas is especially observable in conditions of plain relief in Georgia. This difference is less evident in the mountains because of the vertical (altitudinal) zonality. Among all the components of landscapes—soil is distinguished by the great diversity. According to the conditions of soil formation, each genetic type distinguishes a number of soil subtypes, forms, families, and characterized by certain specificity.

The soil cover of Georgia had always been interesting for foreign specialists. Interesting materials were accumulated for decades and widened the existing views of the genesis, classification, geography, and use of different soils. At present, there are additional data about the use of modern methods to study the soils of Georgia and there are novel views about soil processes introduced making it necessary to create the works dedicated to the modern circumstances. The history of both the origination and development of soils is quite complex. Anthropogenic impact in Georgia, as in an ancient agricultural country, is quite strong. In the twenty-first century, in terms of rapid scientific-technical progress, a human’s impact on soil increased. As a result, the

(a)

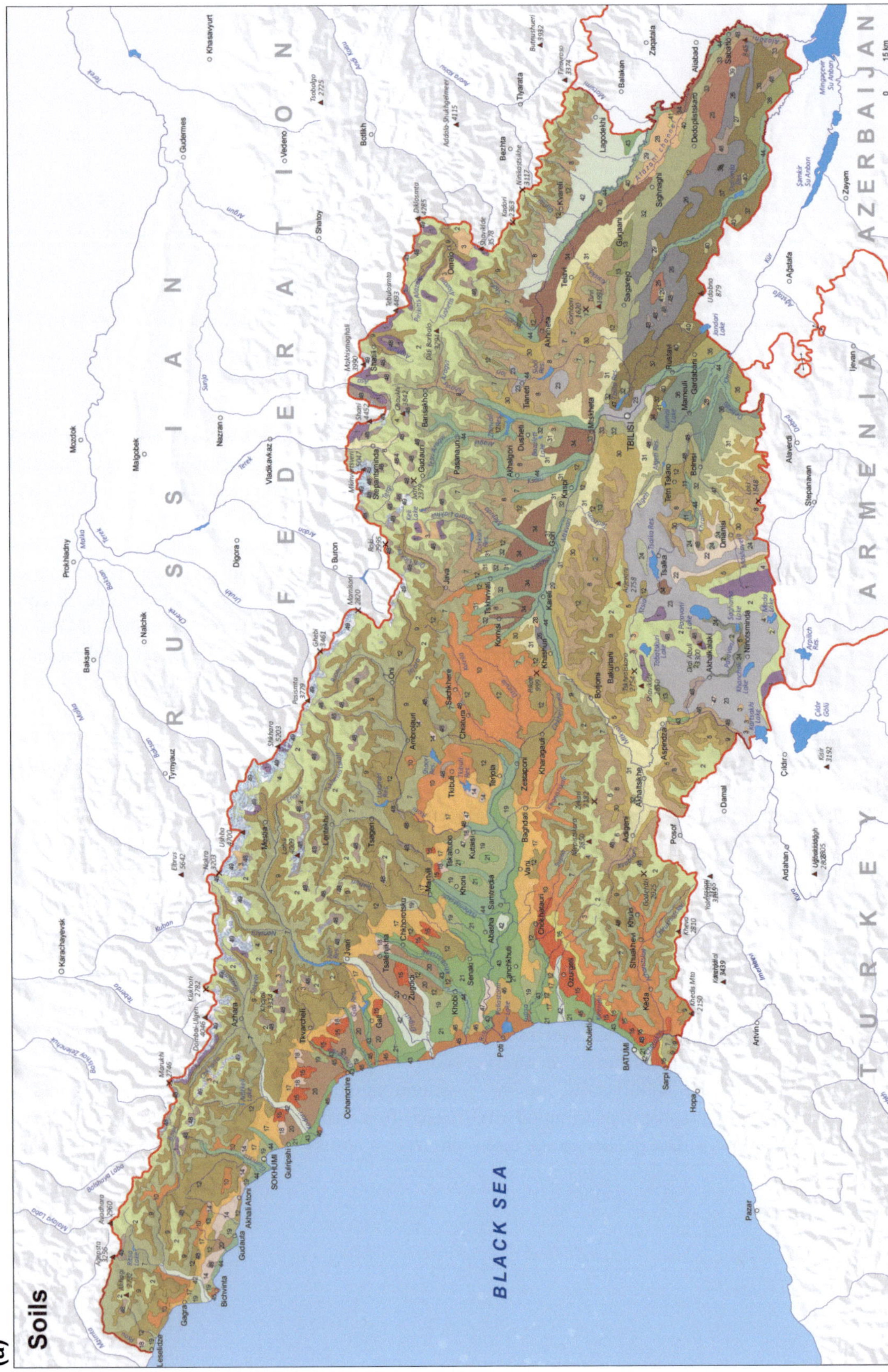

Fig. 1.2 **a** Soil map (T. Urushadze, L. Matchavariani, G. Ghambashidze. National Atlas of Georgia, Stuttgart: Steiner-Verlag, 2018, p. 74). **b** legend of soil map

(b)

Soils Map Legend

1	Mountainous Meadow Primitive Soils (Leptosols Umbric)
2	Mountainous Meadow Turf Soils (Leptosols Umbric)
3	Mountainous Meadow Turf-Peat Soils (Leptosols Umbric)
4	Mountainous Meadow Marsh Soils (Leptosols Umbric)
5	Mountainous Meadow Chernozem like Soils (Leptosols Mollic)
6	Mountainous Meadow Brown Forest Soils (Leptosols Umbric)
7	Brown Forest Soils (Cambisols Eutric & Cambisols Dystric)
8	Brown Forest Unsaturated Soils (Cambisols Eutric)
9	Brown Forest Podsolic Soils (Cambisols Dystric)
10	Yellow Brown Forest Soils (Acrisols Haplic)
11	Brown Forest Black Soils (Chernozems Haplic)
12	Raw Humus Calcareous Soils (Leptosols Rendzic)
13	Raw Humus Calcareous Degraded Soils
14	Raw Humus Red Soils (Leptosols Rendzic)
15	Red Soils (Nitisols Ferralic)
16	Red Podsolic Soils (Nitisols Albic)
17	Yellow Soils (Acrisols Haplic)
18	Yellow Podsolic Soils (Acrisols Albic)
19	Subtropical Podzols (Luvisols Albic)
20	Subtropical Ortshtein Podzols (Luvisols Ferralic)
21	Subtropical Gley Podzolic Soils (Gleysols)
22	Mountain Leached Chernozems
23	Mountain Chernozems
24	Mountain Vertisol Chernozems
25	Black Soils (Vertisols)
26	Chernozems Calcareous (Vertisols)
27	Chernozems Solonetz & Salted
28	Meadow Chernozems
29	Meadow Chernozems Solonetz & Salted
30	Cinnamonic Leached Soils (Cambisol Chromic)
31	Cinnamonic Soils (Cambisol Chromic)
32	Cinnamonic Calcareous Soils (Cambisol Chromic)
33	Cinnamonic light Soils
34	Meadow Cinnamonic Soils (Cambisol Chromic)
35	Meadow Grey Cinnamonic Soils (Cambisol Chromic)
36	Grey Cinnamonic Dark Soils (Cambisol Chromic)
37	Grey Cinnamonic Soils (Cambisol Chromic)
38	Grey Cinnamonic Light Soils (Cambisol Chromic)
39	Raw Humus Sulfate Soils
40	Solonets
41	Salt Soils (Solonetz Humic)
42	Alluvial Acidic (Fluvisols)
43	Alluvial Satisfied (Fluvisols)
44	Alluvial Calcareous Soils (Fluvisols)
45	Marsh Silt Soils (Gleysols)
46	Marsh Peat Soils (Gleysols)
47	Anthropogenic Soils
48	Strong eroded soils and bare rocks
49	Glaciers

Fig. 1.2 (continued)

soil productivity has reduced. The issue of the cadastre and qualitative evaluation of the plots of field was put on the agenda. Soil degradation has led to uncontrolled erosive processes, soil pollution with chemical substances, and other issues, which need a qualified research with modern methods and preventive measures to plan.

History of Soil Survey

2

Lia Matchavariani and Besik Kalandadze

Abstract

This chapter covers the history of soils survey in Georgia starting from Vakhushti Bagrationi (eighteenth century), who divided the country into five botanical–agricultural zones, which are the vertical zones of soil with different fertilities; and Sulkhan-Saba Orbeliani (eighteenth century) which lexicon and encyclopedia contain terms describing features and properties of soils. Later (nineteenth century), V. Dokuchaev used the route method to study the soils in different regions (coastal zone, foothills, and mountainous). Following this trip, he established the regularities of the vertical soil distribution. A number of published textbooks, monographs, and scientific papers on Soil Science and soils of Georgia are chronologically described. The study results of some scientists and groups of researchers, as well as scientific directions of research and educational institutions, are considered in the field of classification and genesis of soils, agrochemical and physical properties, mineral and organic fertilizers, erosion control, soil pollution, mapping, etc.

Keywords

Soil survey • Vakhushti Bagrationi • Sulkhan-Saba Orbeliani • Soil history

Due to the "locked system" of former Soviet Union, previous generation of well-known Georgian researchers, who worked in the last century, published their study results only in Georgian and Russian languages. Therefore, in many cases, the names of some authors and their publications could not reach the international scientific community.

L. Matchavariani (✉) · B. Kalandadze
Faculty of Exact & Natural Sciences, Department of Soil Geography, Ivane Javakhishvili Tbilisi State University, Tbilisi, Georgia
e-mail: lia.matchavariani@tsu.ge

B. Kalandadze
e-mail: besik.kalandadze@tsu.ge

L. Matchavariani (ed.), *The Soils of Georgia*, World Soils Book Series,
https://doi.org/10.1007/978-3-030-18509-1_2

Husbandry of our nation, which dates back in the ancient past, as evidenced by the archeological material monuments and written sources, demonstrates that since the ancient times, a Georgian farmer used organic fertilizers and ash to fertilize meager soil of low productivity, and leguminous and cereal crops rotation, created artificial terraces over the slopes with great inclination (in South Georgia), made irrigation channels on droughty locations, followed irrigation farming on Rustavi Plain (IV c.) and Alazani Plain (XII c.), built water reservoirs to collect and use snow and rainwater (in Davit-Gareji), and used economic "zoning" of (virtually) his own land to grow the field crops, perennial crops, and cattle breeding expediently.

Vakhushti Bagrationi (Fig. 2.1)—Georgian royal prince, son of King Vakhtang VI—geographer, historian, and cartographer, in one of his principal historical and geographic works "Description of the Kingdom of Georgia" (1745) in line with the fields of agriculture, divided our country into five botanical–agricultural zones, which, in fact, are the vertical zones of soil with different fertilities. This work, as well as his "The Geographical Atlas", was inscribed on UNESCO's Memory of the World Register in 2013.

The materials available in Georgia about the development of the fundamentals of soil science are presented in the form of so-called "public" Soil Science. The evidences are also found in one of the important works of Prince Sulkhan-Saba Orbeliani (Fig. 2.2)—a Georgian writer and diplomat "The Georgian Dictionary" (1754), which combines both a lexicon and an encyclopedia, that contains a number of terms describing many features and properties of soils.

A slogan of the vineyard textbook the "Vine and wine", compiled by Tsinamdzgvrishvili (1920) is a smart scientific postulate stating that "The earth is the vein of the lives of all plants and animals". By saying this, the author underlines that the main source of life on our planet is soil (the Earth).

The issue of studying the soils of Georgia is associated with the name of V. Dokuchaev, a famous researcher. At the end of the nineteenth century (in 1898–1899), he was the first to use the route method to study the soils in different

Fig. 2.1 Vakhushti Bagrationi *Source* © Photo Archive of National Parliamentary Library of Georgia

Fig. 2.2 Sulkhan-Saba Orbeliani *Source* © Photo Archive of National Parliamentary Library of Georgia

regions of Georgia, such as the sea coastal zone in Adjara, Borjomi and Bakuriani, Javakheti, Kakheti, and the zone along the Military Road of Georgia. Based on the soil study materials, he delivered a speech in Tbilisi and published articles in the journals of Russia. Following this trip, he also established the regularities of the vertical soil distribution in Georgia.

Following the excursions in the environs of Batumi in 1910, Prof. K. Glinka noted that a special kind of soil-formation process is observed on the Black Sea coast, characterized by mineral weathering—the washout of anions and bases on the one hand, and by accumulation of one and a half oxides on the other hand. It was in that period, when P. Kosovitch traveled in Georgia, who stated that in the environs of Chakvi, soil formation at different depths occurs in terms of different reactions. He proved this opinion by means of chemical analyses of the waters from the river Chakvistskali and wells and of soils.

It should be noted that the researchers found it much appealing and interesting to study the natural conditions of West Georgia, including original Red Soils. Among the researchers studying Red Soils was A. Ostriakov, Professor at Kazan University, who studied the soils in Chakvi (1912–1914) and compiled a monograph-like work about the Red Soils, which he used to attribute to the group of laterite clays.

M. Kalinin worked in the study of soils in respect to development of vine growing in Georgia, and consequently, he plotted a soil map of Kutaisi Province and an explanatory note to it. It should be noted that M. Kalinin spent his life in Georgia and made a certain contribution to the study of soils of Georgia.

Of Dokuchaev's pupils, a merited researcher in Georgia was S.A. Zakharov, who started working on soil problems in 1910 and dedicated most of his life to the study of soils in Georgia and Transcaucasus.

The scientific study of the soils of Georgia is historically associated with the establishment of the chair of Soil Science in 1919 at Tbilisi State University, later named as Ivane Javakhishvili TSU (Fig. 2.3) by Prof. Dimitri Gedevanishvili (Fig. 2.4). Thereafter, Head of Chair was Prof. Mikheil Sabashvili (Fig. 2.5)—later Dean of the Faculty of Geography and Geology at TSU, Academician of Georgian Academy of Sciences, Minister of Agriculture. From 1929, Prof. D. Gedevanishvili supervised the Soil Science Chair in Agricultural Institute of Georgia (Fig. 2.6).

M. Sabashvili was the first to create a scientifically based classification of soils of Georgia and the South Caucasus;

Fig. 2.3 Tbilisi State University, 1918 *Source* © Museum of Ivane Javakhishvili Tbilisi State University

Fig. 2.4 Professor Dimitri Gedevanishvili *Source* © Museum of Ivane Javakhishvili Tbilisi State University

Fig. 2.5 Academician Michail Sabashvili *Source* © Museum of Ivane Javakhishvili Tbilisi State University

clarified the vertical zoning of soils; and created original schemes of geographical zoning of Georgia. Sabashvili was the first to publish several times the textbook "Soil Science" in Georgian (1952, 1970). On the basis of fundamental research, his famous monograph "Soils of Georgia" was published in different years in Georgian and Russian.

Ivane Javakhishvili Tbilisi State University and Georgian Agrarian University, as well as, Mikhail Sabashvili Research

Fig. 2.6 Georgian Agricultural University. Photo by B. Kalandadze

Institute of Soil Science and Agrochemistry, Water Management Institute of Georgia, later named as Tsotne Mirtskhulava Water Management Institute at the Georgian Technical University, Research Institute of Tea and Subtropical Crops (Anaseuli), Georgian Institute of Subtropical Agriculture (Sukhumi), and others are the main Georgian scientific and educational institutions where traditionally develop scientific schools in soil science (Figs. 2.7, 2.8, 2.9 and 2.10).

Prominent work in the Soil Science of Georgia was accomplished by the Georgian soil scientists, such as Gedevanishvili ("The Soil and Landscape Zones of Georgia" 1946), Sabashvili ("The Soils of Georgia" 1948, 1965; "Soil Science" 1941, 1970), Daraselia ("Water Regime of Red Soils" 1939; "Red and Subtropical Podzol Soils and their use for subtropical crops" 1949; "Dynamics of Solutions in Red Soils of Georgia" 1974), Sanikidze ("The Soils of Kakheti" 1940), Daraselia and Kvaratskhelia ("The Production Properties of Red and Gley Soils" 1975), Tarasashvili ("The Mountain Forest and Mountain Meadow Soils of Georgia" 1956; "Soil Science" 1965), Akhvlediani ("The Production Properties of Gypsum Soils" 1973), Talakhadze ("The Black Soils of Georgia" 1962b), Motserelia ("Conversion of Colchis" 1954), Chkhikvishvili ("The Salt Soils of Georgia and Their Melioration" 1970), Klopotkovskyi ("The Soils of Meskheti-Javakheti" 1971), Ambokadze ("The Erosion of the Soils of Georgia" 1973), Baratashvili ("The Soils of South Osetia" 1973), Motserelia and Kostava ("The Methods of Drying Melioration of Colchic Marsh Soils" 1975), Kostava ("Methods for drainage and reclamation of waterlogged soils of Colchis" 1976), Kostava and Ramishvili ("Processes of soil formation and land reclamation of the wetlands of the Colchis lowland" 1987), etc.

In the 1940–1950s, the number of soil scientists studying the agrochemical properties of soils of Georgia and plotting soil maps and cartograms thus rendering a great support to the agriculture, increased a lot.

Following the establishment of various higher educational institutions in Georgia, the demand for new textbooks increased. Here too, the Georgian scientists made a

Fig. 2.7 Mikhail Sabashvili Research Institute of Soil Science and Agrochemistry. Photo by R. Lolishvili

Fig. 2.8 Tsotne Mirtskhulava Water Management Institute, GTU. Photo by G. Gavardashvili

Fig. 2.9 Research Institute of Tea and Subtropical Crops, Anaseuli. Photo by A. Meskhidze

significant contribution to supplying the institutes and vocational schools with textbooks.

For the first time, "The Soil Science" by Samadashvili (1930) was published in the Georgian language; also "A brief course of Soil Science" (1935) for the students of the Institute of Forestry was published. "The Soil Science—husbandry with the principles of Soil Science" by Williams (1939) was published in the Georgian language, and Prof. D. Gedevanishvili was a book editor and author of the Foreword; for the students of higher educational establishments Prof. M. Sabashvili published "Soils of Georgia" (1948, 1965) and "The Soil Science" (1941, 1970); "The course of Soil Science" for agricultural institutions by Gedevanishvili and Talakhadze (1955, 1961) was published; and "The Practical Work in Soil Science" (1962a) and "The Principal Types of Soils in Georgia" (1964) was published by G. Talakhadze as an auxiliary textbooks for the Institute of Agriculture of Georgia. Later, Talakhadze and Mindeli published "Private Soil Science" (1976); and by the group of soils scientists—Talakhadze et al. (1983) was published "The Soils of Georgia".

The first map of soils of Georgia was plotted by Zakharov (1923), followed by D. Gedevanishvili plotting three maps in (1930), (1938) and (1958), respectively. M. Sabashvili plotted two maps of soils of Georgia scaled 1:500 000 and 1:200 000, in (1939) and (1954), respectively. G. Talakhadze plotted a map of the soils of Georgia scaled 1:500 000 (1963), and jointly with I. Anjaparidze, he plotted a map of the soils of Georgia scaled 1:400 000 (1980). Later, the soil map of Georgia scaled 1:500 000 was plotted by the group of soil scientists (edited by Urushadze 1999).

M. Daraselia, with the aim to develop subtropical cultures, explored the Red and Yellow Podzols of Georgia, their physical and hydrophytic properties in particular. In the 1930s, aiming at studying the dynamics of soil solutions, he built a big lysimetric station in Ozurgeti Region. The station is operable to present. Particularly, worthwhile of his works are "The Water Regime of Red Soil" (1939), "Dynamics of the Red Soils of Georgia" (1974), etc.

Motserelia (1954) studied the soils of Colchic Lowland, methods of their melioration and perspectives of their use. Later, a group of soil scientists, geographers, hydrologists, etc. published a collective monograph "Colchis Lowland—Scientific prerequisites for development" (1990).

Nakaidze (1977) explored the soils of East Georgia (Black, Gray-Cinnamonic). Anjaparidze (1979) continued the study of the Cinnamonic soils of Georgia and described their agricultural use.

A number of agrochemical studies were conducted by Prof. O. Zardalishvili at the M. Sabashvili Research Institute of Soil Science and Agrochemistry ("Nitrogen balance in agriculture of Georgia" 1977; "Use of microelements in farming" 1988; "Production and use of organic fertilizers in Georgia" 1990), as well as by Prof. V. Tsanava with colleagues in Research Institute of Tea and Subtropical Crops, Anaseuli, ("Losses of nitrogen by washing out of Red soils

Fig. 2.10 Georgian Institute of Subtropical Agriculture, Sukhumi (Abkhazia). Photo by T. Kacharava

according to the lysimetric research with tea" 1979; "Agrochemistry" 2014; "Main Principles of Ecologization of Fertilization Systems for Subtropical Cultures" 2015), etc.

T. Urushadze dedicated several works to the study of soil in the mountain-forest zone of Georgia. Among them are particularly worth noting "The Soils of Mountain Forests of Georgia" (1987), "The Principal Soils of Georgia" (1997), and others worth mentioning. Later, in (2011), jointly with Austrian professor Winfred Bloom, he published a textbook "The Geography of Soils with the Fundamentals of Soil Science"; and with Prof. L. Matchavariani—"Practicum in Soil Science" (2011), which are of a great help for the students of Ivane Javakhishvili Tbilisi State University and specialists in this field and laboratory works.

Special studies on Georgian soils were carried out by Russian soil scientists, in some cases, together with Georgian colleagues, that were reflected in the following publications: Zakharov ("About the pivotal results and the basic-problems of studying of Georgia's soils" 1924); Gerasimov ("What are the Subtropical Podzols of Abkhazia?" 1966); Gerasimov and Romashkevich ("Soils and weathering crust in genetic profile of Red soils of West Georgia" 1967); Romashkevich ("The Study of Red Soils Microstructure of Western Georgia in Relation to their Genesis" 1966; Soils and weathering crusts in the humid subtropics of West Georgia 1974; Subtropical Pseudo-Podzolic Soils 1979); Gradusov and Urushadze ("Clay Minerals in Mountain-Forest Soils of Georgia" 1968; "Clay Minerals in Soils of Floodplain Forests of

Eastern Georgia" 1972); Zonn and Shonia ("Pseudopodzolization in Subtropical Soils of Western Georgia" 1971); Zonn ("Soil Science and Soils of Subtropics and Tropics" 1974; "Soil Cover and Problems of the USSR Subtropics" 1987); Zonn and Urushadze ("Some issues of Georgia's soils vertical distribution" 1975); Vukolov and Tursina ("Features of Macro- and Microstructure of the Hydromorphic Soils of Colchis Lowland" 1986); Gradusov and Matchavariani ("The Mineralogical Composition of Yellow-Podzolic Soils" 1987); Tursina et al. (Micromorphology of Virgin and Anthropogenic Soils of Humid Subtropics, Georgia 1988); Lezhava et al. ("Micromorphometric Features of Ortshteyn Soils of Western Georgia" 1989); Dobrovolsky and Urushadze ("Soils on red weathering products of Georgia" 1990); Vodyanitskyi and Matchavariani ("The Influence of Hydromorphism of Podzolic-Yellow Soils on the Lepidocrocite Content" 1992); Sokolov and Lezhava ("Actual Problems of Genetic Soil Science in Relation to Georgian Soils" 1997), etc.

Interesting researches on erosion are also reflected by Ambokadze ("Antierosion Measures in Soils of Georgia" 1962; "Development of Erosion Processes in the Eastern Georgia and the Struggle with them" 1968), as well as by Machavariani ("The Results of Studies on Soil Erosion in Georgia" 1976; "Erosion and Protection of Soil" 1988). Afterward, Kereselidze et al. (2013, 2015) studied the issues concerning to allowable erosion rates and quantitative assessment of permissible loads on soils of Georgia.

Detailed studies on erosion control were provided by the Laboratory of Hydraulic Engineering (Fig. 2.11) at the Water Management Institute of Georgia ("Methodological Recommendations…" 1978). A number of publications in this field were published by Mirtskhulava ("Techniques for Calculating and Pre dicting Water Erosion" 1970; "Reliability of Hydroreclamation Constructions" 1974; "Reliability of Drainage Systems" 1985; "Basics of Physics and Mechanics of Channel Erosion" 1988; "Maximum Permissible Erosion" 1989; Environmental Disturbances, Risk Assessment, and Mitigation Measures 1993; "Dangers and Risks of Some Water Management and Other Systems" 2003), Gavardashvili (Prediction of the Erosive Processes in the Corridor of Baku–Tbilisi–Ceyhan Oil Pipeline and Development of Methods to Design the New Engineering Environmental Protection Measures 2014; Predicting of Mountain Slope Erosion in the Catchment Areas of the Laknashera and Lekverari Rivers by Using a Universal Soil Loss Equation of Erosion Processes 2016a; The Forecast of Land Reclamation Risk Factors in Georgia Considering Climate Change 2016b), G. Dokhnadze (erosion by water), V. Nadirashvili (erosion by wind), etc.

Soil erosion problems in river basins of Georgia, as well as quantitative assessment of soil erodibility on the ameliorated lands and arable soils were studied in detail by Gogichaishvili ("Erosive Potential of Rainfalls in Georgia" 2004; "Quantitative Assessment of Soil Erodibility on the Ameliorated Lands in Georgia" 2007; "Erodibility of Arable Soils in Georgia during the Period of Storm Runoff" 2012; "Soil Erosion in River Basins of Georgia" 2016), Gogichaishvili and Sheliya "Annual Variations in Soil Erodibility in Georgia (2006); Gogichaishvili et al. "Testing of hydromechanical predictive model of soil erosion in Georgia" (2014), etc.

From the 1970s, the Georgian works, in addition to the traditional soil studies, gave the descriptions of micromorphological properties, which were descriptive at first (Nakaidze 1966, 1973; Gerasimova and Urushadze 1967; Bobrovitskiy 1973; Beruchashvili et al. 1973; Makeeva 1983; Marshania et al. 1984; Pipia 1986; Varazashvili and Gogoberidze 1986; Jorbenadze and Gogua 1986; Jebisashvili 1986; Iashvili 1986; Iashvili and Makeeva 1986, etc.). Later, from the 1990s, micropedology, following its informative nature, became important in a genetic view. It is

Fig. 2.11 Laboratory of Hydraulic Engineering of the Water Management Institute. Photo by G. Gavardashvili

worthwhile that the most detailed micromorphological studies were accomplished in the region of the humid subtropics of Georgia. By using a system of methods (at macro-, meso-, micro-, and sub-microlevels), the Subtropical Podzolic (Yellow-Podzol) soils and Fe concretions in them were studied in complex (Matchavariani 1987, 2002, 2005, etc.). Later, all principal types of soil were subject to a thorough micropedological study in Georgia (Matchavariani 2008).

B. Kalandadze, together with a group of German scientists for almost a decade, studied soil contamination in the vicinity of the mining enterprise, Bolnisi municipality, Kvemo Kartli Region (Kalandadze et al. 2009; Hanauer et al. 2011; Kalandadze and Felix-Henningsen 2014, etc.).

In different years, the various aspects of soil studies were undertaken by many other scientists: Sanikidze ("The Soils of Kakheti" 1940), Shevardnadze ("The Mountain-and-Forest Soils of Adjara" 1963), Mardaleishvili ("Overview of chernozem soils of the northeastern part of the Iori Upland" 1973), Petriashvili ("The Soils of Meskheti" 1975), Chkheidze ("Raw Humus Calcareous Soils of Georgia" 1977), Samarguliani et al. ("Types of the tea root system structure in Ortstein soils" 1985; "Influence of Heavy Metals on Hydrophysical Properties of Soils" 1994), Iashvili ("Soils of Svaneti" 1987), Kacharava et al. ("Diagnosis of the Nitrogen Fertilization Method Effectiveness" 1987; "Influence of Fractional Fertilization by Nitrogen on Citrus Productivity" 1988), Charkseliani et al. ("Land Cadastre of Georgia" 1988), Mardaleishvili and Pipia ("Cherozems of Iori Upland of Eastern Transcaucasia" 1988), Motserelia ("Soils of the Colchis lowland—an object of melioration" 1989), Kacharava and Tiutiunikov ("Natural and Economic Indicators of Conducting the Agricultural Production in the Abkhazian ASSR" 1990), Mardaleishvili and Tvalvadze ("Agroecological Zoning of Soil Cover in the Intensive Agricultural Zone of Kartli" 1992), Lezhava ("Podzolic Ortshtein Soils (Plintosols) of Georgia" 1998), Palavandishvili ("Red Soils of Adjara and their agroindustrial using" 1987; "The Soils of Georgia" 2002), Mardaleishvili et al. ("Genesis, Diagnostic Indices and Properties of Meadow Solonetzes and Solonchaks of Georgia" 2003; "Ways to Improve the Efficiency of low-fertile Chernozems in Eastern Georgia" 2005), Pipia and Mardaleishvili ("Chernozems of the Iori Plateau of the Eastern Transcaucasia" 1988), Pipia ("Hydrological Regime of the Ordinary Chernozems of Iori Upland" 2008), Lezhava and Pipia ("Determination of Soil Evolution and Age—methodical instructions" 2005), Kirvalidze ("Humus in the highlands of the Central Caucasus" 1993), Lolishvili, Khutsishvili ("Balance of Humus of Southern Carbonate Chernozem under the Crop of Onion" 2007), Lolishvili and Burchuladze ("The Estimation of Fertility of Soils and Ecologically Safe Biotechnologies of its Increase" 2011), Nikolaishvili and Matchavariani ("Humus Reserves and their Distribution in the Landscapes of Georgia" 2010), Gambashidze ("Mineral composition of organically grown tomato" 2014), Gambashidze et al. ("Heavy metals in some soils of Western Georgia" 2014), etc.

References

Akhvlediani GK (1973) The soil cover of Samgori Valley (Samgoris velis niadaguri safari). Sabchota Sakartvelo, Tbilisi, 205 p

Ambokadze A (1973) The erosion of the soils of Georgia (Sakartvelos niadagebis erozia). Sabchota Sakartvelo

Ambokadze VA (1962) Antierosion measures in soils of Georgia (Antieroziuli zoebi saqartvelos niadagebshi). Tbilisi, 50 p

Ambokadze VA, Lobzhanidze VA (1968) Development of erosion processes in the Eastern Georgia and the struggle with them (Eroziuli processebis ganvitareba da mattan brdzols agmosavlet saqartveloshi). In: Transcaucasian scientific conference on the issue of soil erosion and measures to combat it, Tbilisi, pp 53–62

Anjaparidze I (1979) Cinnamonic soils of Georgia (Saqartvelos yavisferi niadagebi). Sabchota Sakartvelo, Tbilisi

Baratashvili I (1973) The soils of South Osetia (Samkhret Osetis niadagebi). Sabchota Sakartvelo

Beruchashvili NL, Maglakelidze RV, Petriashvili RA, Shevardnadze MG (1973) Micromorphological features of the subalpine and mountain forest soils of the Saguramo-Yalon Range. Observations on Markopi physiographic Lab. Tbilisi, fourth quarter

Bobrovitskiy AV (1973) Micromorphology and mineralogical composition of alluvial sediments and Colchis soils. Eurasian Soil Sci (Mikromorfologia i minralogicheskiy Sostav Alluvialnikh Pochv Nanosov i Pochv Kolkhidy. Pochvovedenie) 3:115–128

Charkseliani A, Petrishvili R, Kipiani M (1988) Land cadastre of Georgia (Sakartvelos SSR Mitsis Kadastri). Sabchota Sakartvelo, 735 p

Chkheidze T (1977) Raw humus calcareous soils of Georgia (Sakartvelos kordian-karbonatuli niadagebi). Sabchota Sakartvelo

Chkhikvishvili V (1970) The salt soils of Georgia and their melioration (Sakartvelos damlashebuli niadagebi da mati melioracia). Sabchota Sakartvelo

Chkhikvishvili V (1990) Colchis lowland—scientific prerequisites for development

Daraselia M (1939) Water regime of red soils (Tsitelmitsa niadagebis wylis rezjimi). Metsniereba, Tbilisi

Daraselia MK (1949) Red and subtropical podzol soils and there use for subtropical crops (Tsitelmitsa da subtropikuli etseri niadagebi da mati gamokheneba subtropikuli kulturebis qvesh). VNIIChSK, 445 p

Daraselia M (1974) Dynamics of soil solutions in red soils of Georgia (Niadaguri khsnarebis dinamika saqartvelos tsitelmitsa niadagebshi). Metsniereba, Tbilisi

Daraselia M, Kvaratskhelia N (1975) The production properties of red and gley soils (Tsitelmitsa da galebebuli niadagebis satsarmoo tvisebebi). Sabchota Sakartvelo

Dobrovolsky VV, Urushadze TF (1990) Soils on red weathering products of Georgia. Mecniereba, Tbilisi, 178 p

Gambashidze G (2014) Mineral composition of organically grown tomato. In: EGU general assembly conference abstracts, vol 16

Gambashidze GO, Urushadze TF, Blum WE, Mentler AF (2014) Heavy metals in some soils of Western Georgia. Eurasian Soil Sci 47(8):834–843

Gavardashvili GV (2014) Prediction of the erosive processes in the corridor of Baku-Tbilisi-Ceyhan oil pipeline and development of methods to design the new engineering environmental protection measures. In: 6th international conference on contemporary problems

of architecture and construction, Czech Republic, Ostrava, 24–27 June 2014. VSB-Technical University of Ostrava, pp 478–483
Gavardashvili GV (2016a) Predicting of mountain slope erosion in the catchment areas of the Laknashera and Lekverari rivers by using a universal soil loss equation of erosion processes. Collected papers of Ts. Mirtskhulava Water Management Institute of Georgian Technical University, #71, Tbilisi, 10 p
Gavardashvili GV (2016b) The forecast of land reclamation risk factors in Georgia considering climate change. In: 1st international scientific conference "Waterland-2016", 06–12 June, Kaunas, Lithuania, 14 p
Gedevanishvili D (1930) Map: the soils of Georgia
Gedevanishvili D (1938) Map: the soils of Georgia
Gedevanishvili D (1946) The soils and landscape zones of Georgia (Sakartvelos niadagebi da landshfturi zonebi). Sabchota Sakartvelo
Gedevanishvili D (1958) Map: the soils of Georgia
Gedevanishvili G, Talakhadze G (1955) The course of soil science (Niadagmcodneobis kursi). Metsniereba, Tbilisi
Gedevanishvili G, Talakhadze G (1961) The course of soil science (Niadagmcodneobis kursi). Metsniereba, Tbilisi
Gerasimov IP (1966) What are the subtropical podzols of Abkhazia? (Chto takoye subtropicheskie podzoli Abkhhazii?). Pochvovedenie 11:21–34
Gerasimov IP, Romashkevich AI (1967) Soils and weathering crust in genetic profile of red soils of West Georgia. Pochvovedenye 4:102–113
Gerasimova MI, Urushadze TF (1967) Peculiarities of microstructure of mountain brown forest soils of Carpathians and Georgia. Features of soil formation in the zone of Brown Forest soils. Vladivostok 1967:85–87
Gogichaishvili G (2004) Erosive potential of rainfalls in Georgia. Eurasian Soil Sci 37(5):538–544
Gogichaishvili G (2007) Quantitative assessment of soil erodibility on the ameliorated lands in Georgia. Eurasian Soil Sci 40(11):1227–1232
Gogichaishvili G (2012) Erodibility of arable soils in Georgia during the period of storm runoff. Eurasian Soil Sci 45(2):197–201
Gogichaishvili G (2016) Soil erosion in river basins of Georgia. Eurasian Soil Sci 49(6):696–704
Gogichaishvili G, Sheliya V (2006) Annual variations in soil erodibility in Georgia. Eurasian Soil Sci 39(10):1148–1151
Gogichaishvili G, Kirvalidze D, Gorjomeladze O (2014) Testing of hydromechanical predictive model of soil erosion in Georgia. Eurasian Soil Sci 47(9):1107–1112
Gradusov BP, Urushadze TF (1968) Clay minerals in mountain-forest soils of Georgia (Glinistie minerali v burikh lesnikh pochvakh Gruzii). Pochvovedenie 2:124–132
Gradusov BP, Urushadze TF (1972) Clay minerals in soils of floodplain forests of Eastern Georgia (Glinistie minerali v pochvakh poymrnnilh ledov Vostochnoy Gruzii). Bull Georgia SSR Acad Sci 66 (3):657–660
Gradusov BP, Matchavariani LG (1987) The mineralogical composition of yellow-podzolic soils (Mineralogicheskiy sostav zjelto-podzolistikh pochv). Subtropicheskie kulturi). Subtrop Cult 212:130–136
Hanauer T, Shnell S, Steffens D, Kalandadze B, Navrozashvili L, Felix-Henningsen P (2011) In situ remediation of Cd, Cu and Zn contaminated topsoils by different amendments. In: EGU general assembly, vol 13. XY488 EGU2011-159
Iashvili N (1987) Soils of Svaneti (Svanetis niadagebi). Sabchota Sakartvelo, 225 p
Iashvili N, Makeeva V (1986) Peculiarities of some Georgia's soils microstructure. In: Proceedings of the conference "micromorphological methods of soil studying". Tbilisi University Publishing
Iashvili NN (1986) Micromorphology and Alpine brown soils of Svaneti. In: Proceedings of the conference "micromorphological methods of soil studying". Tbilisi University Publishing
Jebisashvili NV (1986) Features of micromorphological structure of Brown Forest soils under conifers in Eastern Georgia. In: Proceedings of the conference "micromorphological methods of soil studying". Tbilisi University Publishing
Jorbenadze LT, Gogua ZCh (1986) Micromorphological features of soils of the southern part of Kvemo Kartli. In: Proceedings of the conference "micromorphological methods of soil studying". Tbilisi University Publishing
Kacharava ON, Tiutiunikov AI (1990) Natural and economic indicators of conducting the agricultural production in the Abkhazian ASSR. VASHNIL (Prirodno-Ekonomicheskie Pokazateli Vedeniya Selskokhoziaistvennogo Proizvodstva v Abkhazskoy ASSR VASKHNIL, Spravochnik), 197 p
Kacharava ON, Kacharava TO, Gorchakov VV (1987) Diagnosis of the nitrogen fertilization method effectiveness. Bulletin of the V.V. Dokuchaev Soil Institute, issue 41, Moscow (Diagnostika Effektivnosti Sposoba Vneseniya Azotnikh Udobreniy. Bulleten Pochvennogo Instituta imeni V.V. Dokuchaeva, vip 41, Moskva), pp 14–20
Kacharava ON, Kacharava TO, Gorchakov VV (1988) Influence of fractional fertilization by nitrogen on citrus productivity. Bulletin I—activation of functional activity of root systems. J Agrochem (Vliyanie Drobnogo Vneseniya Azotnikh Udobreniy na Produktivnost Tsitrusovikh. Soobshenie I—Aktivatsia Funktsionalnoy Deyatelnosti Kornevikh System. Jurnal Agrokhimia) 9:18–24
Kalandadze B, Felix-Henningsen P (2014) Pollution of irrigation soils and cultural plants with heavy metals in the basin of the river Kvirila due to mining industry (West Georgia). In: Eurasian soil congress abs., Istanbul, Turkey
Kalandadze B, Hanauer T, Felix-Henningsen P, Urushadze T, Narimanidze E, Steffens D (2009) Mining and agriculture in the Mashavera valley (South-East Georgia)—a land use conflict with severe consequences. Biol J Armenia LXI 2:22–29. National Academy of Sciences of the Republic of Armenia
Kereselidze DN, Matchavariani LG, Kalandadze BB, Trapaidze VZ (2013) Allowable soil erosion rates in Georgia. Eurasian Soil Sci 46(4):438–446
Kereselidze D, Matchavariani L, Trapaidze V, Lagidze L, Svanadze D, Bregvadze G, Gulashvili Z (2015) Quantitative assessment of permissible loads on Georgia's soil. Am J Environ Prot, special issue "Applied Ecology: Problems, Innovations", SciencePG 4 (3-1):29–33
Kirvalidze D (1993) Humus in the highlands of the Central Caucasus. Candidate thesis, GSAU, Tbilisi
Klopotkovskiy B (1971) The soils of Meskheti-Javakheti (Meskhet-Javakhetis niadagebi). Sabchota Sakartvelo
Kostava G (1976) Methods for drainage and reclamation of waterlogged soils of Colchis. Sabchota Sakartvelo
Kostava GA, Ramishvili TD (1987) Processes of soil formation and land reclamation of the wetlands of the Colchis lowland. Sabchota Sakartvelo, Tbilisi
Lezhava V (1998) Podzolic ortshtein soils (plintosols) of Georgia. The autoreferat of doctoral dissertation, Tbilisi Agrarian University, 58 p
Lezhava V, Pipia TS (2005) Determination of soil evolution and age (methodical instructions). Centre for Ecology and Pathology, Tbilisi, TSU, Tbilisi
Lezhava VV, Tursina TV, Skvortsova EB, Matchavariani LG (1989) Micromorphometric features of ortshteyn soils of Western Georgia (Mikromorfologicheskie osobennosti ortshteqnovikh pochv Zapadnoy Gruzii. Soobsheniya Akademi Nauk Gruz.SSR). Bull AS GSSR 133(2):373–376

Lolishvili R, Khutsishvili M (2007) Balance of humus of southern carbonate chernozem under the crop of onion. Ann Agrar Sci 5 (2):26–28

Lolishvili R, Burchuladze I (2011) The estimation of fertility of soils and ecologically safe biotechnologies of its increase. In: Proceedings of the conference "biodiversity and agricultural biotechnologies", Agrarian University of Georgia, Tbilisi, 5–6 Dec 2011, pp 95–101

Machavariani V (1976) The results of studies on soil erosion in Georgia. In: Proceedings of the scientific session devoted to the 30th anniversary of Scientific-Research Institute of Soil Science, Agrochemistry and Melioration, Metsniereba, Tbilisi, pp 85–97

Machavariani V (1988) Erosion and protection of soil. Metsniereba, Tbilisi, 250 p

Makeeva V (1983) Micromorphological peculiarities of soil formation of dark cinnamonic soils of Eastern Georgia. In: II conference on soil micromorphology, Tartu

Mardaleishvili R (1973) Overview of chernozem soils of the northeastern part of the Iori upland. In: Proceedings of the Soil Science Institute (Ioris zegnis chrdilo-agmosavleti nacilis shavmica niadagebis dakhasiateba. Niadagmcodneobis institutis shromata krebuli), vol 14, pp 20–35

Mardaleishvili R, Pipia TS (1988) Cherozems of Iori upland of Eastern Transcaucasia (Chernozemi Iorskogo ploskogorya Vostochnogo Zakavkazya). Agropromizdat, Moscow

Mardaleishvili R, Tvalavadze M (1992) Agroecological zoning of soil cover in the intensive agricultural zone of Kartli (Kartlis intensiuri samicatmoqmedo zonis niadaguri saparis agroekologiuri daraioneba. Niadagmcodneobis institutis shromata krebuli)

Mardaleishvili R, Tvalavadze M, Lolishvili R (2003) Genesis, diagnostic indices and properties of meadow solonetzes and solonchaks of Georgia. In: Proceedings of the soil science (Sakartvelo mdelos mlashobi da bicobi niadagebis genezisi, diagnostikuri machveneblebi da tvisebebis genezisi. Niadagmcodneobis institutis shromata krebuli), Tbilisi

Mardaleishvili R, Tvalavadze M, Lolishvili R (2005) Ways to improve the efficiency of low-fertile chernozems in Eastern Georgia (Puti povisheniya effektivnosti nizkoplodorodnikh chernozemov Vostochnoy Gruzii. Izvestiya Agrarnoy Nauki), vol 3(2)

Marshania M, Tursina T, Verba M (1984) Changes of microstructure in Yellow-Podzolic soils of the slopes because of the primary treatments. Subtrop Cult 3:139–148

Matchavariani LG (1987) On the genesis of subtropical podzolic soils (O genezise subtropicheskikh podzolistikh pochv. Trudi TGU, Sovet molodikh uchennikh, seriya yestestvennikh nauk). In: Proceedings of TSU, series of natural sciences, XIII. Tbilisi University Publishing House, pp 165–174

Matchavariani L (2002) The genetic significance of micro-pedological studies relatively to subtropical podzols. Geography of Georgia. Tbilisi University Publishing House, no 1, pp 18–23

Matchavariani LG (2005) Morphogenetic typification of concretions in subtropical podzolic soils of Georgia. Eurasian Soil Sci 38 (11):1161–1172

Matchavariani LG (2008) Geographic paradigms of the microfabric of major soils of Georgia. Universal, Tbilisi, 309 p

Methodological recommendations on the prediction of water (rainstorm) soil erosion (1978) Publishing House VASKHNIL (Metodicheskie rekomendatsii po prognozu vodnoi (dozdevoi) erozii), Moscow, 60 p

Mirtskhulava TsE (1970) Techniques for calculating and predicting water erosion. Kolos, Moscow

Mirtskhulava TsE (1974) Reliability of hydroreclamation constructions. Kolos, Moscow

Mirtskhulava TsE (1985) Reliability of drainage systems. Agropromizdat, Moscow

Mirtskhulava TsE (1988) Basics of physics and mechanics of channel erosion. Gidrometeoizdat, Leningrad

Mirtskhulava TsE (1989) Maximum permissible erosion. Doklady VASKhNIL 2:32–34

Mirtskhulava TsE (1993) Environmental disturbances. Risk assessment and mitigation measures

Mirtskhulava TsE (2003) Dangers and risks of some water management and other systems. Nauka, Tbilisi

Motserelia AV (1954) Conversion of Colchis (Preobrazovanie Kolkhidi). Nauka, Moscow

Motserelia I (1989) Soils of the Colchis lowland—an object of melioration (Pochvi Kolkhidskoy nizmennosti—obyekt melioratsii). Tbilisi

Motserelia A, Kostava G (1975) The methods of drying melioration of Colchic marsh soils (Kolkhetis cjaobiani niadagebis dashrobiti melioraciis metodebi). Sabchota Sakartvelo

Nakaidze EK (1966) Micromorphological features of cinnamonic soils of Georgia. Micromorphological method in the study of soil genesis. Nauka, Moscow, pp 23–41

Nakaidze EK (1973) Features of micromorphology meadow-cinnamonic soils of Georgia. Micromorphology of soils and loose deposits. Nauka, Moscow, 73–77

Nakaidze EK (1977) Cinnamonic and meadow-cinnamonic soils of Georgia (Sakartvelos khavisperi da mdelos-yavisperi niadagebi). Metsniereba, Tbilisi, 303 p

Nikolaishvili DA, Matchavariani LG (2010) Humus reserves and their distribution in the landscapes of Georgia. Eurasian Soil Sci 43 (1):39–48

Palavandishvili Sh (1987) Red soils of Adjara and their agroindustrial using (Adjaris tsitelmitsa niadagebi da mati agroindustriuli gamokheneba). Sabchota Adjara

Palavandishvili Sh (2002) Soil geography with the basics of soil science (Niadagebis geografia niadagmcodneobis safudzvlebit). Adjra Publishing House, Batumi, 350 p

Petrishvili R (1975) The soils of Meskheti (Meskhetis Niadagebi). Sabchota Sakartvelo, 111 p

Pipia TS (1986) Micromorphology of the ordinary chernozems, Shiraki. In: Proceedings of the conference "micromorphological methods of soil studying". Tbilisi University Publishing

Pipia TS (2008) Hydrological regime of the ordinary chernozems of Iori upland. Ann Agrar Sci (Gidrologicheskiy rezjim obiknovennikh chernozemov Iorskogo Ploskogorya. Izvestiya Agrarnoy Nauki) 10(3)

Pipia TS, Mardalelishvili R (1988) Chernozems of the Iori Plateau of the Eastern Transcaucasia (Chernoziomi Iorskogo Ploskogoria Vostochnogo Zakavkaziya). Agropromizdat, Moscow

Romashkevich AI (1966) The study of red soils microstructure of Western Georgia in relation to their genesis. Micromorphological method in the study of soil genesis (Issledovanie mikrostroenia krasnizemov Zapadnoy Gruzii. Mikromorfologicheskiy metod v issledovanii genezisa pochv). Nauka, Moscow, pp 42–57

Romashkevich AI (1974) Soils and weathering crusts in humid subtropics of West Georgia (Pochvi i kori vivetrivania vlazjnikh subtrpikov Zapadnoy Gruzii). Nauka, Moscow, 218 p

Romashkevich AI (1979) Subtropical pseudo-podzolic (eluvial-surface-gley) soils. Genetic soil types of subtropical Transcaucasia (Subtropicheskie psevdopodzolistye (elluvialnj-poverkhnostno-gleevie) pochvi. Geneticheskie tipi pochv Yuzhnoy Gruzii). Nauka, Moscow, pp 52–114

Sabashvili MK (1939) Soil map of Georgia (1:500 000). Tbilisi

Sabashvili MK (1941) Soil science. Metsniereba, Tbilisi

Sabashvili MK (1948) Soils of Georgia (Pochvi Gruzii. Akademiya Nauk GruzSSR). Science Academy of GSSR, Tbilisi

Sabashvili MK (1954) Soil map of Georgia (1:200 000). Tbilisi

Sabashvili MK (1965) Soils of Georgia. Tbilisi, Metsniereba

Sabashvili MK (1970) Soil science. Metsniereba, Tbilisi

Sanikidze AO (1940) The soils of Kakheti. Tbilisi, 150 p

Samadashvili N (1930) The soil science. Ganatleba, Tbilisi

Samadashvili N (1935) A brief course of soil science. Ganatleba, Tbilisi

Samarguliani G, Lezhava V, Totibadze V (1985) Types of the tea root system structure in Ortstein soils. GruzSRISTI (Tipi stroeniya kornevoi sistemi chaiya v ortshteinovikh pochvakh. GruzNIINTI) 6 (70):3–5

Samarguliani G, Zarkua G, Iinasaridze V, Kalandadze B (1994) Influence of heavy metals on hydrophysical properties of soils. In: Proceedings of the Aristotle Greco-Georgian University, pp 43–47

Shevardnadze M (1963) The mountain-and-forest soils of Adjara. TSU Proc 60:89–104

Sokolov IA, Lezhava VV (1997) Actual problems of genetic soil science in relation to Georgian soils. Proc TSU Geogr Geol 322: 95–102

Talakhadze G (1962a) The practical work in soil science. Ganatleba, Tbilisi

Talakhadze G (1962b) The black soils of Georgia. Tbilisi

Talakhadze G (1963) Soils map of Georgia (1:500 000). Tbilisi

Talakhadze G (1964) The principal types of soils in Georgia. Ganatleba, Tbilisi

Talakhadze G, Mindeli K (1976) Private soil science. Tbilisi

Talakhadze G, Anjaparidze I (1980) Map: the soils of Georgia (1:400 000). Tbilisi

Talakhadze G, Anjaparidze I, Lataria M, Kirvalidze R, Mindeli K, Nakashidze L, Mindeli M (1983) The soils of Georgia. Tbilisi

Tarasashvili GM (1956) Mountainous-forest and mountainous-meadow soils of East Georgia. AS GSSR, Tbilisi

Tarasashvili G (1965) Soil science. Ganatleba, Tbilisi

Tsanava V, Tsanava N, Meskhidze A (1979) Losses of nitrogen by washing out of red soils according to the lysimetric research with tea (Poteri azota putiom vimivaniya v krasnozemakx po dannim lizimetricheskikh issledovaniy s chaem). Agrochemistry 6:18–25

Tsanava V, Lominadze Sh, Bajelidze A (2014) Agrochemistry. Batumi Shota Rustaveli State University, 380 p

Tsanava V, Meskhidze A, Lominadze Sh, Tsanava E (2015) Main principles of ecologization of fertilization systems for subtropical cultures. In: Proceedings of the international conference "applied ecology: problems, innovations", Tbilisi-Batumi, pp 138–141

Tsinamdzgvrishvili I (1920) Vine breeding (Vazis mosheneba). Vine and wine. Literasi

Tursina TV, Lezhava VV, Matchavariani LG (1988) Micromorphology of virgin and anthropogenic soils of humid subtropics, Georgia (Mikromorfologiya celinnikg i antropogennikh pochv vlaikh subtropikov Gruzii). In: Proceedings of V. Dokuchaev Soil Institute "Optimizing of use of soil resources in subtropical areas", Moscow, pp 17–25

Urushadze T (1987) The soils of mountain forests of Georgia. Metsniereba, Tbilisi, 243 p (Pochvi Gornikh Lesov Gruzii, 243 c)

Urushadze T (1997) The principal soils of Georgia (Sakartvelo Dziritdi Nidagebi). Metzniereba, Tbilisi, 268 p

Urushadze TF (ed) (1999) Soil map of Georgia. Scale 1:500000. Tbilisi

Urushadze T, Blum W (2011) Soil geography with the basics of soil science. Tbilisi University Publishing House, 425 p

Urushadze TF, Matchavariani LG (2011) Practicum in soil science. Tbilisi University Publishing House, 126 p

Varazashvili N, Gogoberidze I (1986) Micromorphology of steppe and meadow soils of Georgia and its change during reclamation. In: Proceedings of the conference "micromorphological methods of soil studying". Tbilisi University Publishing

Vodyanitskyi UN, Matchavariani LG (1992) The influence of hydromorphism of podzolic-yellow soils on the lepidocrocite content (Vliyanie gidromorfizma podzolisto-zheltozemnikg pochv na soderzjanie lepidokrokita). Eurasian Soil Sci 12:79–90

Vukolov NG, Tursina TV (1986) Features of macro- and microstructure of the hydromorphic soils of Colchis lowland (Osobennosti makro- i mikrostroyeniya gidromorfnikh pochv Kolkhidskoy nizmennosti. Pochvovedenie). Eurasian Soil Sci 2:94–103

Williams W (1939) The soil science—husbandry with the principles of soil science. Ganatleba, Tbilisi

Zakharov SA (1923) Map: the soils of Georgia (Pochvi Gruzii). Tbilisi

Zakharov SA (1924) About the pivotal results and the basic-problems of studying of Georgia's soils. Polytechnic Institute, Tiflis, issue 3

Zardalishvili O (1977) Nitrogen balance in agriculture of Georgia (Azotis balansi saqartvelos motsatmoqmedebashi). Sabchota Sakartvelo, Tbilisi, 164 p

Zardalishvili O (1988) Use of microelements in farming (Mikroelementebis gamokheneba mitsatmoqmedebashi). Sabchota Sakartvelo, Tbilisi, 128 p

Zardalishvili O (1990) Production and use of organic fertilizers in Georgia (Organuli sasuqebis tsarmoeba da gamokheneba saqartveloshi). Tbilisi, Georgia, 133 p

Zonn SV (1974) Soil science and soils of subtropics and tropics (Pochvovedenie i pochvi subtropikov i tropikov). Moscow

Zonn SV (1987) Soil cover and problems of transformation the nature and economy of the USSR subtropics (Pochvenniy pokrov i problemi preobrazovania prirodi i khoziaystva subtropikov SSSR). Nauka, Moscow

Zonn SV, Shonia NK (1971) Pseudopodzolizationn in subtropical soils of Western Georgia. Pochvovedenie, pp 103–115

Zonn SV, Urushadze TF (1975) Some issues of Georgia's soils vertical distribution (Nekotorie voprosi vetikslyno-poyasnogo raspredeleniya pochv Gruzii. Geographia v Gruzinskoy SSR). Geography in Georgia SSR, vol 2. Metsniereba, Tbilisi, pp 101–104

Soil-Forming Factors

3

Lia Matchavariani

Abstract

The chapter discusses all major natural factors, influenced soil formation, such as parent material, climate, relief, organisms, and time. Due to considerable influence of humans on soils, it is also considered as a specific factor of soil formation. Considering the significant role of each of these factors, their descriptions for Georgia are given. In particular, geological structure, orographic indicators, climatic features, including soil climate, specificity of vegetation cover and biodiversity, landscape diversity, importance of time factor and relict parameters in soils of Georgia, etc. are described in detail.

Keywords

Soil formation • Parent material • Relief • Climate • Living organisms • Time factor

3.1 Introduction

The soil-forming factors (geographical components) such as parent material, relief, climate, and living organisms appear in some way or another in any definition of soil as a natural and historical body. The soil is a natural body due to the fact that it is formed in nature as a result of the interaction of natural factors; and the historical body due to the fact that its formation needs a certain time. The age of the place, where the soil is located, is also of an utmost importance. It is known that one of the main concepts of soil science is the following—soil formation is influenced by five natural factors: parent material, climate, topography, organisms, and time. All the soil-forming components are equally important and play an equal part in its formation. Therefore, it is necessary to have full knowledge of all the abovementioned components to study the soil genesis.

In its turn, humans have considerable influence on soils, but this influence for thousands of years has increased from the last century. Now, large areas of soils worldwide are heavily altered. Because of their huge impact on soils, humans are also considered as a specific factor of soil formation. "Conservation of the soil resource and its continued use to perform ecosystem functions to support the ever-increasing global population depends on understanding the properties of and processes occurring in the soil at any point in the landscape" (West et al. 2017).

In addition to the basic soil-forming factors (parent material, relief, climate, organisms, and time), sometimes number of local factors are identified. For example, the groundwater with their regime and chemical composition often determine the processes going on in the soil. Sometimes the surface waters of floods periodically swamping the groves and deltas are also considered as the independent local factors of soil formation. Also, the sea waters can also be considered as local factors of soil formation in the deltas of the rivers flowing into the seas and oceans. Sometimes a volcanic factor is also considered to be a local factor of soil formation. The periodic volcanic eruptions in the regions of the active volcanoes are accompanied by an eruption of volcanic ash in the atmosphere and their precipitation on the soil surface.

All the abovementioned local factors have a specific impact on soil formation. Within certain areas, they can determine the orientation of soil formation. But some local factors such as volcanism, earthquakes, underground waters, etc. can be considered as the principal factors like parent material, climate, etc. All soil-forming factors have a specific impact on soil formation, and they are equally important.

In the article of Bockheim et al. (2014), the past and present roles of soil-forming factors are analyzed in USDA *Soil Taxonomy* and it is believed that the factorial and genetic approaches are clearly present there. There is an imbalance in the utilization of the soil-forming factors in *Soil Taxonomy* (Bockheim et al. 2014).

L. Matchavariani (✉)
Faculty of Exact & Natural Sciences, Department of Soil Geography, Ivane Javakhishvili Tbilisi State University, Tbilisi, Georgia
e-mail: lia.matchavariani@tsu.ge

L. Matchavariani (ed.), *The Soils of Georgia*, World Soils Book Series,
https://doi.org/10.1007/978-3-030-18509-1_3

3.2 Factor of Parent Material in Soil Formation

As is known, the parent material is the substrate, from which the soil is formed. It determines most of the properties of soil: the mineral and chemical contents of soil depend on the composition of the parent material; the density and porosity of parent materials determine the nature of the soil structure; the soil depth, its composition in vertical and horizontal directions, water content, thermal and physical properties, etc., also depend on the parent materials. Thus, the parent material is considered as a substrate from which soils form because the parent material provides the geochemical foundation of the soil.

In the process of soil formation, the original features of the parent material change in different ways. Some are almost unchanged in the soil while others experience significant transformation. Young soils with their composition and structure are close to those of the parent materials. The older the soil is, the longer the process of soil formation and exhaustion is going on and the greater the difference between the soil and their parent materials.

The diversity in petrographic composition of parent materials causes the diversity of soils in general. Soils formed in the same bioclimatic and geomorphological conditions might be different when they are formed of different soil-forming parent materials. The minerals that compose the parent material are the sources of elements that serve as nutrients for plants (Graham and Indorante 2017).

Analogously, the parent material limits the textural range of the soils derived from it. As noted by the authors of one Chapter "The Soil of USA" (Graham and Indorante 2017), soil textures are typically finer than the grain sizes of the original parent material because weathering reduces the size of the original grains and precipitates clay-size material, thus reducing the grain size overall.

It is known, out of three basic types of parent materials (magmatic, metamorphic, and sedimentary), mainly the sedimentary parent materials of different genetic types of the Quaternary Period sedimentary parent materials (eluvia, diluvia, proluvia, alluvia, lacustrine, moraine, aeolian, marine or fluvio-glacial sediments, loess, etc.) take part in soil formation.

The existence of the primary (quartz, spars, amphiboles, pyroxenes, and mica) and secondary minerals (hydroxide and oxide minerals, clay minerals) in soils is determined by the soil-forming parent materials.

As the loose sedimentary ran a long life cycle on the earth's surface, experienced weathering, and multiple re-sedimentation, its composition is characterized by a relatively low content of the primary and high content of the secondary minerals.

According to literary sources (Bockheim et al. 2014), parent material is used to fully define two soil orders (Histosols and Andisols), and partially to define the suborders in the Entisol order (Fluvents, Psamments), whereas relief and time are not used in defining taxa in *Soil Taxonomy*.

3.2.1 Geological Structure

The territory of Georgia consists of the different age and composition rocks (Fig. 3.1). Mostly, the young Mesozoic and Cenozoic parent materials are widely spread, while the old Paleozoic and Precambrian parent materials are relatively localized (Gamkrelidze 2018).

The ancient, Paleozoic parent materials are predominantly spread along the axial line of the Caucasus (in its western and central part). They are represented by metamorphic and crystalline parent materials—granites, gneisses, crystalline slates.

Mesozoic parent materials, especially the Jurassic and Cretaceous rocks, are extensively spread. Jurassic sediments are predominantly distributed on the southern slopes of the Caucasus and in the Eastern Georgia along the axial line of the Caucasus and in the northern slopes. They are represented by clay slates, sandstones, and conglomerates with porphyritic, tuff, and tuffogenic parent materials in some places.

Cretaceous sediments are almost continuously distributed in the southern slopes of the Caucasus and its foothills. In Western Georgia, they are mainly represented by limestones, marls, and dolomites. In the region of their distribution, the karst processes are developed.

The sediments of the Tertiary Period are mostly characteristic of the Lesser Caucasus and intermountainous lowland of Georgia represented mostly by clays, sandstones, conglomerates, and limestones.

The whole Adjara–Trialeti Range mostly consists of the Palaeogenic system, while its certain areas are Neogenic. There are extensively distributed porphyrites, tuffs, and tuffogenic parent materials. There are neogene volcanic systems on the Erusheti Mountain, a major part of the Samsari and Javakheti Ranges, and the bottom of the Akhalkalaki Plateau represented by volcanic lavas—basalts, andesites, andesite-basalts, and tuffs (Fig. 3.2).

The Quaternary sediments are the youngest and most diverse. They are divided into continental and sea deposits. The continental sediments are very diverse. According to their origin, they are volcanic, potamogenic, lakustrine, glacial, cryptaline, etc. Volcanic deposits are mostly characteristic of the South Georgia Pass (Fig. 3.3), Mount Kazbek massif, and Jvari Pass. They are relatively localized on the Lesser Caucasus and are mainly represented by lava

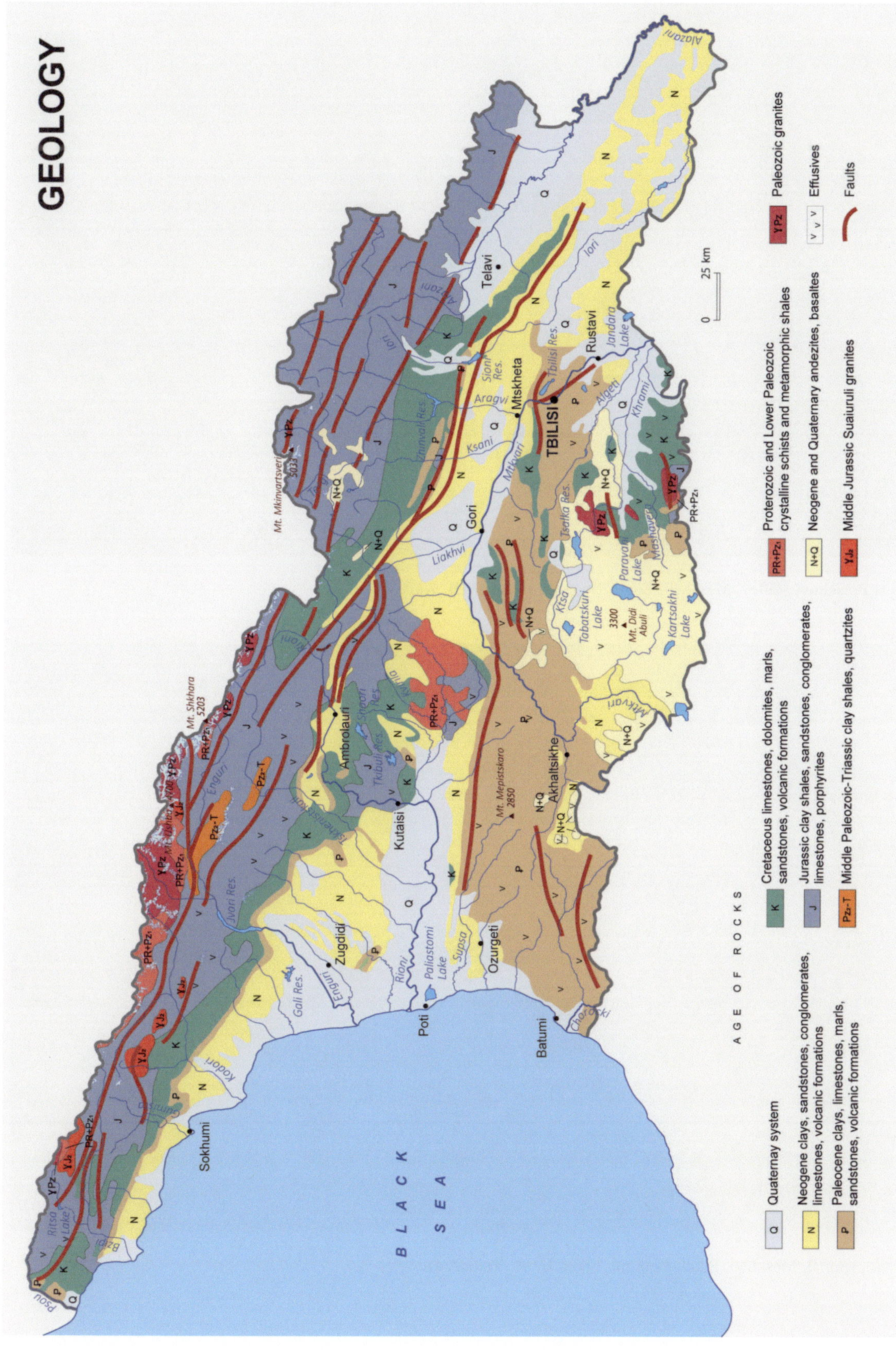

Fig. 3.1 Geologic map (E. Gamkrelidze). *Source* Copyright © Vakhushti Bagrationi Institute of Geography 2008. All Rights Reserved

Fig. 3.2 Volcanic rocks at Potskhovi Valley, Meskheti, South Georgia. Photo by B. Kalandadze

Fig. 3.3 Weathered parent material, Meskheti, South Georgia. Photo by B. Kalandadze

sheet—dolerites, andesites. The potamogenic, lakustrine deposits are mostly distributed in the intermountain lowlands of Georgia. The Colchis Lowland, Shida Kartli, Kvemo Kartli, and Alazani Valley are covered with those deposits. Fluvial deposits depth is rather great and is defined by tens and/or hundreds of meters. The potamogenic deposits are rather intensive defined by tens and sometimes by hundreds of meters. It is represented by riverside cobblestones, sands, and clay soils. Glacial deposits are mainly characteristics of the current glaciers of the Caucasus, though they occur even farther in high-mountainous regions.

On the territory of Georgia, there are two kinds of sea deposits—the Black Sea deposits and the Caspian Sea deposit. The Black Sea deposits are distributed in a narrow line along the coast of Abkhazia, Guria, and Adjara. In the Colchis Lowland, they are covered by the potamogenic deposits. The Caspian Sea deposits are distributed in the Kartli Plain and Iori Plateau.

3.3 Factor of Relief in Soil Formation

The relief as an important factor in soil formation and geographical distribution plays a leading part in the distribution of weathering, heat, moisture, and soil-forming products on the earth's surface. Relief is the "legislator" of soil structure and the basis of soil cartography.

Solar energy and atmospheric precipitations are distributed according to the relief form. Radiation energy is conditioned by the existence of the slopes of different inclination and exposition. In any latitude of the North hemisphere, the northern slopes receive the least radiation in any season of the year and are the coldest. Usually, the temperature of the northern and southern slopes is considerably different. On the southern slopes, the solar radiation increases compared to the horizontal surface, especially in winter. In summer, the steep slopes of southern exposition are less favorable compared to the slightly inclined slopes because the noontime sunrays fall on the earth's surface at an obtuse angle and start sliding across the slope.

On the slopes of different expositions, the peculiarities of soil thermal regime affect the water regime and the character of plants that can cause serious changes in soils in relation to the relief conditions. The soils on the southern slopes are characterized by relatively less humidity and more contrastive temperature regime, while there are colder and moisturous soils formed on the northern slopes. This is especially noticeable in the mountains. The soils on the southern slopes are underdeveloped, often carbonated, and the soils on the northern slopes are more stony, better developed, and deeper.

The climate of different reliefs is different due to the peculiarities of the air mass flows. As a result of the cold air flowing down from the higher places, the climate on the lowland is usually characterized by more contrasts in heat and moisture compared to the slopes and crest. Consequently, the soils are absolutely different there.

The uneven reliefs affect the surface water runoff playing a great role in the distribution of the atmospheric moisture on the earth's surface. The precipitated water flows down the slope from the higher elements to the lower places, so the high watersheds of the relief lose a part of atmospheric moisture of precipitations, while the soils in the lower places receive additional moisture of the water downflowing from the higher places.

As the horizontal surface is never ideally smooth consisting of ups and downs, it distributes the moisture of atmospheric precipitation that is finally reflected in the character of the soil cover. Even the slightest changes in the relief heights can cause a sharp difference in moisturization and the formation of soil that will be clearly reflected in the zones of excess or insufficient moisture.

In addition to the mentioned factors, an absolute altitude of the location is also very important. The height change causes the change of all climate factors: pressure, insulation, temperature, air humidity, precipitation amount, etc. The greater the height the greater the solar radiation and the flow of the biologically active ultraviolet rays. Therefore, in the mountains the temperature decreases by 0.5 °C per 100 m of altitude, on average.

In the mountains, more complex changes in the amount of precipitations are observed. The zone of the most precipitations quite often coincides with the forest belt or the lower part of the subalpine belt characterized by intensive clouds, heavy fogs, and excess rains.

Along with the increase of the height, the climate change also causes vertical differentiation of plants and soils, i.e., the formation of the vertical natural zones. The soil and plant zones successively change each other, forming vertical soil structures. The relief exposition and inclination are the basic distributors of the solar radiation and precipitation that influences the soil water, heat, and oxidation and salination regimes.

The direction of the soil formation is determined by all relief forms such as macro- (plain, plateau, mountain systems) and mesoreliefs (hills, fields, terraces, and their elements) as well as the small forms of the micro-relief occupying insignificant areas. Therefore, in scientific literature, relief is often called an "arbiter of the soils' fate".

The relief, the surface water runoffs, and the depth of the groundwater determine formation of the automorphic soils (when the free water runoffs on the even surface or slope and the groundwaters are at the depth of more than 6 m); semi-hydromorphic soils (when the short-time surface waters or the groundwater are at the depth of 3–6 m, the capillary line can reach the roots of plants); and hydromorphic soils (when the surface longstanding waters or the groundwater are at the depth of less than 3 m, capillary line can reach the soil surface).

In addition, the relief can be a determining factor in the evolution of plants and soils. When the riverbed gradually flows in the floodplain terrace, it transforms into an upper terrace causing the change in hydro-regime and, consequently, the development of soils in automorphic and not hydromorphic or semi-hydromorphic conditions.

3.3.1 Relief

From the orographic point of view, Georgia is quite diverse. There are high ridges, hills, plateaus, flat lowlands, and plains at different levels, deep gorges, and depressions (Fig. 3.4). Absolute altitude of its surface is up to 5068 m asl (Mount Shkhara). In Western Georgia, the lowest point is 1.5–2.3 m below sea level, which is the bottom of the wetlands located between the town of Poti and the village of Kulevi. In Eastern Georgia, the lowest point is at 91.5 m above sea level located in the southeastern part of the Eldari Lowland, at the northwest coast of the Mingechauri water reservoir.

The altitudinal zones of relief are not equally distributed on the territory of Georgia with 20% of the total area of the country falling on high-mountain relief characterized by the intensively fragmented surface and rocky massifs, though steep slopes (with the inclination of more than 35°) are often changed by plain surfaces. About 34% of the total area of Georgia fall on the zone of the middle mountains with the height of mainly 1000–2000 m, where the slopes of steep and medium inclination (20–30°) dominate. The rivers cause deep erosion creating deep V-shaped gorges. Less than 23% of the total area of the country falls on the low-mountain relief, which is within the vertical zone beginning from 500–800 m up to 1000 m with dominating steep slopes (because the transitional zones from mountain to plain are characterized by accumulative processes and well-developed trains). The river gorges are broad and terraced.

The flatlands and the hills of the foothills occupy more than 23% of the total area of the country, represented in the western part of the intermountainous lowland of Georgia at 200–600 m asl and in the eastern part at 400–1000 m asl. The river gorges are very broad. In Georgia, an average height of the surface is 1508 m asl, which is 1691 m asl in Eastern, and 1314 m asl in western Georgia.

According to the relief, the Caucasus is particularly distinguished. While the distance between the peaks and the settlements of the river gorges is just about 9–16 km, their relative height is often more than 3300 m.

The territory of Georgia is divided into four major orographic units: the Caucasus Mountains, the intermountainous lowland of Georgia, the Lesser Caucasus Mountains, and the volcanic upland of Southern Georgia (Gobejishvili 2010; Geomorphology of Georgia 2018; Tielidze et al. 2019).

The Caucasus Mountains are located in the northern part of Georgia, occupying about 1/3 of the total area of the country. The meridians of Elbrus and Mount Kazbek divide the Caucasus into three parts: Western, Central, and Eastern Caucasus. The highest peak in the Western Caucasus is Dombay-Ulgen (4046 m), in the Central Caucasus it is Shkhara (5068 m) and in the Eastern Caucasus Mount Kazbek (5033 m) (Fig. 3.5). Among those parts of the Caucasus, the Central Caucasus is the highest, which is characterized by volcanic masses and an almost continuous line of current glaciation. From the orographic point of view, the Great Caucasus is divided into three parts: the main watershed ridge of the Caucasus and its north and south slopes. Among them, the core unit is the watershed ridge of the Great Caucasus stretching out from the northwest to the southeast. The highest point of the ridge is the Bezeng Wall with Shkhara, the highest peak of Georgia (5068 m). Shkhara is the second highest mountain in the Caucasus (5648 m) after Elbrus. Other peaks to be worth mentioning are Janga (5049 m), Katyn-Tau (4970 m), Shota Rustaveli Peak (4960 m), Histola (4860 m), and Lalveri (4350 m). From the Bezengi Wall, the height of the ridge is gradually decreasing to the west and to the east.

The western part of the main watershed ridge of the Caucasus is the watershed for the rivers of the Sea of Azov (river Kuban) and the Black Sea, while the middle part is the watershed of the Black Sea and the Caspian Sea. A part of the Caucasus from the mount Zekari to mount Tinovroso is the watershed for the rivers flowing into the Caspian Sea in the north of the Absheron Peninsula (Tergi, Sulak) and in its south (Mtkvari). According to the natural peculiarities, the Caucasus is divided into the Caucasus of Abkhazia, Svaneti, and Racha.

A significant section of the northern slope of the Caucasus includes a part of a parallel ridge extending northwards, as well as hollows and gorges surrounded by the main watershed and northern ridges (Fig. 3.6).

Among the depressions, the Tusheti depression is worth mentioning, which latitudinally stretches over 42 km. It is one of the largest depressions in the Caucasus covering the area of almost 800 km^2. The Truso depression is also large located in the upper part of the river Tergi gorge.

The southern slope of the Caucasus is different from the northern slope. There are no similar orographic units of the lateral ridge. The ridges are mainly the branches of the main watershed of the Caucasus. The western part of the Caucasus is mostly characterized by latitudinal ridges, while the eastern part is characterized by the ridges of longitudinal orientation. They are connected to the principal ridge with

Fig. 3.4 Physical map. This map is created by D. Svanadze

Fig. 3.5 Sunrise on Peak Kazbegi (Mkinvartsveri), Stepantsminda, North Georgia. Photo by B. Kalandadze

Fig. 3.6 North slope of the East Caucasus Range, Gergeti. Photo by B. Kalandadze

Fig. 3.7 Dariali Gorge, Tergi River Basin, Greater Caucasus. Photo by G. Dvalashvili

one end. One more orographic difference between the southern and the northern slopes is that the depressions on the southern slope occupy larger area.

The rivers of the Caucasus flow through the gorges of narrow and steep slopes where they often form the ghats and narrows. The most important of them is Dariali Gate in the Tergi River basin (Fig. 3.7), the Enguri narrows on the river Enguri and others.

The intermountain lowland of Georgia is located between the Caucasus and the Lesser Caucasus. From the orographic point of view, it is diverse, where the plains and lowlands, hilly foothills, low-mountain reliefs, and some middle-mountain reliefs occur. The lowland is the territory located at 500 m above sea level, though the middle ridges of Likhi and Gombori are considered with it due to the central location of the mentioned ridges in the intermountain lowland of Georgia. However, their landscapes are more similar to the Caucasus and the Lesser Caucasus.

Orographically, the intermountain lowland of Georgia is divided into western, central, and eastern parts.

The western part or the Colchis Lowland is sloped to the Black Sea. It comprises a hilly line of Colchis and seaside lowlands. The central part of the Colchis Lowland is triangular with its bottom rested on the sea. Hypsometrically, the lowest is the line of the coastal lakes and wetlands, the bottom of which is below sea level. The western and the central parts of the lowland are quite flat, slightly sloped to the west, while the surface of the outer part is more sloped. Also, there are some small hills in some places. The large rivers (Enguri, Rioni, Tskhenistskali, Kvirila) formed the lowland. The absolute height of the relief is 100–150 m, on average.

On the extension of the Colchis Lowland, there are coastal plains along the seaside. In the north, there is the Pitsunda plain and in the south the Kakhaberi and Gonio plains. Near the Colchis Lowland there is stretched out a hilly line of the foothill. The northern hilly line is relatively wide. The central part of the lowland of Georgia is small. It is represented by Imereti Upland, which is asymmetric: the western part is wider than the east one. The highest is the

Likhi (Surami) Ridge, which is the watershed of the Black Sea and the Caspian Sea rivers. It is of medium height (1500–1600 m) and intensively the relief is cut by gorges and canyons.

The eastern part of the intermountain lowland of Georgia (Eastern Georgia) is the largest area and orographically more versatile. There are several important orographic units: Shida Kartli, Kvemo Kartli and Alazani Valleys, Iori Upland, Eldari Lowland, and Tsiv-Gombori Ridge.

On the territory of Georgia, the Lesser Caucasus is represented by its northwestern part. It is quite different from the Caucasus as it does not have a shape of a uniform watershed. Orographically, it is divided into several parts: Meskheti, Trialeti, Eastern Ponto, Shavsheti, Arsiani, Lokhi Ridges, the Shuakhrami Mountain Massif, Akhaltsikhe, and Adjara depressions (Fig. 3.8).

The upland of Southern Georgia represents a northern part of the Transcaucasus Upland. In geologic and geomorphological terms, it greatly differs from the Lesser Caucasus, because there is an upland relief developed in the past geological epochs in the result of intensive volcanic eruptions, but from the orographic point of view it is quite diverse. The central part of the upland is a plateau, and the periphery is represented by the outermost ridges. Like the Lesser Caucasus, it does not have the shape of a uniform watershed. Orographically, it contains several units: Javakheti Plateau, Erushti Highland, Samsara, Javakheti, and Nialisquri Ridges.

As for the impact of the relief forms on the soil surface temperature, it is subject to A. Voevikov Law, implying that positive relief forms increase the annual and daily temperature amplitudes, while negative forms reduce them. It is established that in autumn and winter, in Georgia, the plateaus and sea coastal plains are warmer than mountain slopes, basins, and intermontane plain. During the spring, the basins get warm swiftly and the temperature of the soil surface is higher than that of other relief forms, summer temperatures are somewhat higher over the intermontane and sea coastal plains and are minimal over the mountain slopes and plateaus (Elizbarashvili 2017). The differences in the soil surface temperatures are also seen with the slopes with different expositions. The distribution of air temperature in different morphographic conditions is subject to the same regularity. The maximum annual amplitude of the soil surface temperature is typical to the basins and intermontane plain and it is 26 °C on average, with its minimum of 23 °C typical to the soils of the sea coastal plain. The minimum temperature amplitude on the sea coastal plains can be explained by the soil and climatic conditions of Colchis Lowland, where the soils contain large amount of water and

Fig. 3.8 The Alpine Zone, Kvabliani Gorge, Meskheti, South Georgia. Photo by B. Kalandadze

have great inertia as a result. Due to this, their warming and cooling are slower processes compared to the soils found in other morphographic conditions.

According to Bockheim et al. (2014), the most important factor to define two soil orders (Aridisols and Gelisols), used at the highest level in Soil Taxonomy, is soil climate.

3.4 Factor of Climate in Soil Formation

Usually, identification of the most common regularity of soil geography is related to climate, because the climate directly affects soil formation controlling all its phenomena. Solar radiation is the main source of energy for a geographic area, including the soil cover.

The influence of climate on soil formation is largely through the combined effects of water and temperature, although wind and solar radiations also play important roles (Graham and Indorante 2017).

Atmospheric climate implies the average state of the atmosphere in the given area, characterized by the average indices (temperature, precipitation, air humidity, etc.) and greatest indices of meteorological elements providing day-night, seasonal, and annual fluctuation amplitudes.

Since the rhythms of solar energy reaching the earth's surface are of different duration (daily, seasonal, annual, perennial), the soil experiences alteration of the processes of warming and cooling, freezing and melting. Different combinations of such processes together with their specific features determine the thermal regime of soils.

Thus, the climate as a soil-forming factor is very important in various aspects. It plays a significant role in the development of biological and biochemical processes. Certain combinations of the temperature and moisture determine the type of vegetation, the rate of formation and decomposition of organic substances, the composition and intensity of activity of soil microflora and fauna. Through the character and composition of soil, the atmospheric climate greatly affects the air–water, temperature, and oxidation regimes in soil. The processes of transformation of mineral compounds in soil (the direction and rate of weathering, accumulation of soil-forming products, etc.) are closely related to climatic conditions. The climate has a great impact on the processes of soil erosion caused by wind and water.

Identification of the major thermal groups of climates is based on the sum of average daily temperatures (sum of active temperatures) of the vegetation period.

Distribution of sediments by seasons, intensity of precipitations, relative humidity of air, and the rate of airflow plays an important role in soil formation. All these phenomena affect the peculiarities of the soil processes and determine the development of the erosion of soils by water and wind.

The climate has a direct and indirect impact on soil formation. The direct effect is the soil moistening, heating, cooling, while the indirect effect—flora and fauna.

3.4.1 Climate

A lot of publications are devoted to the study of the climatic parameters assessment of Georgia, in particular, temperature regime, zoning, ecological monitoring, climate change, anthropogenic influence, environment transformation, soil-atmospheric system, etc. (Kordzakhia 1961; Javakhishvili 1977, 1992, 2000; Elizbarashvili et al. 1992; Elizbarashvili and Sulkhanishvili 2002; Elizbarashvili 2007, 2017; Begalishvili et al. 2009; Beritashvili et al. 2010; Gunia 2011; Lagidze et al. 2017a, b and others).

As a rule, there are many factors influencing the climate formation: the latitudinal location of the area, the hypsometry, the circulation of the atmosphere, the substrate's surface, etc.

Georgia is situated in the extreme northern part of the subtropical zone (the border between the subtropical and moderate belts passes through the main watershed ridge of the Caucasus). Georgia is in such an area of the Earth's surface, where the western winds (latitudinal) dominate in lower layers of the troposphere almost all the year round.

It is known that climate change impacts the whole world, including Georgia (GSNC to the UNFCCC 2009; Matchavariani and Lagidze 2012). Problem of soil degradation, desertification, salinization, erosion, and chemical pollution is the main results of climate change. A number of vulnerable regions and economic sectors are identified. Adaptation of critical ecosystems is a priority for Georgia (GSNC to the UNFCCC 2009).

The three most vulnerable to climate change ecosystems has been revealed in Georgia: the Black Sea coastline—zone of tourism development; high-mountainous region—identified as a vulnerable area to disastrous events as snow avalanches, landslides, mud torrents, erosion, which is damaging forests and agriculture; desertification zone, one of the main agricultural areas—identified under the threat of degradation (Beritashvili et al. 2010).

The best indicator of climate change in the Caucasus Mountains is increasing of glaciers number that is connected with their partitions and retreating back because of their thawing (GSNC 2009; Tielidze et al. 2018). In addition, relative elevation of sea level on the eastern coast of the Black Sea, provoking the flooding processes of the lowland territories, is caused by extensive melting of glaciers in the mountain regions.

Air masses enter the territory of Georgia from two different sides. When they enter from the west the air temperature falls, the weather becomes cloudy and rainy.

Crossing the Black Sea, it gains a large amount of moisture causing large atmospheric precipitations mainly in Western Georgia. While in Eastern Georgia (after crossing the Likhi ridge) those air masses are depleted of moisture. The air masses enter from the west all the year round, especially in the warm period of the year.

The influence of air masses invaded from the east is mainly observed in Eastern Georgia marked with the rainy weather. The invasion of the air masses from the east is more common in the cold period of year. Invasion of air masses from two sides occurs when a strong anticyclone is developed on the Russian Plain and a low pressure in the South Caucasus and Asia Minor. At that moment, the air masses simultaneously move around the Caucasus Ridge from two sides—from the west and east. They meet mainly in Eastern Georgia. Such a circulation of the atmosphere is not frequent; it occurs once or twice a year mostly in winter.

The orographic barriers also play a significant role in determining the climatic peculiarities. In this regard, the mountainous region of the Caucasus is worth mentioning, which hinders the penetration of the northern cold air masses in the territory of Georgia (and generally in the South Caucasus). Also, it changes the direction of the air masses penetrated in the territory of the country from other sides, bypass of the Great Caucasus Range. However, sometimes the cold air masses reach Georgia through some river gorges of the Caucasus, though their influence is not great. The airflow coming from the north crosses the Black Sea region and the warm surface of the East Transcaucasia Lowland and spreads to the territory of Georgia in a relatively warm form. It is the Caucasus that determines that the territory in the south of the main watershed is within the subtropical belt, and the territory in the north of the Caucasus is in the moderate belt. If there was not that orographic barrier, the climate would be much colder in Georgia, which would be particularly felt during the cold period.

In Georgia, the other important orographic barriers are also Likhi and Arsiani Ridges hindering the distribution of the humid air masses coming from the West. Therefore, there are more precipitations in the West Georgia than in the Eastern Georgia. The border between the wet and dry subtropics passes right across the spine of the Likhi and Arsiani Ridges.

The Lesser Caucasus (Anticaucasus) is also a rather significant orographic barrier. It somehow retards the impact of the hot air masses coming from the south. If there was not such an orographic barrier, it would be relatively hotter and drier climate in Georgia, especially during the warm period of the year.

The Black Sea has a significant influence on the climate of Georgia. The location of the country on the eastern coast of the Black Sea determines a warm and humid climate in the western part of the country. Moving over the sea surface to the territory of Georgia, the air masses get saturated with moisture (even in the winter, since the sea does not freeze and the water is intensely evaporated).

The upland of the Transcaucasia, which represents the so-called "Coldness factory", makes the climate of Georgia colder to some extent—in the cold period of year the cold air masses move from the upland to the direction of the lowlands of Georgia.

The adjacent dry valleys, deserts, and semideserts of the Mtkvari-Araksi Lowland, Asia Minor, and Middle Asia are reflected on the climate of the lowlands of eastern Georgia, which is characterized by the surface overheating in summer.

The Transcaucasian intermountainous lowland is favorable for free motion of air masses and ventilation; therefore, it is called the "Transcaucasian Corridor".

In Georgia, an average annual sunshine duration is 1300–2500 h, which is the longest in the Shiraki Valley and Gardabani Plain, and relatively less on the Colchis Lowland due to frequent clouds, and the least in Adjara and in the mountainous highlands of the Caucasus. Quite a different picture is observed in the mountainous regions of southern part, where the sunshine duration is longer.

The greater the absolute height is, the greater the air transparency and the total radiation. In the lowland, it is 120–130 kcal/cm^2 and in the mountain—150–155 kcal/cm^2. The radiation balance in the lowland of Georgia is 45–60 kcal/cm^2 and 10–20 kcal/cm^2 in the mountain. Radiation balance is positive throughout the year except just some of the mountainous regions. According to the radiation rate, the Caucasus, Lesser Caucasus, and Javakheti Plateau are slightly different from each other.

On the territory of Georgia, the air temperature changes in accordance with the increase of the absolute height. The temperature is falling by 0.2–0.9 °C per 100 m, on average. Annual air temperature amplitude is minimum on the Black Sea coast (16–17 °C), and maximum on the Eldari Lowland (26–27°). In the high mountains of Georgia, an average annual temperature of the air does not exceed 2–5 °C. In mountainous uninhabited areas, it is even lower.

An average annual air temperature (Fig. 3.9) is the highest on the Black Sea coast (+14, 15 °C). In the intermountain lowland, this figure is gradually decreasing from the west to the east. It is +13, 14 °C in the hilly line of Colchis, and 9–13 °C on the Shida Kartli Valley.

Winter in the lowland of Western Georgian is relatively warm. At 600–700 m above sea level, an average temperature never goes down below 0° in January. It is particularly warm on the Colchis Lowland and adjacent hilly line, where an average temperature of the coldest month (January) is from +4 to +5 °C and it is from +5 to +6 °C on the Black Sea coastline. Therefore, warm, sunny weather lasts long enough in winter. That determines the widespread of the humid subtropical cultures are. The farther we go from the

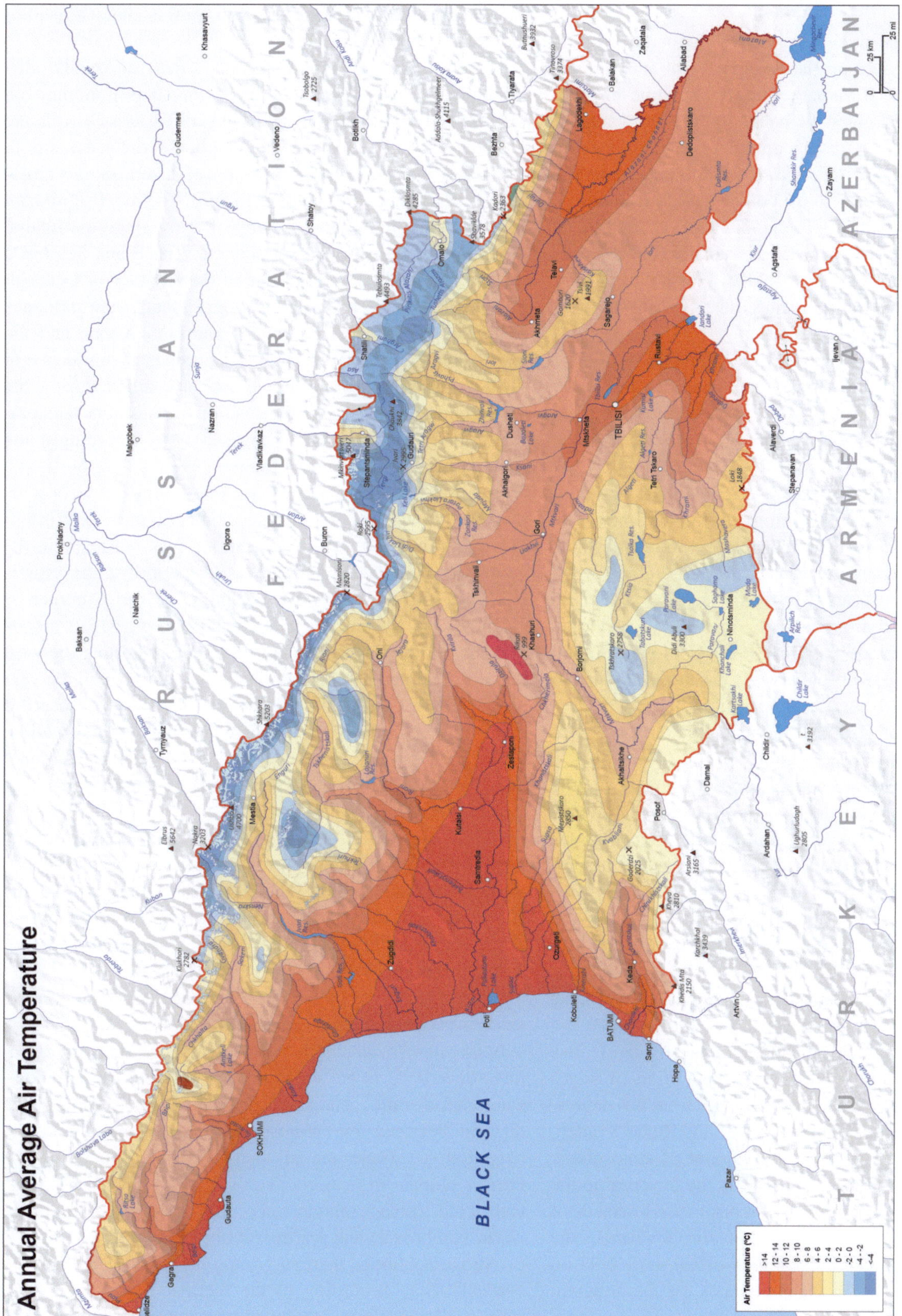

Fig. 3.9 Annual average air temperature (D. Mumladze, G. Gagua). *Source* National Atlas of Georgia. Stuttgart: Steiner-Verlag, 2018

Black Sea coast the lower the average temperature of January.

In Eastern Georgia, an average highest temperature of the coldest month is marked on the Kvemo Kartli and Alazani Valleys/Plains, where it varies around 0 °C while in other areas of eastern Georgia it is below 0 °C. Compared to western Georgia an average temperature of January is about 1–2 °C lower in eastern Georgia at a similar height. The lowest average temperature in the coldest month is observed on the Iori Upland (−3.8 °C). In the lowlands of Georgia, an average temperature of January is negative just in its outermost eastern part. On the Javakheti Plateau, the winter is characterized by cold and severe climate, where an average temperature of January is −8 °C.

The warm and humid summers are characteristic of the Colchis Lowland and the adjacent hilly line. On the Black Sea coast, an average temperature is from +22 to +23 °C in July, but in case of invasion of the hot air masses from the south it reaches even 40 °C.

In the warmest month, an average temperature is gradually rising in the east direction. Summer is hot and dry on the lowland of eastern Georgia. With the increase of the absolute height, the air temperature decreases. At 2000 m above sea level an average temperature is from −6 °C to −8 °C in January and in July from +10 °C to +12 °C, while at 3500 m above sea level, it is from −14 °C to −16 °C and from +3 °C to +5 °C, respectively. At the mountain station of Stepantsminda (Kazbegi) it is −15 °C.

On the territory of Georgia, the annual amount of atmospheric precipitations varies within 400–4500 mm. It is a minimum on the Eldari Lowland and maximum on the mountain Mtirala (on the slope of the Chakvi Ridge overlooking the sea), where the atmospheric precipitations exceed even 5000 mm. Western Georgia is richer in precipitations compared to the eastern Georgia. The abundance of precipitation is characteristic of the Colchis Lowland and the adjacent hilly line, where its amount is no less than 1000 mm in any place. The Abkhazian Caucasus is considered as another pole of abundant precipitation, where its amount is over 3000 mm. In western Georgia, there are places, where atmospheric precipitations are relatively less (less than 1200) due to the presence of the barriers of ridges forming the so-called "Rain shadows", such as the depressions of Adjara, Racha, Kvemo Svaneti, and Zemo Svaneti.

On the territory of Georgia, the amount of atmospheric precipitations decreases from the west in the direction to the east. An average amount of precipitations in the lowlands of eastern Georgia is 400–500 mm. In the lowlands, the greatest amount of precipitations (800–1000 mm) falls on the Alazani Valley and in the mountains on the southern slopes of the Kakheti Caucasus.

With the increase of the altitude, the amount of atmospheric precipitation increases (1500–2000 m) at first and then decreases. Relatively less precipitation falls on the Javakheti Plateau and the least (400 mm) is characteristic of the Akhaltsikhe depression.

Atmospheric precipitations are often unequally distributed in terms of seasons. In Spring, it is minimum on Colchis Lowland, while the lowlands of eastern Georgia are characterized by abundant precipitations in the same period.

As for the wind rates in the lower layers of the atmosphere, it depends on plenty of factors. The most significant of them is the ununiform heating of the sea and the inland, the nature of the subfield surface, etc. The Black Sea has a great influence on the wind rate in the territory of Georgia.

In winter, the territory of Georgia is under the influence of the western branch of the Siberian anticyclone and the anticyclones developed in the West Europe. The influence of the Siberian anticyclones is no longer marked during the warm period of the year. Because of the unstable temperature regime of winter, the air pressure (and accordingly, the wind) varies more than in summer.

According to the wind pattern, the territory of Georgia is divided into western, eastern, and southern parts. In Western Georgia, the east winds of descending type dominate in the cold period of years. And the warm period of the year is characterized by the west ascending wind. Summer is characterized by the sea winds—the breeze, which moves into the depth of 130–135 km. In eastern Georgia, the west and north winds of descending type blow almost all the year round. South Georgia is characterized by the low speed of wind. In the winter, there is mostly an anticyclone situation, where the west and southwest winds prevail, and in summer there is a cyclone situation and winds of opposite direction develop.

Wind speed is different in regions of Georgia. It is a maximum in high mountains. The highest average speed of wind is recorded in the highlands of the Caucasus in the sub-nival and nival belts. Winds of minor strength are observed in the lowlands, deep gorges, and depressions of eastern Georgia.

As a rule, in the climate description, a particular importance is given to thermal groups, so called the sum of active temperatures (meaning the sum of average daily temperatures of >10 °C in the vegetation period) and precipitation–evaporation ratio, so-called aridity index (meaning the relation between the atmospheric precipitations and the evaporation). There are all major thermal groups of the Earth's climate in Georgia (Table 3.1), except the hot tropical: cold (polar); moderately cold (boreal); moderately warm (subboreal) and warm (subtropical).

Abovementioned thermal groups are distributed along the vertical belts. According to the thermal groups, the main soils of Georgia are distributed as follows: cold (polar)—Leptosols Umbric, Leptosols Molic; moderately cold (boreal)—Cambisols Dystric, Chernozems; moderately warm

Table 3.1 Climate thermal groups

Climatic groups	Sum of air temperature >10 °C
Cold (polar)	<600°
Moderately cold (boreal)	600–2000°
Moderately warm (subboreal)	2000–3800°
Warm (subtropical)	3800–8000°

Table 3.2 Aridity index groups

Climatic group	Humidity coefficient
Extra-humid	>1.33
Humid	1.33–1
Semi-humid	1–0.55
Semiarid	0.55–0.33
Arid	0.33–0.11

(subboreal)—Acrisols Haplic, Vertisols, Cambisol Chromic soils); warm (subtropical)—Nitisols Ferralic, Acrisols Haplic, Luvisols Albic, Kastanozems.

As for the humidity factor, i.e., precipitation–evaporation ratio (aridity index) of the main six climatic groups on the Earth, mostly extra-humid, humid, semi-humid, and semiarid groups are spread in Georgia (Table 3.2). An arid group is spread over small areas in the southeastern part of the country (Fig. 3.10), and there is no extra-arid group in the territory of Georgia. Figures 3.10 and 3.11 show the aridity index and the annual average evaporation in Georgia.

According to the humidity coefficients, the major part of soils in Georgia are distributed as follows: the extra humid climatic group consists of Nitisols Ferralic, Acrisols Haplic, Luvisols Albic, Leptosols Umbric, Leptosols Molic; the humid climate group—Acrisols Haplic, Cambisols Dystric; the semi-humid group—Vertisols and Chernozems; and the semiarid climatic group—Cambisol Chromic, Kastanozems.

The humidity coefficient with respect to one is significant. If the humidity coefficient >1, the soil is acidic, and if it is <1, then the soil is alkaline, carbonate, and sometimes saline.

The thermal regime of soil is mostly determined by the radiation and thermal balance values of the soil surface. The radiation absorbed from the surface transforms into the thermal energy, warms the soil surface, and the heat from the soil surface is transferred to the depth of soil and adjacent atmosphere layers.

Soil thermal properties depend on a number of factors: altitude of the Sun, atmospheric transparency, altitude of the location above sea level, soil texture and color, vegetation cover, etc. However, the major factors are thermal capacity and thermal conductivity of the soil.

As for the heat exchange in the soil-atmospheric system, usually, the solar radiation acting on the soil surface determines the thermal regime of the underlying surface. The heat from the warmed earth's surface is transferred to the atmosphere. Heat is transferred from the soil to the atmosphere by means of molecular thermal conductivity, turbulent mixing, thermal convection, radiation thermal conductivity, moisture evaporation, and further condensation, with the major role played by turbulent mixing and thermal convection. It is mainly under the influence of these processes, the ratio between the soil and the atmospheric temperatures is formed.

Overall, on the territory of Georgia, for the most of the year, when the radiation balance is positive, soil is warmer than air, while in winter, due to radiation, soil loses heat and it cools more than air.

The soil temperature in Tbilisi approximately from mid-February to mid-November exceeds the air temperature. An opposite picture is observed in other seasons of the year: the air temperature is higher than the soil temperature, i.e., negative thermal exchange takes place. An almost similar state is observed in Batumi and Dedoplistskaro. Across Jvari Pass (East Caucasus), the air warms up for a short period of the year, from the end of May to the end of October. For most of the year, negative thermal exchange dominates. Such a difference in the formation of the thermal regime of the soil-atmospheric system is the result of the different thermal properties of soils.

A decisive role in the occurrence of the periods of positive and negative thermal exchanges between the soil and the atmosphere is played not by the altitude of the location, but by the soil type. In particular, in the mountainous zone of Adjara, the period of positive thermal exchange lasts for 8 months, from March through October, while on the Black Sea coast (Chakva), it lasts for 7 months. A long thermal exchange period in Khulo can be explained by a high value of the soil-warming coefficient. A soil-warming coefficient is

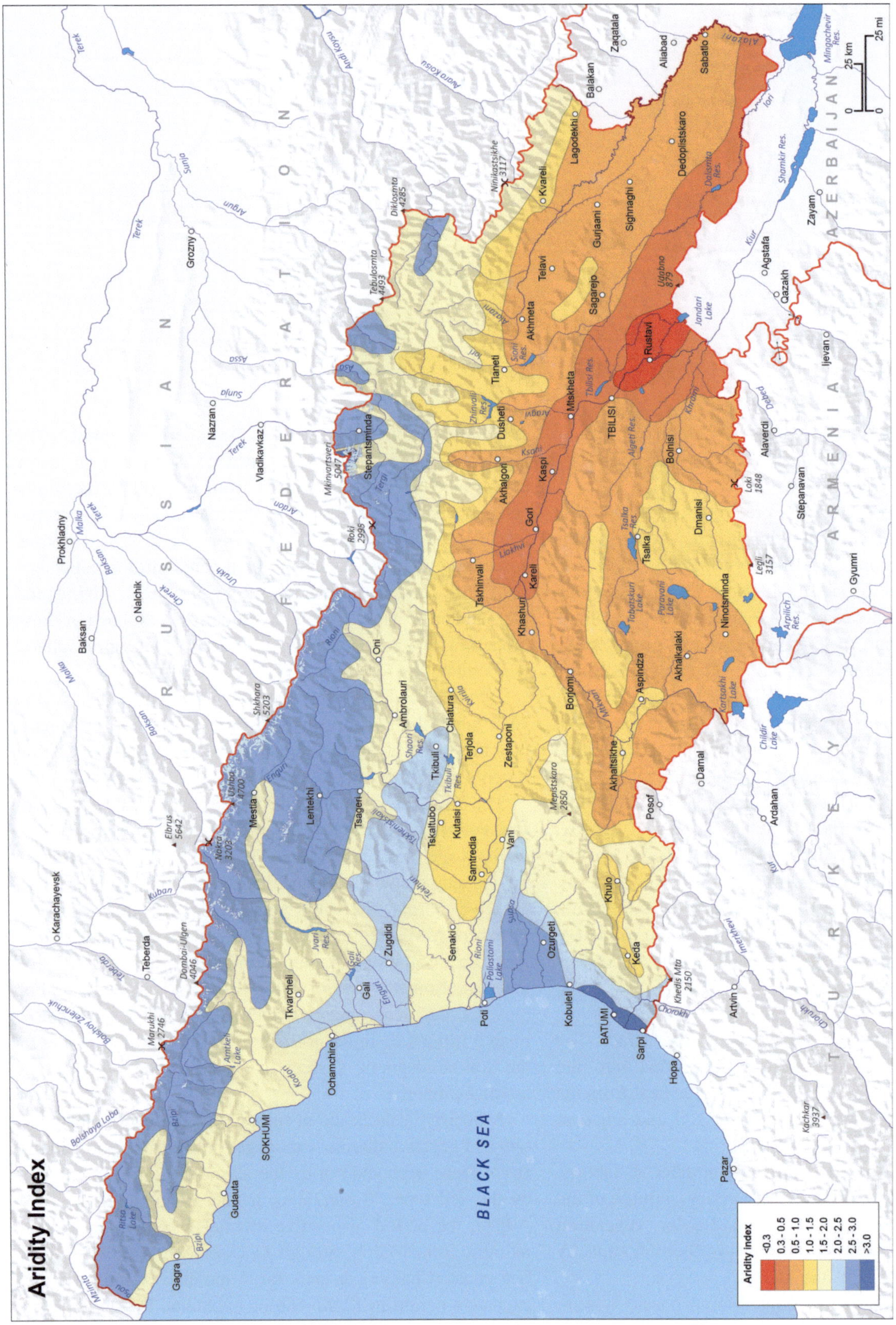

Fig. 3.10 Aridity index in the territory (G. Gogichaishvili). *Source* National Atlas of Georgia. Stuttgart: Steiner-Verlag, 2018

Fig. 3.11 Annual average evaporation in the territory (G. Gogichaishvili). *Source* National Atlas of Georgia. Stuttgart: Steiner-Verlag, 2018

the ratio between the duration of a frost-free period on the soil surface and the duration of the same period in the air.

The duration of the frost-free period on the soil surface is less than it is in the air. Their ratio on the territory of Georgia varies between 0.65 and 0.95. It can be used to describe the soil-warming velocity and it is therefore, called soil-warming coefficient (Elizbarashvili 2007).

The soil-warming coefficient on the Black Sea coast is approximately 0.9 and it is slightly higher on Kolkheti Lowland, reaching its maximum in the steppes and semi-deserts of East Georgia (>0.95). The soil-warming coefficient in the mountains decreases and is less than 0.8 in the Caucasus Range. It is minimum (>0.7) in the most humid regions of West Georgia and glacial-nival zone. The soil-warming coefficient is a good indicator of the main climatic peculiarities of the area, and based on the soil-warming coefficient, the highest soil surface temperatures in summer are fixed in East Georgian plain and on Colchis Lowland.

According to Elizbarashvili (2017), in January, the highest soil surface temperature (>4 °C) is fixed on the Black Sea coast and Colchis Lowland. Average temperature in January in Shida Kartli and Kvemo Kartli, as well as on Kakheti Plain varies between +2 and −2 °C and falls to −4 °C on Alazani Valley. In the mountains of South Georgia, the soil surface temperature falls to −12 °C and it falls to −16 °C in the high-mountainous zone of the Caucasus. In July, the warmest are the soils in Kvemo Kartli, on Alazani Valley and on the Black Sea coast. In July, average monthly soil surface temperature in these regions varies within the range of 26–30 °C. In the mountains of South Georgia and over Meskheti Ridge, the soil temperature falls to 16 °C and it falls to 10–14 °C in the high-mountainous zone of the Caucasus.

Depending on the thermal properties of soils spread on the territory of Georgia, there are four geothermal regions identified: very warm, moderately warm, moderate, and cold soil regions (Elizbarashvili 2017). Very warm soils occupy almost all the area of Colchis Lowland and part of its piedmont, as well as Alazani Plain, Iori Plateau and lowland areas of Kvemo Kartli. The temperature of soil surface here during the warm period of the year (April–October) is 22 °C and it is more than 20 °C at the depth of 20 cm. The majority of the territory of Georgia, including the Caucasus and piedmont of the mountainous area of South Georgia, Imereti Plateau, Shida Kartli Plain, etc., is covered with the very warm soils. In these regions, the soil surface temperature is 15–20 °C and it is more than 20 °C at the depth of 20 cm. Warm soils are spread within a narrow strip of the Caucasus Range and in a vast area of the mountainous region of South Georgia, where the temperature in the warm period of the year both, at the soil surface and at the depth of 20 cm, is 10–15 °C. A large area of the Great Caucasus Range is covered with moderately warm and cold soils, where the soil surface temperature is <10 °C and it is 0–10 °C at the depth of 20 cm.

A dependence of soil surface temperature on the altitude and morphographic properties of a site is of no less interest. The soil surface temperature, like air temperature, decreases regularly at greater altitudes. The vertical gradient of the soil surface temperature essentially depends on the soil type and season of the year. The greatest gradient for Nitisols Ferralic, Cambisols Dystric, and Rendzinas is fixed in January (0.8–1.0 °C for every 100 m altitude), and the greatest gradient for Chernozems and Cambisols Chromic is fixed in July (0.7–1.0 °C). The minimum gradients of soil surface temperature in January are typical to the Vertisols and Cambisols Chromic (0.5 °C), while in July, the minimum gradients are typical to Nitisols Ferralic, Cambisols Dystric, and Rendzinas soils (0.5 °C for every 100 m altitude).

Such a nature of the soil surface temperature variation is in good compliance with the regularities of the air temperature variations and can be explained by the climatic peculiarities of Georgia (Elizbarashvili 2007). In particular, in winter months, in the areas of Cambisols Chromic soils and Chernozems in East Georgia, in terms of anticyclone state of the atmosphere, efficient radiation and temperature inversions become stronger and reach such intensity that the air temperature in the layer at the height of 1000–1200 m exceeds that in the lower layers even in the long-term regime. Due to the negative thermal exchange between the soil and the atmosphere, what is typical to East Georgia (Elizbarashvili et al. 2007), the soil surface temperature reduces as well. In the area of Nitisols and Rendzic Leptosols in West Georgia, the inversion occurs in the warm period of the year, the reason for which is the transfer of warm air masses formed at the sea during the winter.

As regards the surface temperature regime of the different types of soils in Georgia, it is known that in summer, up to 500 m altitude, it is Acrisols Haplic soils and Chernozems warming best of all; at 500–1500 m altitude, particularly warm are Cambisol Chromic and Rendzinas, but Nitisols Ferralic and Cambisols Dystric soils are colder; above 1500 m, the warmest is Leptosols Rendzic and the coldest is Vertisols. In winter, up to 500 m altitude, the warmest are Nitisols Ferralic and Rendzinas soils; Cambisol Chromic soil is the warmest and Cambisols Dystric is the coldest at 500–1500 m altitude; above 1500 m altitude, the warmest is Chernozems and the coldest is Rendzinas and Leptosols Umbric soils.

A decisive role in the formation of the soil surface temperature is played by the climate. The impact of the climate on the soil temperature is determined by the nature and distribution of precipitations and snow cover, as well as wind anisotropy and evaporation heterogeneity. The Nitisols Ferralic soils, which are spread in the humid subtropical zone of West Georgia, and which have high moisture content, warm less than the Cambisol Chromic soils of the continental part of Eastern Georgia. In winter, due to the continental nature of the climate, the Cambisol Chromic cools faster than the soils in the subtropical zone.

Heat propagation deep in the soil depends on the soil surface temperature, humidity, properties of the snow cover, and thermal conductivity, thermal capacity and structure of soil, etc. The temperature of the upper soil layers (up to 20 cm) correlates with the soil surface temperature (Elizbarashvili et al. 2007). 1 °C change of the soil surface temperature results in the temperature change even at a depth of 20 cm. This is namely true with the temperature change of the soils of the subtropical zone of West Georgia by 0.5 °C and by 1.5 °C of dry soils in East Georgia.

Heat propagation through the soil is characterized by a deep temperature gradient. The least temperature gradient for the soils of the intermountain depression of Georgia is typical to Nitisols and it is 0.5–1.1 °C for every 10 cm evidencing its good thermal conductivity. This is the result of the fact that this type of soil is formed in terms of high moisture content, hot summer and warm winter and maintains the warmth during the cold period of the year. The maximum gradients in summer are fixed with Vertisols in East Georgia and they are 1.0–1.3 °C, as a result of the soil being formed in the continental climate and being dry. In the mountainous regions, the least gradients of soil temperature in summer at the depth of 10 cm are typical to Cambisols Chromic (0.4–0.9 °C), while maximum gradients are typical to Cambisols Dystric (0.5–1.3 °C). The temperature gradients are almost the same in Rendzinas (both, in West and East Georgia) despite their different climatic conditions. Thus, the temperature regime of the upper soil layers, unlike the soil surface temperature regime, depends on the type of the soil, and the climate peculiarities of the location play only a secondary role.

The daily amplitudes of soil temperature have an annual course, both on the surface and at any depth. In Tbilisi, the maximum amplitude at the surface is fixed in the summer (37.5 °C) and the minimum amplitude is fixed in winter (14.7 °C). As the depth increases, the amplitude of soil temperature decreases. As compared to the soil surface, the minimum temperature is retarded by 1 h at every 5 cm depth, while the maximum temperature is retarded by 1.5 h (Kotaria 1992).

In the cloudy and rainy weather, the daily amplitudes of the soil surface and its adjoining layers as compared to the clear days decrease significantly. This is particularly true with summer months (10–20 °C). At the depth of 40 cm and more, the daily variation of winter and summer temperatures is almost the same and there is no difference between the amplitudes at the depth of 70–80 cm.

The daily temperature conditions and amplitudes of soil are influenced by the soil structure and slope exposition. The looser the soil, the less it conducts the heat, as more air accumulates in its pores. This is why the daily amplitude of the surface temperature of loose soil exceeds that of the leveled and compacted soil.

The average annual temperature of the bare soil surface in Tbilisi is 17.5 °C what is 4.8 °C more the average annual temperature. This difference is the least in December (0.4 °C) and is the greatest in July (9 °C). As per Kordzakhia (1961), a similar situation is observed in other areas of East Georgia. In the lowland areas of West Georgia, the difference between the average annual temperatures of soil and air is 0.5–1.0 °C. In terms of humid climate, the difference between the soil and air average annual temperatures is less than it is in terms of dry climate. The annual amplitude of average soil temperature in Tbilisi is 7.5 °C higher than that of air.

The thermal conditions of soil are basically affected by the natural cover (grass, forest, snow, etc.). The vegetation shades the soil surface from the immediate action of solar radiation and protects it against overheating. During the day, much heat is consumed during the evaporation from the vegetation cover surface, while at night, plants reduce the overcooling of soil surface, as in this case, the object of radiation is the vegetation cover itself. As the vegetation cover uses more heat for evaporation than a bare surface, its thermal balance is less. Therefore, the average annual temperature of a bare sandy-clay soil surface in Tbilisi is higher than that of the soil covered with vegetation.

Thermal conditions of soil are particularly strongly influenced by a snow cover. Due to the low thermal conductivity coefficient of snow, the latter protects the soil against sharp temperature changes and deep freezing. Despite the fact that snow albedo is quite great, it anyway absorbs a certain amount of solar radiation. The radiation penetrates the snow deeply and if the snow cover is not very high, it may even reach down the soil surface. Such state of affairs supports the warming of the snow cover and its layers at different depths. In all winter months, in Tbilisi, the annual, monthly temperature of the layers at different depths of soil with no snow on it is higher than or equals to the temperature of the soil covered with the vegetation. However, at locations where in the cold period of the year, the soil is mostly covered with snow, the average temperatures at different depths are 4–8 °C higher than those of the bare soil at the same depths. Soil temperature is strongly influenced by forests as well, which reduces the soil overcooling in winter and soil overheating in summer.

As for soil freezing, it depends on a number of factors: frost intensity and duration, soil thermal capacity and thermal conductivity, vegetation cover and height of the snow cover, duration of winter, etc. Soil freezing in Tbilisi is insignificant, with the temperatures below 0 °C at the depth of 15 cm occurring quite frequently, but very rarely at greater depths. The maximum depth with a negative temperature is 40 cm. The depth of soil freezing increases with the increase of the altitude of the location above sea level. The temperature of the soil surface at the Jvari Pass (North Georgia, 2395 m asl) is −13 °C, it is +1.5 °C at the depth of 40 cm and +3.4 °C at the depth of 160 cm; the soil surface temperature at Paravani (Javakheti, South Georgia, 2100 m asl) is −8 °C, it is −1.6 °C at the depth of 40 cm and +3.5 ° C at the depth of 160 cm. On Jvari Cross located 300 higher than Paravani, no negative temperature is fixed at any soil depth, as the thickness of snow cover is quite great (up to 150 cm), while it is only 3 cm at the Paravani weather station.

There are 12 soil-climatic zones identified on the territory of Georgia (Elizbarashvili 2017): extremely warm, intensely water-infused soils; extremely warm, moderately water-infused soils; extremely warm, slightly water-infused soils; extremely warm soils with capillary humidification; extremely warm, full spring-moistened soils; warm slightly water-infused soils; warm soils with capillary humidification; warm, full spring-moistened soils; moderately warm, slightly water-infused soils; moderately warm soils with capillary humidification; moderate and cold, slightly water-infused soils; and moderate and cold soils with capillary humidification.

3.5 Biological Factor in Soil Formation

A biological factor (living organisms) is an integral part of the soil-formation process. No soil exists without biological influence. Many living organisms and their products are immediate components of soil. Their unity, despite the minor amount in relation to the mass of our planet, guides the geochemical processes and is the major factor in the formation of the Earth landscapes. In the course of vital activity of the organisms, the major chains of soil formations are realized: synthesis and degradation of organic substances, degradation of minerals, migration, accumulation, and other phenomena being the essence of soil formation and determining the principal property of soil—fertility.

The soil formation may occur under the influence of a biological factor on the parent materials. The geography of plant communities has much determined the geography of soils. It is the living organisms engaging the solar radiation energy in the process of soil forming by transforming it into potential energy and later into kinetic energy of geochemical processes. This is why the biological factor of soil formation is often described as "leading". Three groups of organisms take part in soil formation: green plants, microorganisms, and animals forming complex land communities.

Soil humus accumulates energy, which is assimilated in plants in the process of photosynthesis. The primary and secondary minerals of soils are degraded under the relevant impact, and organic mineral substances are formed. Owing to humus compounds, the individual particles of soil stick together into structural aggregates.

Almost 99% of the living organisms on Earth is made up of an organic mass of the organisms. Consequently, the nature of the biological cycle of substances is first of all determined by the vital activity of green plants. The principal characteristics of the types and quantitative properties of a biological cycle are as follows: phytomass (biomass), dead organic matter, litter, intensity of the degradation of the vegetation residue, and ash content. Phytomass is the total amount of living organic matter in the above- and underground plant communities; dead organic matter is the amount of organic matter under the dead cover; litter is the amount of organic matter accumulated per unit area annually; intensity of the degradation of organic matter is the ratio between the dead cover and the litter; and ash content is the content of ash elements (%) in plants.

The greatest amount of phytomass is found in forest vegetation (4000–5000 centner/ha) and the minimum amount of biomass is found in polar and tropical deserts (<50 centner/ha).

The animals living in the soil influence the soil in many ways: they accelerate the degradation of organic residue, loosen the soil, and contribute to the formation of a soil zoogenic structure. The soil is the home of many thousand animal species, much different in size, types of vital activity, and their impact on soil. These animals range from nanofauna, the simplest organisms living in humid environment; microfauna, the smallest insects (mites and the like); mezofauna (potworms, spiders, pauropods, rotifers); and macrofauna consisting of worms, crawfish, rodents, etc.

As a rule, plants, animals, and microbes' habitats in soils, strongly impact the soils. Plant roots have direct impacts on soil physical properties as well. They create pores and promote soil aggregation (Graham and Indorante 2017).

Thus, living organisms influence the chemical and mineralogical structure of soil, its physical properties, thermal and water regime, etc.

3.5.1 Vegetation Cover and Biodiversity

The diversity of the natural conditions, periodic climate changes in the past geological epochs, and location on the brink of different floristic areas has resulted in the rich and diverse vegetation cover of Georgia (Gegechkori

2007, 2019). The total number of plant species exceeds 4500, including over 300 species of trees and bushes. The vegetation covers of Georgia and Caucasus, as well as their geography and productivity were studied by a number of scientists at different times (Grossgeim 1948; Ketckhoveli 1959; Gulisashvili et al. 1975; Nakhutsrishrvili et al. 1980; Gigauri et al. 1987; Dolukhanov 1989; Gagnidze 1996 and others).

The vegetation cover of Georgia is rich in relict species. On the territory of the Caucasus, Colchis is the habitat of the survived tertiary flora. In the past, this flora covered quite a large territory and formed a single area. Such ancient species as *Rhododendron ponticum, Castanea Sativa, Pinus pithyusa, Laurocerasus officinalis, Quercus pontica, Phillyrea medwedewii, Buxus colchica, Zelcova carnifolia, Taxus baccata, Ruscus hypophyllum, Dioscorea batatas*, etc. are found in Colchis. They are thermophytes and have survived in the areas with a mostly warm climate for the most of the year. Such areas are mainly deep gorges of the Abkhazia hilly zone. Besides Colchis, the elements of the Tertiary flora are found in other parts of Georgia, such as Kakheti and Kartli, though as fragments only. At the same time, the Caucasus of Kakheti is the only region in the Caucasus presenting the elements of Colchis and Hyrcanian flora at the same time.

Georgia houses not only individual relict species, but their ecosystems as well. They are protected in Batsara, Lagodekhi and Tusheti Reserves, in the gorges of the rivers Kodori and Kintrishi. The ancient relict, the yew forest survived in Batsara Reserve is the only such plantation in the world, while in the past geological epoch, it, together with sequoia, occupied a huge area.

There are many endemic species in Georgia, typical to our country only, amounting to approximately 400, including 7 oak species, as well as *Trapa colchica, Alchimilla, Paeonia abchasica, Pinus pithyusa, Pinus eldarica, Crataegus colchica, Acer sosnowsky, Juniperus pygmea, Pyrus sachokiana, Salix Kazbekensis, Betula megrelica, Amygdalus georgica, Ulmus georgica, Inula magnifica, Aquilegia gegica, A. colchica, Draba migrelica, D. Ossetica, D. imeretika, D. meckhetika*, etc.

Recently, the species from foreign countries (advent, introduced species) to Georgia have occupied an important place in the vegetation cover of the country. A particularly great number of such plants is found along the Black Sea coastline, piedmont and low mountains, and areas of Alpine pastures in high mountains. The areas with their least number are the middle-mountainous forest area as a result of complex orographic conditions. Some introduced species (tree shea, fig tree, false acacia, etc.) have become wild by now.

According to the genesis, there are following species growing on the territory of Georgia: Colchis, Mediterranean-Turgai, Boreal, and other types of vegetation.

A Colchis-type forest is characterized by poly-dominancy. There are five dominant species here: oak, beech, hornbeam, chestnut, and ash tree. However, other tree plants, such as elm, box elder, maple, lime, etc. also grow at many places. Forest composition is particularly diversified in the areas under a human influence. Another peculiarity of Colchis forest is evergreen sub-forest and lianas (Fig. 3.12).

The species of Mediterranean-Turgai flora were introduced to the territory of Georgia from North Caucasus in the Tertiary period, to the mountains first and to the plains later: oak, maple, elm Zelkova, Circassian walnut, Japanese persimmon, etc.

Following the Quaternary glaciation, the vegetation cover of Georgia was enriched with different plant species, in particular, psychrophilic (boreal) deciduous plants, which spread more commonly in the northwestern part of Georgia, in the high mountains of Caucasus and in the environs of lakes and marshes.

Out of the vegetation cover, a forest is very important, which grows both in the lowlands and the mountains of Georgia up to 1900–2000 m asl, occupying 38% of the total territory of Georgia. Most common are hardwood forests (Fig. 3.13), while the coniferous forests, with some exceptions, grow in the mountains only.

The major forest-forming species in the forests growing on Colchis Lowland are oak, beech, hornbeam, and ash tree. These forests have typical lianas. They are mostly degraded and so-called secondary forests (ash, hornbeam) grow in their place. A marsh forest with dominant ash and wing nut also grows in Colchis Lowland.

Tugai or floodplain forest growing as a narrow strip along the gorges of the major rivers of Georgia contain ash, willow, weeping willow, white aspen, oak, elm tree, mulberry, etc. The floodplain forest is mostly degraded, with secondary herbaceous or bushy plants, or agricultural plots of field occupying their place.

In the environs of Bichvinta Cape, there grows a plantation of Bichvinta pine, the Tertiary relict. In the prehistoric epoch, it was quite widely spread, but has survived as a plantation to present. The area with pines diminishes gradually promoted by the coastline washout and fires. There is another pine species known in Georgia called Turkish Pine. The Turkish Pine forest plantation is found at the Georgian–Azerbaijani border on Iori Plateau.

Forest vegetation covers larger areas in the mountainous amounting to 95% of the territory of Georgia covered with forest. Lower mountains are covered with a hardwood (poly-dominant) forest with oak and hornbeam as major

Fig. 3.12 Colchis forest of humid subtropics, Adjara, Southeast Georgia. Photo by B. Kalandadze

Fig. 3.13 Hardwood forest, Gombori Ridge, East Georgia. Photo by B. Kalandadze

Fig. 3.14 Dark coniferous forests, river Kvabliani Gorge. Photo by B. Kalandadze

species, with oriental hornbeam, chestnut, lime, ash tree, and others growing in some locations. Great areas of the forests are demolished, with settled areas and agricultural plots of field occupying their place at present. The middle mountains are dominated by beech forests, with hornbeam, chestnut, elm, and maple growing in some locations. In some locations of the middle mountains, deciduous plant species form the plantations, such as yew forest, elm Zelkova forest, cherry laurel forest, etc.

As the absolute height increases, the forests in West Georgia are changed by beech and dark coniferous forests with spruce and fir as dominant species (Fig. 3.14). The lower border of such forests in West Georgia is presented as fragments. There are elfin and crooked trees and plants above the areas of beech and dark coniferous forests (from 180 m asl): birch, maple, beech, and Caucasian oak. In some regions (in Tusheti), there grows quite a strong forest.

Bushes are evergreen or deciduous. Evergreen bushes are more common in the mountain forests and high-mountainous regions of West Georgia, while deciduous bushes are mostly met in East Georgia. Evergreen bushes, such as Rhododendron, ilex, and cherry laurel, are one of the most important features of a Colchis Forest found in the deep gorges of a hilly zone of Colchis. Here, the function of a sub-forest is sometimes played by a willow or yew. Evergreen bushes in the high-mountainous areas are typical to subalps and above 1800–2000 m asl. Caucasian rhododendron, red bilberries, juniper, etc. are spread here. Deciduous bushes are presented by shibliak or Christ's Thorns.

In Georgia, along the Black Sea coastline in Apkhazeti, there grow Mediterranean plants—the low xeric trees and bushes with coriaceous or thorny leaves.

There are numerous herbaceous plant species found in Georgia: meadow, steppe (plain), marsh, semidesert, and mountain. There are no pure deserts on the territory of Georgia, but semidesert plants grow in the extreme south-eastern part of East Georgia. Out of steppe plants, drought- and frost-resistant perennial grasses dominate. Beard grass, which is a transient species between the semidesert and steppe vegetation, is dominant. Meadows in the high mountains are spread at 1800–3500 m asl; however, over the mountain cuts, they are spread even at lower altitudes forming secondary meadows. In high mountains, meadows are common in subalpine and alpine zones. Meadows in the subalps are presented as a tall herbaceous cover with a dense grass with its height reaching 3–5 m. Due to its height, they are often called "mammoth flora". A significant area of subalpine meadows is modified. It has almost lost its original natural appearance. Therefore, tall herbaceous cover is rare in the subalps. Alpine meadows occupy larger areas and are presented as a low herbaceous cover (with the height of up to 20–30 cm). However, due to intense grazing, it has lost its original natural appearance. Alpine meadows can be of two kinds: pure meadow and a carpet. Pure meadows are formed by Gramineae + Carex meinshauseniana, while a carpet is created by herbs. Alpine carpet has typical vivid-color, flowering herbs: red camomile, bellflower, gentian, etc., while feather grass, fescue, etc. are more common for pure meadows.

Marsh vegetation is widely spread in the central and western parts of Colchis Lowland, with a reed, deer grass, sedge, mosses, etc. Such vegetation is present also in the environs of the lakes of Georgia, on the banks of water reservoirs, and on the location of the cut-down forests.

On the plateau of West Georgia and over the adjacent slopes, mountain xerophytes dominate. They are presented as perennial herbs and bushes, so-called phrygana.

Cliff and stone-fill herbs form a peculiar type of vegetation, which are most common for high-mountainous cliffy areas. In Georgia, they do not form a single area, but are presented as fragments.

The variation of the absolute height of a site in the territory leads the formation of different vegetation covers at different altitudes and different zoning consequently. The lowest locations are occupied by marsh vegetation and subtropical Abkhazia forests, while sub-nival vegetation grows at the highest altitudes.

The human impact is most clearly seen on the vegetation cover of the components of nature. The quality and productivity of the forests have deteriorated a lot leading to the diminution, or loss of the water-regulation, soil-protection, and recreation functions of the forest.

The economic activity is an important factor, which changed the vegetation cover of Georgia. Forests were cut down at many places and secondary meadows, herbs, and agricultural plots of field (orchards, vineyards, cucurbitaceous, plantations) have occupied their place. This is particularly true with the Black Sea coastline, plains and lowlands, and piedmonts. As a result of predatory exploitation, many oak, willow, chestnut, and beech plantations were destroyed.

Some species of trees and plants have survived only at hardly accessible locations, on the mountain slopes, in the reserve areas, etc. The plants less typical to this region were spread on the cuttings. A cut-down forest is usually changed by xerophytic bushes or meadows. Such plants grow not only in the east, but in western Georgia as well, giving a certain peculiar appearance to the humid subtropical landscape. A negative human impact on the vegetation cover is evidenced not only by the changing composition of flora, but at some locations, the soil and vegetation cover are completely destroyed showing a bare ground surface. If the anthropogenic impact on the vegetation cover occurs once, following such an impact, an inverse process may develop depending on its scales, such as the self-restoration of the vegetation cover, and there are many examples of this process in Georgia.

The lowland and hilly zone of West Georgia (landscapes with Mediterranean climate) are particularly distinguished for biodiversity (a set of living organisms and forms of the Earth). There are 4225 plant species in Georgia making the country rank the 60th in the world and the 5th in Europe after Italy, Spain, Greece, and France. Out of 4629 mammals known in the world, 97 species are spread in Georgia making the country rank the 89th in the world and the 1st in Europe. The total number of birds in Georgia is 322, and with this indicator the country ranks the 73rd in the world and is one of the leading countries in Europe. With the number of reptiles, Georgia ranks the 65th in the world with 51 species and the 3rd in Europe (after Spain and Azerbaijan). With the number of amphibians, Georgia ranks the 74th in the world, and it ranks the 39th in the world with the number of freshwater fish species (84 species).

One of the principal means to preserve the biodiversity is the establishment and development of protected areas. The area of protected areas in Georgia is small. Biological and landscape diversity is one of the major resources of our country, which, if used rationally, can be made an efficient source of the well-being of the country citizens.

3.5.2 Landscapes

Georgia is a very interesting country in respect to the biological and landscape diversity. As a part of the Caucasus, the country is on the lists of 25 biologically richest and endangered “hot spots” of the world (CI, CEPF); 200 sensitive and vulnerable eco-regions of the world; locations of endemic bird habitats (BirdLife International); one of the world centers of agro-biodiversity; “hot spots” of large herbivores (WWF) (Biological and Landscape Diversity… 2000; Biodiversity of the Caucasus… 2001; An Ecoregional conservation… 2006). This list can be made longer if considering such factors as the well-preserved diversity of species and ecosystems in the country, richness of Georgia in endemic, relict, medicinal, and decorative plant species, forests occupying over 40% of the territory of the country on the one hand and the environment of the country not subject to major changes like many regions of the world on the other hand.

Therefore, in an environmental respect, Georgia looks a much “cleaner” region in the world (Beruchashvili et al. 2002; Nikolaishvili 2009). Besides, Georgia is one of the outstanding countries in the world in respect to landscape diversity (Fig. 3.15). In addition, with the rich biodiversity, Georgia is ahead of many countries (Beruchashvili 2000). These natural values are still less studied and the ecological

(a)

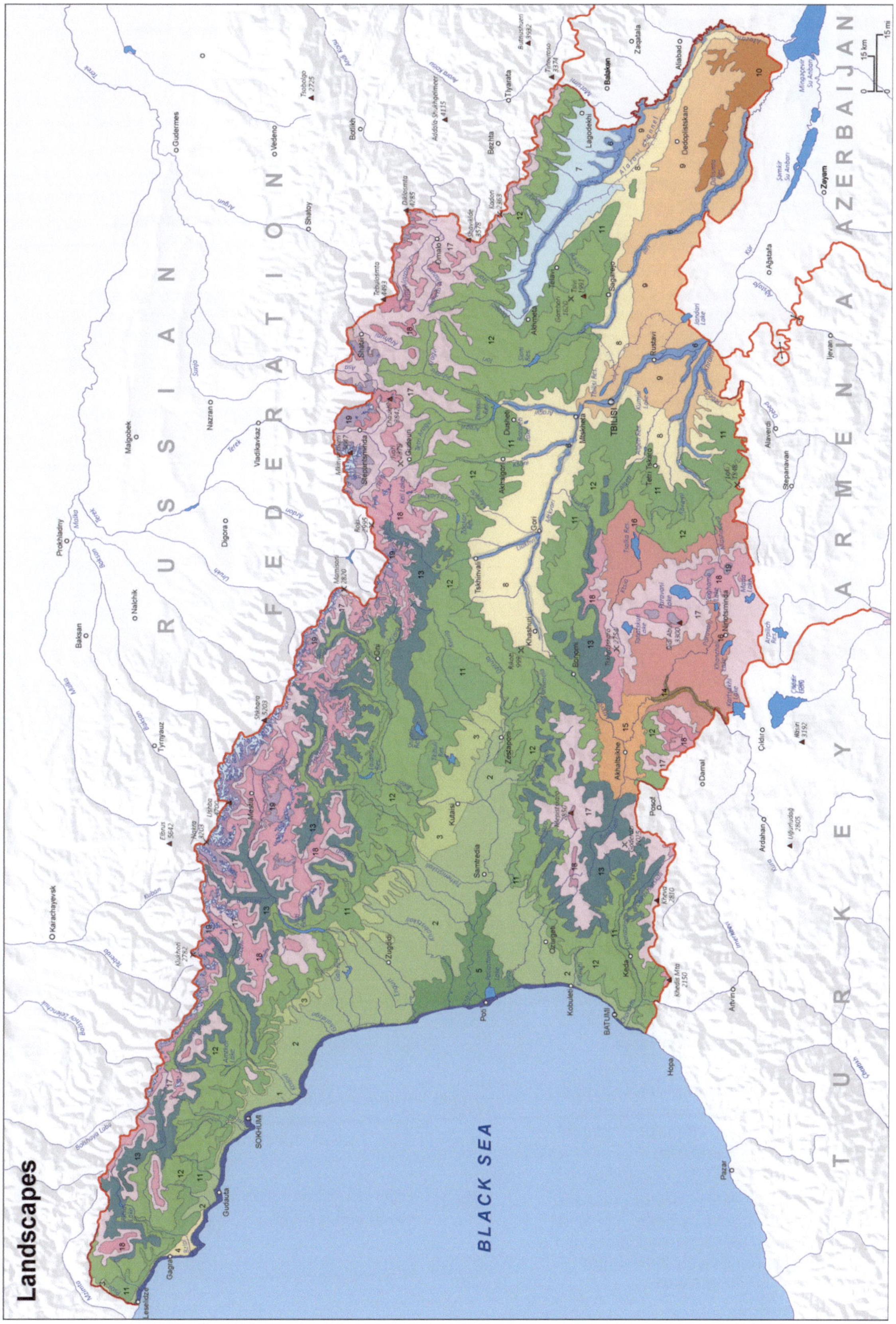

Fig. 3.15 **a** Landscapes map (D. Nikolaishvili). *Source* National Atlas of Georgia, Stuttgart: Steiner-Verlag, 2018. **b** Legend of landscapes map

(b)

Landscapes Map Legend

I. HILLY LANDSCAPES OF PLAINS AND FOOTHILLS

I a. Humid subtropical

1 — Coastal beaches and seaside dunes zone with psammophyte (growing on sandy soils) vegetation, with strongly transformed settlements and coastal pollution in some areas

2 — Plains and lowlands with Alders and Oak forests in the Podzolic and Gley Podzolic Soils, with strongly transformed melioration systems and farm lands

3 — The foothill hillocks with polydominant forests in the Red Soils and Yellow Soils with strongly transformed farm lands

I b. From Semi-humid to Mediterranean transitive

4 — Plain-lowlands and foothill hillocks, fragments of maquis on the coastal cliffs, Pitsundian pine in the plain and polydominant deciduous forest in the gorges; partially transformed

I c. Excessive humid

5 — The plain-lowlands with alders or sedge, peat moss and reed marshes on Slime-Marsh Soils; almost untouched

6 — River terraces and floodplains with tugai forests and meadow vegetation on Alluvial Soils, with strongly transformed farm lands

I d. Moderately warm semi-humid

7 — Forests of Kakheti Plain (oaks, oak-zelkova woods), shrubs and meadow vegetation, with strongly transformed settlements and farm lands

I e. Moderately warm semi-arid

8 — Hillocks with arid light forests, thistle-thorns and steppe vegetation on Meadow Cinnamonic Soils with strongly transformed farm lands and melioration systems

9 — Hillocks with big bluestem, needle grass and buckthorn steppes, in some areas with forest elements on the Grey Cinnamonic Soils and Black soils, with significantly transformed farm lands and winter pastures

I f. Transitional to moderately warm

10 — Plain-hills with semi-desert vegetation, partially transformed farm lands and winter pastures

II. MOUNTAIN LANDSCAPES

II a. Moderately warm and moderately cold humid

11 — Lower mountains with hornbeam-oaks and oak trees in the Brown Forests Soils and Raw Humus Calcareous Soils; substantially transformed in some areas

12 — Middle Mountains with excess of beeches on the Brown Forest Soils; slightly transformed

13 — Middle Mountains with excess of dark coniferous and dark coniferous-beech forests on the Brown Forest Soils, slightly transformed

II b. Moderately warm and dry

14 — Canyon-shaped cavities with thistle-thorn and meadow-steppe vegetation on Cinnamonic Soils; significantly transformed

15 — Middle mountains with arid light forests, buckthorns and thistle-thorns, and semi-desert vegetation in some areas

16 — High mountains with plateau-steppe and meadow-steppe vegetations on Black Soils with significantly transformed farm lands

II c. Cold

17 — High-mountains subalpine forests-shrubs and meadows on Mountain-Forest-Meadow Soils; partially transformed

18 — High mountains alpine meadows and rhododendrons on the Mountain-Meadow Soils; partially transformed

19 — High mountains subnival, with distribution of lichen and mosses

20 — High mountains nival-glacial

Fig. 3.15 (continued)

function Georgia can play on a global scale is not thoroughly realized yet.

According to Beruchashvili's (1979), landscape map of the Caucasus, there are 2 classes, 20 types, 40 subtypes, and 71 genera of landscapes in Georgia. Mountain and plane landscapes on the territory of the country are distributed very unevenly. Mountain landscapes occupy 53.1 thousand km^2 making 76% of the country area. Even according to the altitudinal zoning, the landscapes are distributed too unevenly in the mountains: the low-mountain landscapes occupy 3% of the total territory of Georgia, mountain depressions occupy 1%, lower mountain landscapes occupy 12%, middle-mountain landscapes occupy 24%, upper mountain landscapes occupy 7%, and high-mountain sub-nival and nival landscapes occupy 1% of the total territory of Georgia. The landscapes of meadows and meadows-and-steppes occupy almost equal areas in the planes-and-lowlands and mountains. These landscapes are most common in the mountain depressions of the Caucasus and high plateaus of South Georgia. Karst and volcanic landscapes occupy only 8% and 6% of the total territory of Georgia (Beruchashvili 1979).

The distribution of the landscapes in different regions of Georgia is also extremely uneven (Seperteladze 2001). Plane landscapes occupy largest areas in the region Kakheti (5.4 thousand km^2—over 40% of the total territory of the region). It is also outstanding with the areas of low-mountain and upper mountain forest landscapes. Mtskheta-Mtianeti region falls little back with the area of the upper mountain forest landscapes. The middle-mountain forest landscapes occupy the largest area on the territory of Apkhazeti amounting to almost 30% of the region's total territory. The high-mountain subalpine landscapes occupy the largest areas in Imereti and Samegrelo-Zemo Svaneti regions, while high-mountain alpine landscapes occupy the largest area in Mtskheta-Mtianeti. Middle-mountain forest landscapes and high-mountain subalpine landscapes occupy larger areas in West Georgia, while lower and upper mountain landscapes are mostly spread in the east of the country. As for the alpine landscapes, they occupy almost equal areas in the both regions of Georgia (Nikolaishvili 2009).

"Every corner of Georgia and nature of this country generally, is the most beautiful on earth",—wrote Arthur Leist, a German writer, publicists, and translator. The nature of Georgia is indeed rich and diversified, and it is fairly called the country of contrasts. The Colchis Forests extremely rich in species, high-productive fir and pine forests, intact forest massifs, bulk of mineral sources, rare, endemic and relict flora and fauna species, fruitful soils of eastern Georgia, unique Nitisols Ferralic (in national classification called as "Red soils"), buried rare monuments of flora and fauna, natural nutrition areas, high hydroelectric potential, great biological, and landscape diversity—are an incomplete list of the natural riches of Georgia, which are so much in the country. It should also be noted that the specific weight of the areas, which are under the intense anthropogenic impact, is little. The ecological problems of Georgia have a more local nature and with their severity fall much back many regions of the world. This is why they say that Georgia has high environmental potential, which can be used to further develop recreation and tourism in the country (Nikolaishvili 2009).

3.6 Time Factor in Soil Formation

The environmental factors of parent material, relief, climate, and organisms interact with each other for some time to produce soil. The longer these factors are able to act together, the more developed and differentiated will become soils (Graham and Indorante 2017). Soil character reflects how long the soil-forming factors have exerted their influence.

Soil, like any natural-historical body, has an age: absolute age, showing the time lapse from the onset of its formation to present, and relative age, showing the rate of alternation of the stages of soil-formation process and development. Absolute age varies from some years to a million years. The oldest are the soils in the tropical areas that have not been subject to various destructions (erosion by water, deflation, etc.). As for the relative age, it depends on the rate and direction of the soil-formation process, as well as changes of relief properties and parent material composition.

The soil-formation process usually takes place during a certain time. Every new cycle of soil formation (seasonal, annual, plurannual) makes certain changes to the transformation of the organic and mineral substances of the soil profile. Therefore, time factor is very important in the formation and development of soil. In the Quaternary Age (Antropogene) when the Earth had an active transformation history, only a small part of the land could maintain its former soil cover. Old soils destructed, got washed down, and were buried. The origination of the new soils became possible only in the Holocene and has continued for the last 12 thousand years to present. The Quaternary Age, the modern stage of the Earth history, started 2588 million years ago and ended with the Holocene. This was the shortest geological period, but nevertheless, most of the modern relief forms were formed and many important phenomena on the Earth, like glaciation and appearance of a man, occurred during it.

The main postulate of Soil Science is a famous opinion suggesting that "soil is a mirror of landscape", as soil reflects the modern conditions of the environment, in which it really exists and in which it was formed. However, soil is the

Fig. 3.16 Poligenetic soil profile; Mashavera river valley, South Georgia. Photo by B. Kalandadze

mirror of not only modern landscapes, in which it exists at present, but also of the landscapes, in which it existed in the past. Consequently, soil, in its immediate sense, not only gives a "mirror reflection". Rather, the properties formed in the past did not vanish utterly, but survive for some time. So, the soil evolution is a widely spread phenomenon and is a unit not only varying in space, but is also quite unstable in time (Karpachevsky 1997).

Soil profiles are not always adequate to contemporary conditions. Sometimes, they maintain residual properties from the past stages of development, "remember" all phenomena of the landscape life. Consequently, soil is not only a "mirror of landscape" but also a "memory of landscape" maintaining the paleogeographic and relict properties (Targulyan 2008; Targulian and Arnold 2008). It is this feature distinguishing the soil from other landscape components.

As a rule, soil evolution takes place with landscape evolution, although, some soil properties formed in the previous environment do not disappear without a trace. They are preserved for a long time. The modern evaluation of soils identifies several indicators: soil types age, soil profiles age, and soil horizons age. With their structure, soil profiles are divided into monogenetic and polygenetic. Soil is monogenetic, if its profile has the same age, horizons are syngenetic and were formed at the same time; but if soil layers' age is different, the soil profile is polygenetic (Paleopedology glossary 1997).

In scientific sources (Targulyan and Sokolov 1997; Sokolov 1984; Makeev 2002) are mentioned three main categories of soil properties: modern features, which were formed in the current environment; relict features, which were inherited from the early stages of pedogenesis; and lithogenic features, which were inherited from the parent materials.

Presence of buried soils or horizons allows restoring past conditions of the environment and is the base for paleogeographic reconstructions (Fig. 3.16).

3.6.1 Relict Features in Soils[1]

The relict properties of the microstructure of Georgia's soils incorporate both the buried horizons and separate properties not corresponding to the present environment. Usually, relict

[1]Updated Version from an Article Originally Published in the Journal of Environmental Biology;

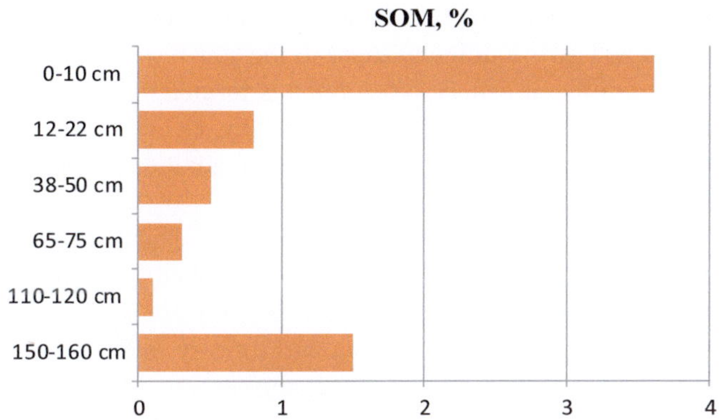

Fig. 3.17 Distribution of soil organic matter in profile of Luvisols Albic with buried horizon; Pit-15M, 150–160 cm. Created by L. Matchavariani

soils are surface non-buried soils, which have the features formed in the medium different from the current one. Its development started in a preexisting landscape and is ongoing to date as it was never buried (Paleopedology glossary). As for paleosols, they are either buried or surface. Sometimes the stability of the relict properties in the soils determines the regimes of present soils (Makeev 2009).

Micropedology is the most informative instrument to diagnose the modern and relic features of soils. It was used by us for the study of various soils of Georgia and our micromorphological studies duly presented (Matchavariani 2008).

In order to show the micropedological diagnosing of buried or a surface paleosols, we have cited an example of Luvisols Albic—so-called as Subtropical Podzolic soils, describing the microfeatures of buried layer. Rather deep second humus horizons on 1–1.5 m have been identified in micro-depressions. The micromorphological data of burial horizons demonstrate the following specific nature of its microstructure (Matchavariani 2008). Generally, a microstructure of a buried horizon is similar to the surface accumulative horizons. Quite mixed and different materials with a large number of organic microzones and individual clots are present there. This fact is evidenced that the conditions of soil organic matter formation were more favorable when the second organic layer was on the surface. The soil matrix of this horizon is characterized by a brown color and is distributed throughout an extremely dissimilar matrix. The matrix is also characterized by large quantities of skeletal grains, with their number much higher than of those in a surface accumulative horizon. The microzones enriched with iron in the form of spotted flaky accumulations are also typical to them.

One more common feature for the buried organic layer is the great number of weakly decomposed, sometimes charred plant residues with a well-preserved cell structure, which are not seen in the surface horizon (Matchavariani 2008, 2012). The abundance of phytoliths of different forms and sizes at the depth of 150–170 cm is of a special importance and is a certain evidence of the second organic horizon being a former surface layer buried during the sedimentation. This is confirmed by the content of soil organic matter in the profile. Figure 3.17 shows the presence of the second maximum of SOM in the buried horizon.

So, the presence of numerous accumulations of soil organic matter in presumable burial layers, as well as incorporated phytoliths and conserved vegetation tissues in the mixed aggregated material (Fig. 3.18), with some chemical analysis data allow diagnosing the polygenetic and heterogeneous nature of the profiles. The mechanism of formation of the textural-differentiated profiles of abovementioned soil is primarily associated with the lithological dissimilarity of parent materials.

Distribution of clay in profiles of Luvisols Albic with an intense increase in lower horizons has resulted from the

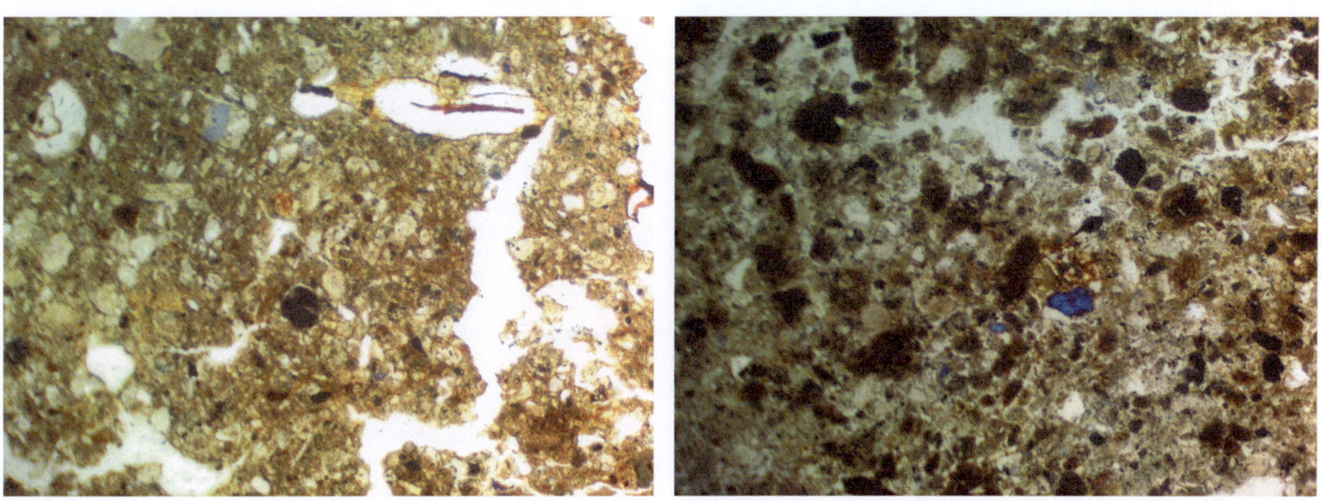

Fig. 3.18 Microstructure of the buried horizon of Luvisols Albic soil; Pit-15M, 150–160 cm. Photos by L. Matchavariani

lithological dissimilarity of the parent material of the Holocene Age and only partially, from the Quaternary Period (Janelidze 1980).

The relict properties of these soils are also seen in Fe-nodules (Matchavariani 2005). The nodules can be grouped as pedogenic (modern and relict) and lithogenic. Selection criteria for the relict features in soils are the disagreement of nodules composition, structure, and enclosing matrix with the modern one.

The characteristic features of lithorelicts are compact fabric, a clear separation from the matrix, heterogeneity of iron impregnation of a large size (about 10 mm in thin sections) and admixtures of different materials (clay, Mn), irregular-round forms, etc. Besides, the concentric forms of pellets are named lithorelicts as well.

As noted in our previous publication (Matchavariani 2005), "The presence of an ortshtein (plintic or petroplintic) horizon in the profiles of Luvisols Albic does not correspond to modern environmental conditions. Notwithstanding the humid climate, which acts in favor of intensive vertical migration of substances, the iron content in the upper horizons is not sufficient to explain the origin of Ortshtein layer by means of iron leaching from the upper horizons, and its accumulation in the iron-cemented layer cannot be the outcome of the exclusively vertical migration of substances".

Micropedological study was used to reveal some other examples of relict features in Georgia's soils. Sometimes, carbonate impregnation of plasma and crypto-grain calcite is observed in Luvisols Albic (so-called Podzolic-Gley soils), at the depth 120–150 cm. The presence of carbonates in the acid soils does not comply with the modern climate. Besides, the ferrum accumulations in the lower horizons of Vertisols, which are formed in terms of semiarid subtropical climate with high seasonal contrasts evidence the climate being wetter in the past.

In the mountainous regions, the main soil-forming factor is age. In Georgia, at the altitudes higher than 1000 m asl soils are of the Holocene Epoch, while soils located up to 1000 m are older. Fluvisols (the floodplain alluvial soils) are the youngest. Mountain soils correspond to the ecological conditions more than those in the plain regions and they can be viewed as a mirror of landscapes as one of the most important postulates of the soils. Otherwise, we are dealing with a distorted mirror of the landscape science (Matchavariani 2012).

3.7 Humans Factor in Soil Formation

A human's factor (man's economic activity), as a soil-forming factor, was added to the soil definition at a later stage. It can be considered a factor only if the anthropogenic or human's economic impact on soil is so strong that it can change the course of the natural development of soil formation and give a different trend to the soil formation. As a result, a different genetic type of soil is formed than it would have been formed naturally. Consequently, an anthropogenic (human's) factor can be considered as an additional (and not major) soil-forming factor.

However, a man's economic activity has changed from a local factor influencing only on the cultivated areas into a strong global factor. This is associated with the general chemization of the rural farming and forestry, realization of large-scale projects of irrigation and drying large areas, development of industry and transport, and increased general technogenic load. The regions distanced from the centers of economic activity, including reserves, cannot be isolated from the impact of technogenic chemical substances occurring from the atmosphere caused by global and regional transfers of air masses.

One of the outcomes of this factor is soil cultivation, man's production activity (cultivation, fertilization, melioration, etc..), which is a strong specific factor of the impact on the soil and whole soil-forming complex (vegetation, climate elements, hydrology). Purposeful actions on soil (liming, plastering, fertilizing, drying or irrigation melioration, etc.) lead to the change of the soil properties and regimes at much higher rates than in the case of natural soil-formation process. A man accomplishes production operations on vast land areas, and it is a decisive factor in improving soil fertility. In addition, the nature and importance of the soil depend on the social–economic relations and the level of the development of science and techniques. On the territories, which are under a strong industrial impact, soils are strongly modified, and so-called anthropogenic soils are formed.

The anthropogenic soils in Georgia occupy over 80 km^2 (0.1% of the territory of the country). The upper section of the profile is under anthropogenic impact and has lost its natural structure. The hazard of erosion by water and/or wind to develop with the anthropogenic soils is real. A failure to consider the properties and terms of development of soils, wrong use of the scientifically proved recommendations does not really contribute to the improvement of the soil fertility and may essentially deteriorate them (development of erosion, secondary salinization, bogging, soil pollution, etc.).

References

An Ecoregional Conservation Plan for the Caucasia (2006)

Begalishvili N, Tsintsadze T, Gzirishvili M, Arutinian N, Makhmudov N (2009) Regional classification of Georgian territory on the basis of evaluating of droughts, vol 115. Works of Hydrometeorology Institute, 308 p

Beritashvili B, Kapanadze N, Chogovadze I (2010) Assessment of climate response in Georgia to global warming, 174 p
Beruchashvili N (1979) Landscape map of Caucasus (1: 1 000 000). Tbilisi, TSU
Beruchashvili N (2000) Potentially possible intact landscapes of Georgia. In: Biological and landscape diversity of Georgia. Tbilisi, Georgia's Office of World Wildlife Fund (WWF), pp 203–220
Beruchashvili N, Shotadze M, Nikolaishvili D, Melikidze V (2002) Caucasus environmental outlook (CEo 2002). UNEP, Grid Tbilisi, p 100
Biodiversity of the Caucasus Ecoregion (2001) An analysis of biodiversity and current threats and initial investment portfolio. Baku-Erevan-Glard-Moscow-Tbilisi
Biological and Landscape Diversity of Georgia (2000) Proceedings of the First National Conference. Tbilisi, Georgia, 312 p
Bockheim JG, Gennadiyev AN, Hartemink AE, Brevik EC (2014) Soil-forming factors and soil taxonomy. Geoderma 226–227:231–237
Dolukhanov A (1989) Rastitelnost Gruzii, tom 1. Lesnaya rastitelnost Gruzii (Vegetation of Georgia, vol. 1. Forest vegetation of Georgia). Tbilisi
Elizbarashvili E (2007) Sakartvelos klimaturi resursebi (Climatic recourses of Georgia). Tbilisi
Elizbarashvili E (2017) Sakartvelos hava. Sakartvelos teqnikuri universitetis hidrometeorologiis instituti (Climate of Georgia. Institute of Hydrometeorology at the Technical University of Georgia). Tbilisi, 360 p
Elizbarashvili E, Sulkhanishvili N (2002) Climatic zoning of Georgia. In: Sakartvelos niadagebis klimaturi daraioneba. Sakartvelos mecnierebata akademiis hidrometeorologiis institutis shromebi (Proceedings of the Institute of Hydrometeorology of the Georgian Academy of Sciences), vol 105
Elizbarashvili E, Kheladze T, Chavchanidze Z, Sulkhanishvili N (1992) To the question of thermal problems of soil-atmospheric system. Bull Georg Sci Acad 145(1)
Elizbarashvili E, Elizbarashvili M, Maglakelidze R, Sulkhanishvili N, Elizbarashvili S (2007) Specific features of soil temperature regimes in Georgia. Eurasian Soil Sci 40(17):761–765
Gagnidze R (1996) Geography of vegetation (mcenareta geografia). Tbilisi, 232 p
Gamkrelidze E (2018) Map: geology. Vakhushti Bagrationi Institute of Geography
Gegechkori A (2007) Caucasus—a treasure of nature. CEPF, WWF, Meridiani, Tbilisi, p 168
Gegechkori A (2019) The biomes of the Caucasus: comprehensive analysis. Springer (in press)
Geomorphology of Georgia (Ed. Tielidze L) (2018) Springer, 289 p
Gigauri G, Makhatadze L, Urushadze T (1987) Sosnovie lesa Kavkaza (Pine forests of the Caucasus). Tbilisi
Gobejishvili R (2010) Relief of Georgia. Universal, Tbilisi
Gogichaishvili G (2018a) Map: aridity index. In: Bolashvili N, Dittmann A, King L, Neidze V (eds) National atlas of Georgia. Stuttgart: Steiner-Verlag, p 51
Gogichaishvili G (2018b) Map: annual average evaporation. In: Bolashvili N, Dittmann A, King L, Neidze V (eds) National atlas of Georgia. Stuttgart: Steiner-Verlag, p 53
Graham RC, Indorante SJ (2017) Chapter 2, Concepts of soil formation and soil survey. In: West LT, Singer MJ, Hartemink AE (eds) The soils of the USA. Springer, pp 9–27, 394 p
Grossgeim AA (1948) Rastitelniy pokrov Kavkaza (The vegetation cover of the Caucasus). Moscow, 267 p
GSNC (2009) Georgia's Second National Communication to the UNFCCC
Gulisashvili V, Makhatadze L, Prilipko L (1975) Rastitelnost Kavkaza (Vegetation of the Caucasus). Moscow, 234 p
Gunia G (2011) About ecological monitoring in Georgia the anthropogenous factors influencing climate change. In: Contemporary problems in geography (Collection of papers of international conference). Tbilisi, pp 145–152
Janelidze C (1980) Paleogeography of Georgia in Holocene. Tbilisi, Mecniereba, p 178
Javakhishvili S (1977) Sakartvelos klimatografia (Climatography of Georgia). Tbilisi University Publishing House
Javakhishvili S (1992) Klimatologia. Tbilisis universitetis gamomcemloba (Climatology. Tbilisi University Publishing House), 187 p
Javakhishvili S (2000) Meteorologia da klimatologia. Tbilisis universitetis gamomcemloba (Meteorology and climatology. Tbilisi University Publishing House), 303 p
Karpachevsky LO (1997) Dynamics of soil properties. GEOS, Moscow, p 170
Ketckhoveli N (1959) Sakartvelos mcenareuli afari (The vegetative cover of Georgia). Tbilisi
Kordzakhia M (1961) Sakartvelos hava. Sakartvelos ssr mecnierebata akademiis gamomcemloba (Climate of Georgia. Publishing House of Georgian Academy of Sciences), 246 p
Kotaria A (1992) Meteorologiis sapudzvlebi. Tbilisis universitetis gamomcemloba (Basics of meteorology. Tbilisi University Publishing House), 446 p
Lagidze L, Matchavariani L, Metreveli G, Tsivtsivadze N, Paichadze N, Gulashvili Z, Svanadze D (2017a) Evaluation of climate parameters of Black Sea coastline in the background of climate change. J Environ Biol. Specia Issue Environment and Ecology 38(5):1069–1077
Lagidze L, Matchavariani L, Paichadze N (2017b) The influence of circular processes on change in precipitation in the scope of climate change. Int J Geomate, vol 13(39):213–219
Makeev AO (2002) Paleo soil science: condition and prospects. Eurasian Soil Sci 4:398–409
Makeev AO (2009) Pedogenic alteration of Aeolian sediments in the upper loess mantles of the Russian plain. Quatern Int 209:79–94
Matchavariani L (2005) Morphogenetic typification of concretions in subtropical podzolic soils of Georgia. Eurasian Soil Sci 38:1161–1171
Matchavariani L (2008) Geographic paradigms of the microfabrics of major soils of Georgia. Universal, Tbilisi, p 309
Matchavariani L (2012) Time factor in soils of Georgia—Mirror or memory of landscapes? J Environ Biol, Special Issue of JEB "Environment & Geography in the Mediterranean 33(2):393–400
Matchavariani L, Lagidze L (2012) Chapter 35, Environment transformation in Georgia as a result of climate change. In: Environment and ecology in the mediterranean region. Cambridge Scholars Publishing, pp 379–393
Mumladze D, Gagua G (2018) Map: annual average air temperature. In: Bolashvili N, Dittmann A, King L, Neidze V (eds) National atlas of Georgia. Stuttgart: Steiner-Verlag, p 45
Nakhutsrishrvili GS, Chkhikvadze AK, Khetsuriani LD (1980) Productivity of high mountain grassland communities of Central Caucasus. Metscniereba, Tbilisi, 159 p
National Atlas of Georgia (2018) Steiner-Verlag (Ed. Bolashvili N, Dittmann A, King L, Neidze V)
Nikolaishvili D (2009) Spatial-temporal analysis of Georgia's landscapes. Tbilisi University Publishing House, 431 p

Nikolaishvili D (2018) Map: landscapes. In: Bolashvili N, Dittmann A, King L, Neidze V (eds) National atlas of Georgia. Stuttgart: Steiner-Verlag, p 77

Paleopedology Glossary (1997) Amsterdam, London

Seperteladze Z (2001) Landscape differentiation and physical-geographical zoning. Tbilisi University Publishing House

Sokolov IA (1984) Soil formation and time: polyclimax and polygenetics of soils. Soil Sci 2, 102–113

Targulyan V (2008) Soil memory: general approaches to problem. Soil memory. LKI, Moscow, pp 23–57

Targulyan V, Sokolov I (1997) Vzaimodeystvie pochvi i sredy: pochva – pamyat i pochva – moment. Izuchenie i osvoyeniy prirodnoy sredi (Soil and environment interaction: soil—memory and soil—moment). Studying and Environment Development, Moscow, pp 150–164

Targulian VO, Arnold RW (2008) Pedosphere: evolution, memory and ecological functions. Elsevier, encyclopedia of ecology, global ecology, vol 4, pp 2665–2670

Tielidze L, Gobejishvili R, Tutberidze B, Maruashvili L, Astakhov N, Wheate R (2018) Chapter 1, Long-term geomorphic history of Georgia. In: Geomorphology of Georgia. Springer, pp 3–14

Tielidze L, Trapaidze V, Matchavariani L, Wheate R (2019) Chapter 2, Climate, hydrography and soils of georgia. In: "Geomorphology of Georgia", geography of the physical environment. Springer, pp 15–34

West LT, Singer MJ, Hartemink AE (2017) The soils of the USA. Springer, 394 p

Soil Processes

4

Lia Matchavariani

Abstract

The micromorphological diagnosis, typification, and assessment of main soil-forming processes have been conducted for soils of Georgia: humification, argillization, gleization, podzolization, lessivage, ferrugination, and carbonization. Morphotypes of soil organic matter were determined in main soils—five groups were identified: raw, raw-moder or moder-raw, moder, moder-mull or mull-moder, mull. The process of argillization is diagnosed as in situ weathering and depending on the clayzation intensity, three main grades were identified: intense, medium, and weak. Lessivage is diagnosed by the presence of the optically oriented clay cutans in transit pores and is visible as clay cutans, silty cutans, and complex cutans in the form of films on the walls of the pores and mineral grains. The diagnostic sign of gleization is the contrast in the distribution of ferrum hydroxides. It was grouped according to the degree of intensity as strong, medium, and weak. Ferrugination processes take place mostly in humid zone. Two main forms are distinguished: concretions and ferruginated plasma. The gradation of ferrugination was done by the intensity of process. Carbonization is diagnosed according to genetic-morphological groups of carbonates and the sizes of calcite crystals. It is divided into concretions and carbonized plasma; with the intensity of strong, average, and weakly calcareous.

Keywords

Argillization • Podzolization • Lessivage • Gleization • Ferrugination • Carbonization

L. Matchavariani (✉)
Faculty of Exact & Natural Sciences, Department of Soil Geography, Ivane Javakhishvili Tbilisi State University, Tbilisi, Georgia
e-mail: lia.matchavariani@tsu.ge

L. Matchavariani (ed.), *The Soils of Georgia*, World Soils Book Series,
https://doi.org/10.1007/978-3-030-18509-1_4

4.1 Introduction

One of the major methodological principles of genetic soil science is the concept of the soil-forming process as a complex set of elementary soil processes, which are the result of the interaction between the transformation and the migration of organic and mineral substances.

As a rule, the soil-forming processes are defined as a set of phenomena of transformations and movement of substances within the Earth's pedosphere. The processes constituting the soil formation in general were named by Rode (1948) as common soil-forming processes, since they take place in any soils in different qualities and quantities and various combinations. A specific manifestation of general processes depending on the factors and conditions of soil formation are called private soil-forming processes. All soil-forming processes are divided into macro-processes concerning the entire soil profile, and micro-processes, the mineral and organic transformations within the local sections of the profile.

The private soil-forming macro-processes were proposed to call Elementary Soil Processes (ESP) by Gerasimov (1973). In the early works (Gerasimov and Glazovskaya 1960), these processes were called as elementary soil-forming processes. The ESP plan was later updated by Rozanov (1975).

The elementary soil processes are rather complex with their significance and nature; virtually, they determine the formation of the genetic profile and are by no means elementary in the meaning of the word *elementary* itself.

As a result, the elementary soil processes incorporate the natural and anthropogenic processes, which are specific only for soils and form specific soil horizons in the profile, and they determine the structure of the profile, composition, and ratio of the system of genetic horizons, and their various combinations take place in several types of soils. In other words, the major profile-forming processes belong to the elementary soil processes.

According to Bockheim and Gennadiyev (2000), Soil Taxonomy (ST) and the World Reference Base (WRB) on soil resources do not adequately assess the role of soil processes in soil taxonomic systems, despite the fact that these modern systems are based on genetic principles. The authors believe that the consideration of soil processes is important for understanding the genetic basis of modern soil taxonomic systems and the development of quantitative models of pedogenic systems. Therefore, the effects of soil-formation processes on current and future soil classification systems and pedogenic models are examined (Bockheim and Gennadiyev 2000).

The most successful means to diagnose ESP in pedology is the micromorphological analysis of soil thin sections. Consequently, special importance was given to the micropedological research in the diagnosis of the processes, forming the genetic soil profiles of Georgia.

Thus, based on the existing manuals and guidelines on micropedology (Kubiena 1938, 1970; Brewer 1964; Ball 1973; Bullock et al. 1975, 1985; Dobrovolsky 1977; Jongerius 1981; FitzPatrick 1984; Gerasimova et al. 1992; Tursina 2002; Stoops 2003), the micromorphological diagnosis, typification, and assessment of the level of main soil-forming processes manifestation have been conducted for soils of Georgia, in particular: humification, argilization, gleization, podzolization, lessivage, ferrugination, and carbonization.

4.2 Diagnosing and Distribution of Profile-Forming Processes

The options of formation and accumulation of soil organic matter were described by a set of micromorphological features reflecting the transformation of organic matter; argillization (weathering in situ)—by the nature of microstructure and the optical orientation of plasma; gleization—by decolorization of the basic mass (due to the loss of ferrum) and segregation of Fe-hydroxides; podsolization—by the presence of signs of movement of weathering products of the primary minerals; lessivage (illimerization, desilting) is described by the properties of the mechanical movement of the mobile clay material, i.e., by the nature of the sinter deposits forms; ferrugination (laterization, ferrallitization, and other processes associated with the movement, accumulation, and transformation of ferrum)—by the presence of various forms of ferruginous formations, the nature of ferruginous micro-zones, or zones of impregnation of plasma with a ferrous substance; carbonization—by the presence of calcite grains in the skeleton, microforms of carbonates and impregnations, or calcite crystals of various dimensions scattered in plasma material.

4.2.1 Distribution of Soil Organic Matter Morphotypes in the Soils

For the purpose of diagnosing the morphotypes soil organic matter, a thorough registration of micromorphological indices of all organic components in the organic profiles of the soils of Georgia was done: vegetation remnants, soil fauna, their metabolic by-products, and end products of humification (Fig. 4.1). On the basis of developing the relevant criteria, the main morphotypes of soil organic matter (Kubiena 1953; Müller 1987) were determined in the studied organic profiles of soils depending on the roughness and/or dispersion of the organic matter of soils. As a result, five groups of SOM in the soils of Georgia were identified (Matchavariani 2008): I—raw, II—raw-moder or moder-raw, III—moder, IV—moder-mull or mull-moder, and V—mull (Fig. 4.2). Of the studied soils, mainly: Gleysols and Solonetz correspond to the raw soil organic matter group; Rendzinas, Luvisols Albic, Acrisols Haplic, and Fluvisols correspond to group moder (medium dispersed soil organic matter); Vertisols, Chernozems, and, in some cases, Cambisol Chromic correspond to group mull (dispersed soil organic matter); Leptosols Umbric and Luvisols Albic correspond to group raw-moder and/or moder-raw (a transitional form from the dispersed soil organic matter to the medium one); Nitisols Ferralic, Kastanozems, Cambisols Dystric, and Cambisol Chromic correspond to group moder-mull and/or mull-moder (a transitional form from the medium soil organic matter to the dispersed one).

4.2.2 Distribution of Argillization in the Soils

Depending on the conditions of soil formation, there are different kinds of a microstructure of clay plasma. The process of argillization at a microlevel is diagnosed as in situ weathering, which is manifested in the soils of Georgia less or more (Fig. 4.3). Depending on the intensity of

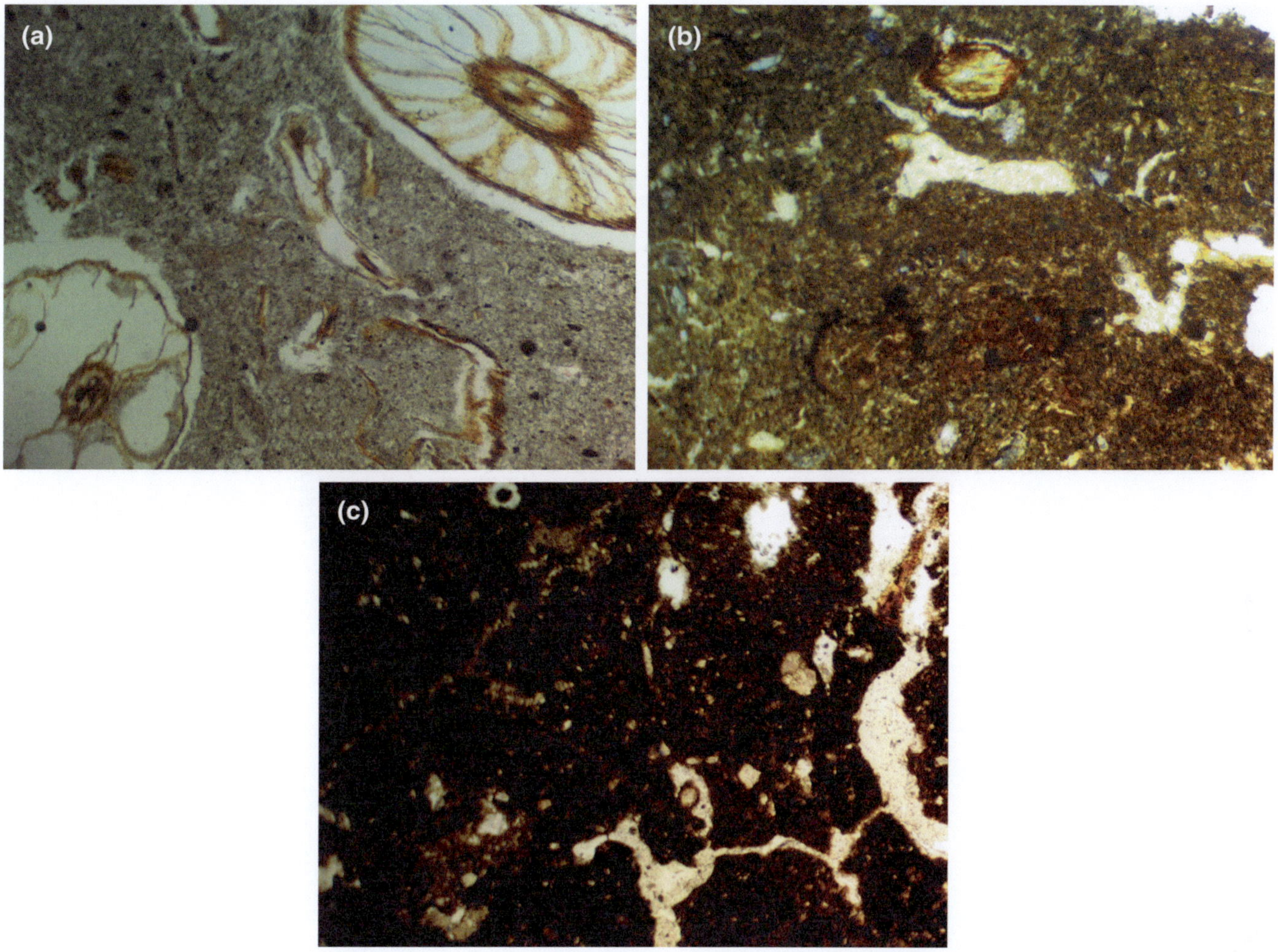

Fig. 4.1 Soil organic matter morphotypes: **a**—raw type (Luvisols Albic, nic.II); **b**—moder type (Fluvisols, nic.II); **c**—mull type (Vertisols, nic.+). Photos by L. Matchavariani

manifestation of the process of claying, three main grades were identified for the soils of Georgia (Matchavariani 2008): intense, medium, and weak (Fig. 4.4). This process is manifested most intensely in the following types of soils: Cambisols Dystric, Vertisols, Chernozems, and Cambisol Chromic. The mean values of the manifestation of the process of claying are identified in Kastanozems, Leptosols Rendzic, Leptosols Umbric, Leptosols Molic, partially Fluvisols. The process of argillization is weaker in Solonetz, Nitisols Ferralic, Acrisols Haplic, and Luvisols Albic. With Gleysols, the process of argillization as ESP is not virtually diagnosed in a micromorphological respect.

4.2.3 Distribution of Podzolization in the Soils

Usually, a diagnostic index of podzolization is the presence of the signs of destruction of primary minerals in soil surface horizons and transfer of the products of their chemical transformation down the profile. Podzolization, as the major profiling process, with the types of the soils of Georgia, including Luvisols Albic (in national classifications called as Subtropical-Podzolic/Yellow-Podzolic, and Gley-Podzolic soils), is not clearly diagnosed (Matchavariani 2008) despite the fact that this term, according to national classifications and FAO-UNESCO, even appears in the titles.

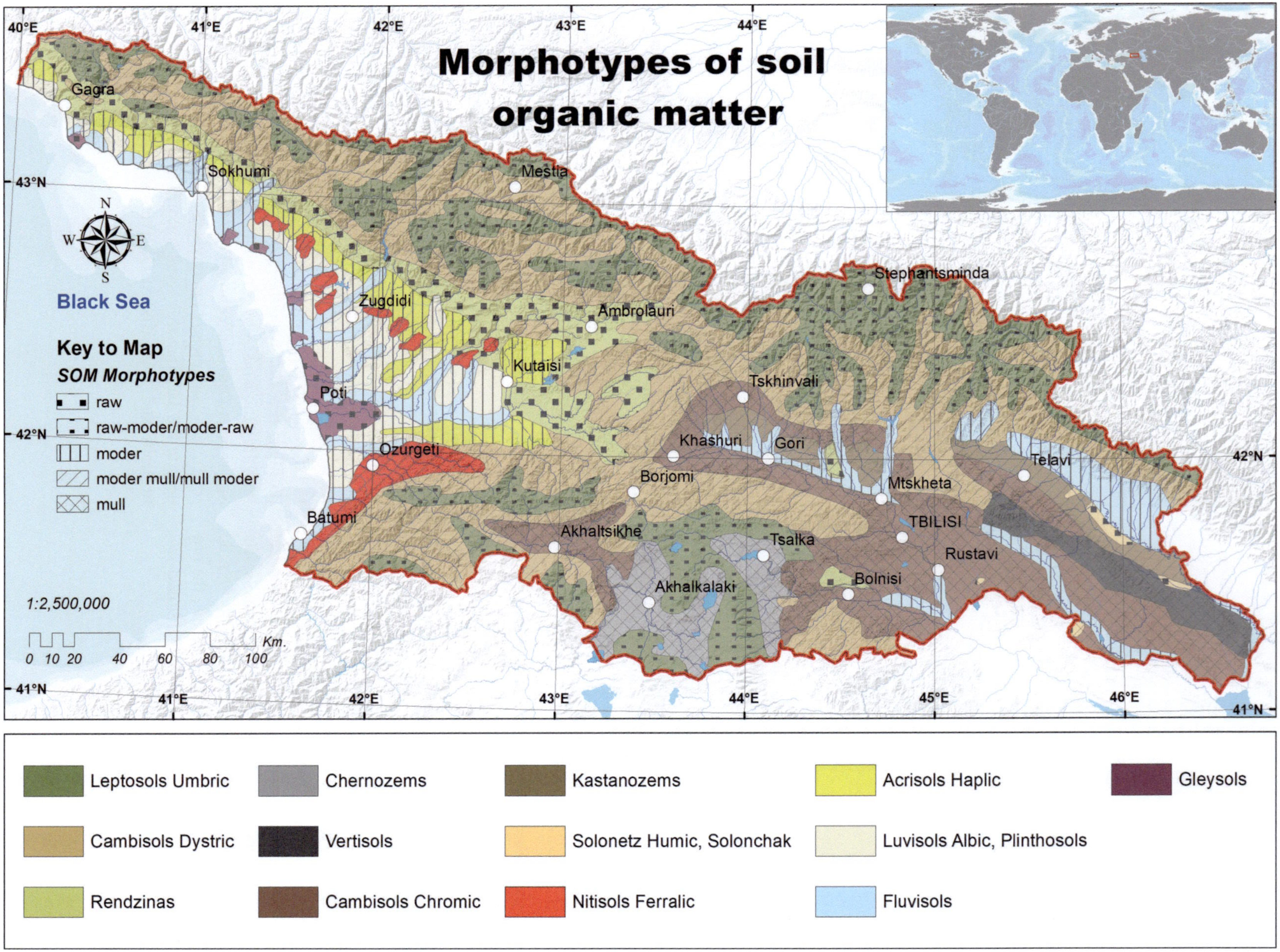

Fig. 4.2 Morphotypes of soil organic matter in soils. This map is created by D. Svanadze, based on data of L. Matchavariani

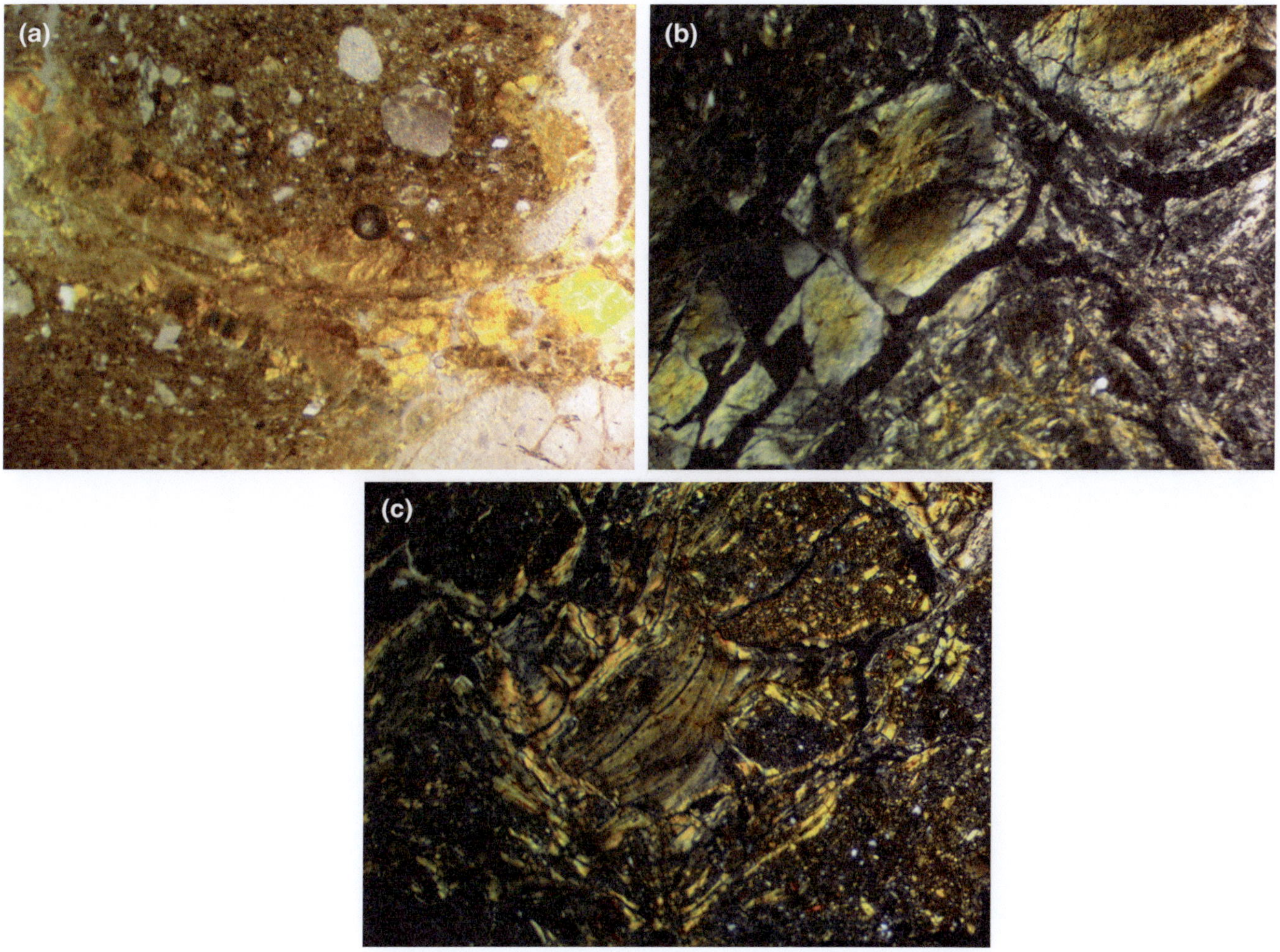

Fig. 4.3 Appearance of argillization process in soils on a microlevel; **a**, **b** and **c**—nic.II. Photos by L. Matchavariani

In theory, humid subtropical climate should promote intense degradation and movement of substances through a vertical profile. However, based on micromorphological studies, some signs of the presence of the products of chemically modified substances in the transit pores are fixed only locally. Therefore, the process of podzolization in these soils sometimes has only an intra-horizon, not an intra-profile value.

4.2.4 Distribution of Lessivage in the Soils

Micromorphologically, lessivage is diagnosed by the presence of the optically oriented clay cutans (sinter deposits, clay flows) in vertical, transit pores (Fig. 4.5). In the soils of Georgia (Matchavariani 2008), this process is manifested as following three forms (Fig. 4.6): clay cutans, as a sign of the actual lessivage and typical to the soils in humid regions—Nitisols Ferralic, Acrisols Haplic, and Luvisols Albic; silty cutans (sometimes clayey silty or sandy silty), typical to the soils of mountain regions—Cambisols Dystric, Acrisols Haplic, Rendzinas, and Leptosols Umbric; and complex cutans in the form of films on the walls of the pores and mineral grains (clay particles move short distances in different directions), typical to the soils of East Georgia—Cambisol Chromic, Kastanozems, and partially Vertisols.

4.2.5 Distribution of Gleization in the Soils

The diagnostic sign of the process of gleization at a microlevel is the contrast in the distribution of ferrum hydroxides (Fig. 4.7). This process, as a profile-forming one, in Georgia is manifested primarily in humid subtropical soils (Matchavariani 2008). In addition, as local signs, gleization occurs in low, depressed areas of different regions, partly in

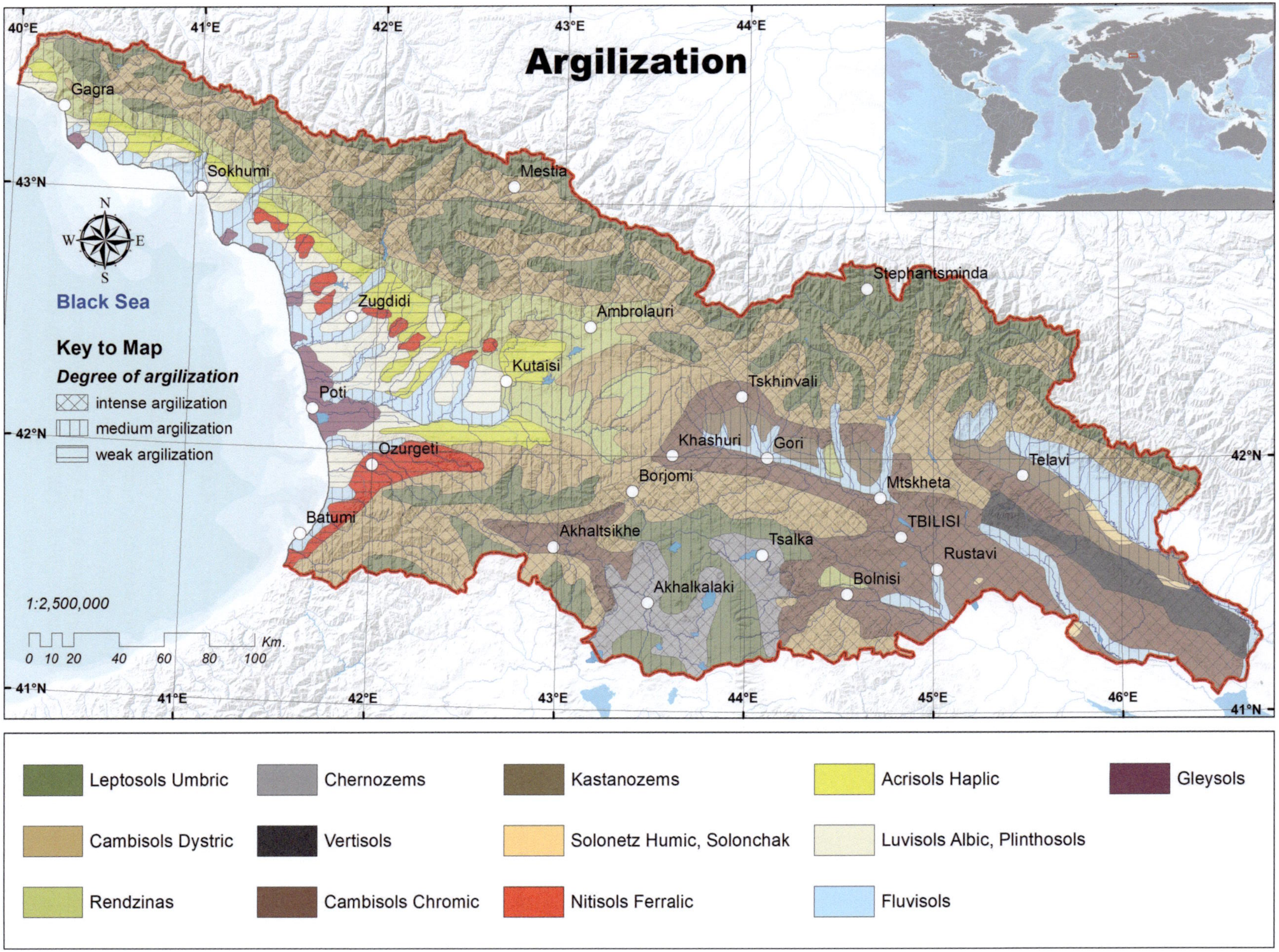

Fig. 4.4 Intensity of argillization in soils. This map is created by D. Svanadze, based on data of L. Matchavariani

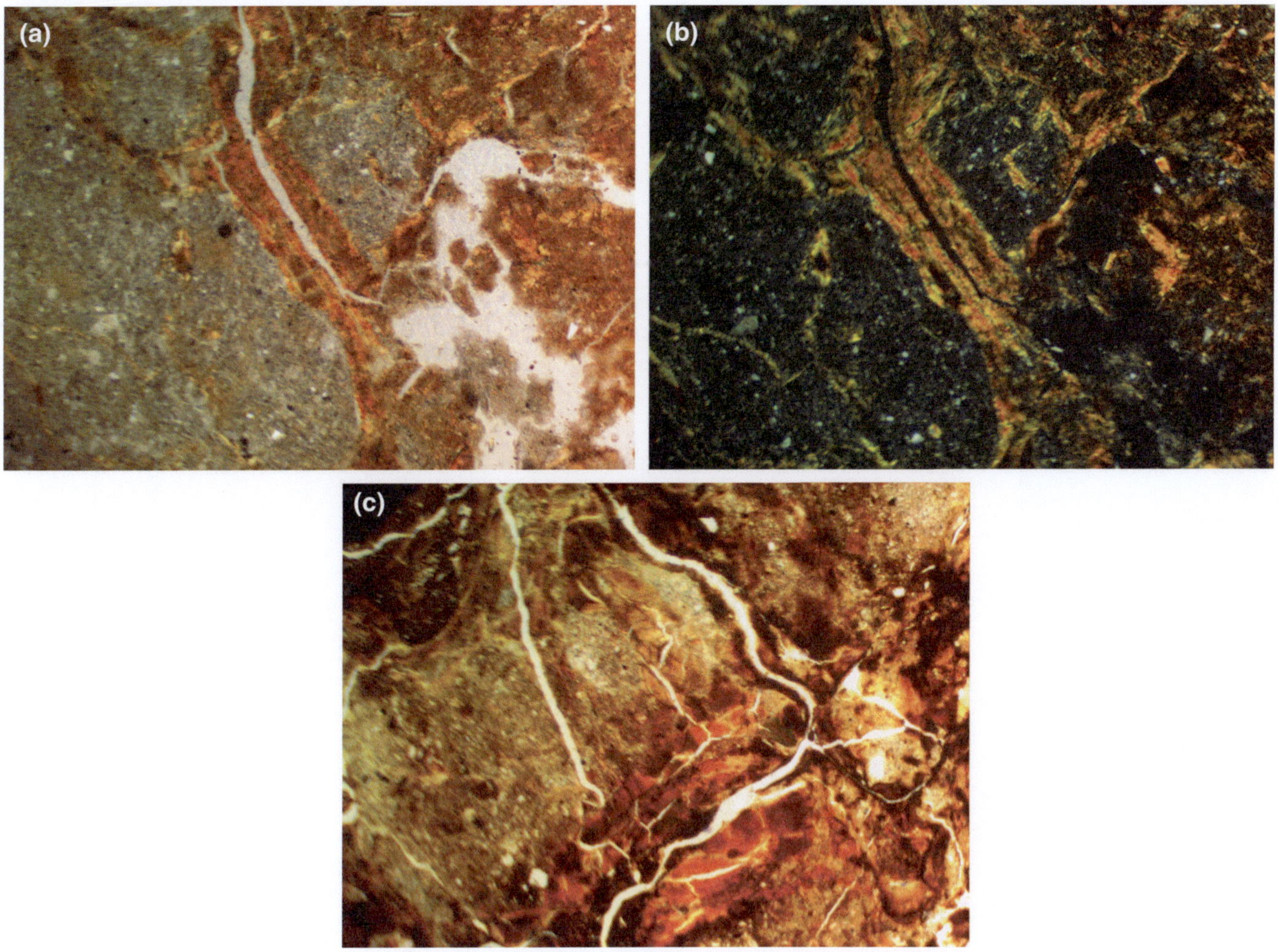

Fig. 4.5 The manifestation of the lessivage process at a microlevel—the sinters of optically oriented clay; **a** and **c**—nic.II; **b**—nic.+. Photos by L. Matchavariani

the mountain-meadow zone. Therefore, this process in the soils of Georgia is grouped according to the degree of intensity of the signs of gleization, as strong, medium, and weak (Fig. 4.8). The most intense process of surface gleization, like ESP, is manifested in the following types of soils: Acrisols Haplic, Luvisols Albic, Gleysols, and Fluvisols; the gleization of a medium intensity occurs in Nitisols Ferralic, Leptosols Umbric, and Solonetz Humic; weaker signs of gleization are also diagnosed in Cambisol Chromic, Kastanozems, and Chernozems.

4.2.6 Distribution of Ferrugination in the Soils

The processes associated with sedimentation, movement, and transformation of ferrum take place in many different kinds of soils of Georgia (Matchavariani 2005, 2008), and in humid subtropics first of all: Nitisols Ferralic, Acrisols Haplic, Luvisols Albic, Gleysols, and Fluvisols; the signs of ferrugination as small micro-zones of segregation and impregnation of plasma are manifested in Cambisols Dystric. Partially ferruginous secretions are also noted in arid regions. In Solonetz Humic, ferrum secretions are sometimes observed as large concretions with complex structure, being the result of the past stages of soil formation. Depending on the kind of manifestation of ferrugination in soils, two forms can be easily distinguished: concretion and plasma ferrugination, although with many humid subtropical soils, ferrugination takes place in fact, in both forms, both in the form of concretion formations and in the form of ferruginous zones impregnated in the basic mass (Fig. 4.9). As a result, the gradation of ferrugination was done by considering the intensity of the process manifestation (Fig. 4.10).

4.2.7 Distribution of Carbonization in the Soils

The carbonization, at the microlevel, is diagnosed according to a number of characteristics, taking into account the genetic-morphological groups of carbonates and the sizes of

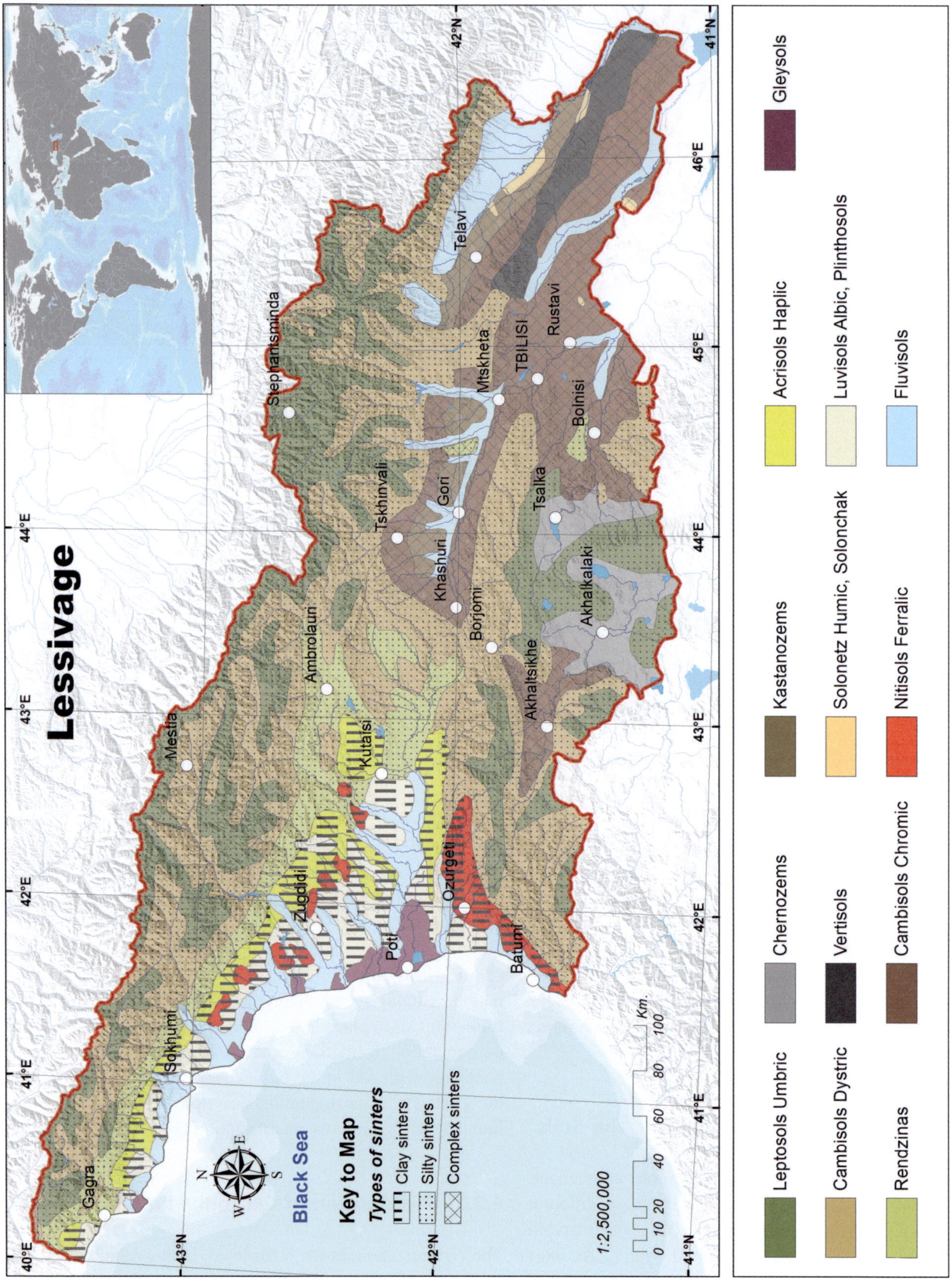

Fig. 4.6 The types of sinter deposits in soils. This map is created by D. Svanadze, based on data of L. Matchavariani

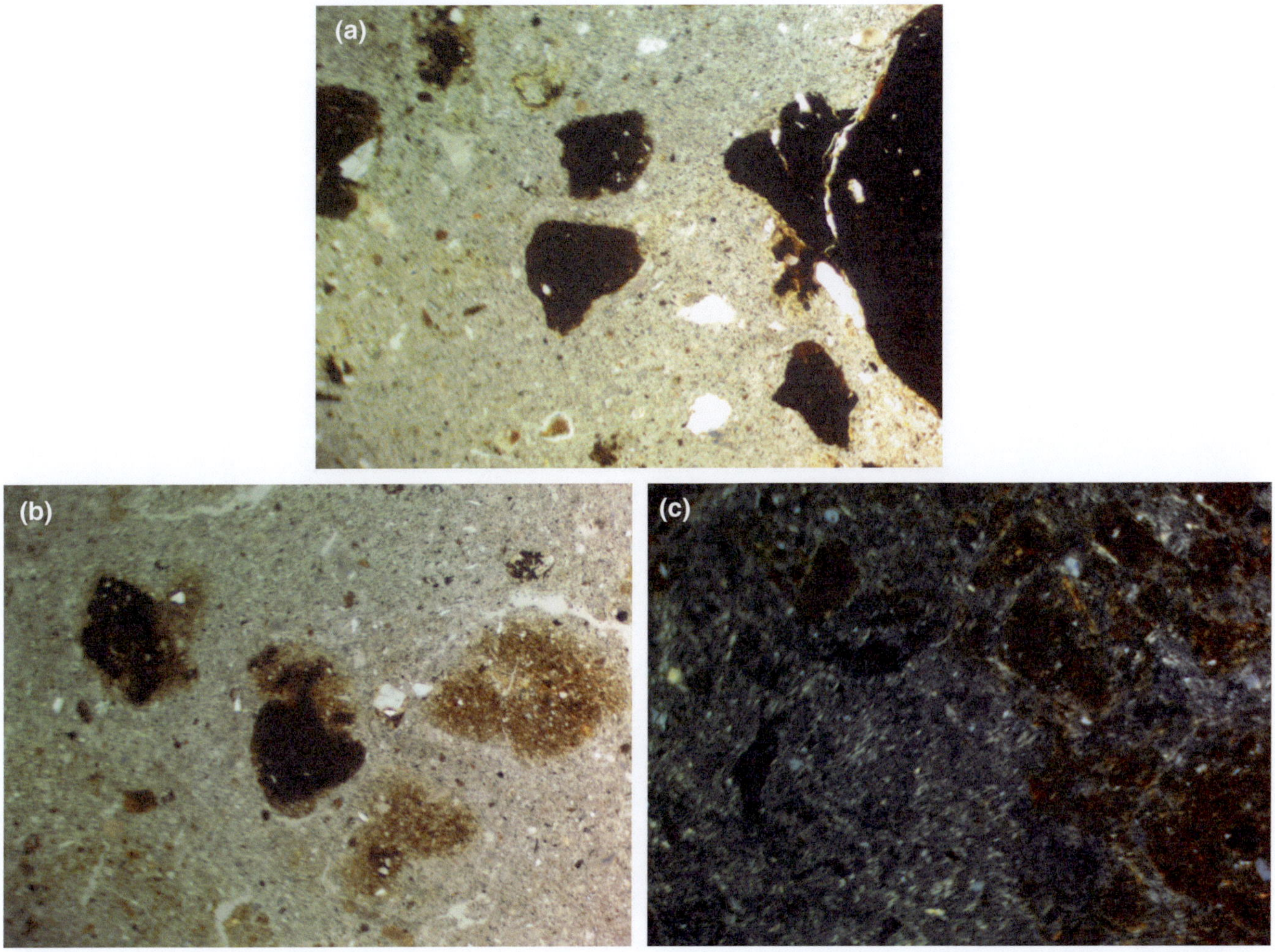

Fig. 4.7 The manifestation of the gleization process in soils at a microlevel; **a** and **c**—nic.II; **b**—nic.+. Photos by L. Matchavariani

calcite crystals (Fig. 4.11). This process, as a major profile-forming one, is manifested in the following soils of Georgia (Matchavariani 2008): Rendzinas, Vertisols, Cambisol Chromic, Kastanozems, partly in Leptosols Molic, Solonetz Humic and Solonchak, and depending on the region of formation, in the Fluvisols. With its form of expression, the carbonization is divided into two main types: concretion and plasma; with the intensity of manifestation, it is divided into strong, average, and weakly calcareous. In soils where this process is the most intense and leading, practically both groups of carbonizations are fixed (Fig. 4.12).

4.2.8 Conclusion

Based on the micromorphological diagnosis of the above-mentioned ESP, the groups of processes most typical to the specific soil types were identified. A map showing the distribution of the main profile-forming processes in the soils of Georgia (Matchavariani 2008) is compiled (Fig. 4.13). In order to establish the geographic features of soils, an attempt has been made to correlate this material with the landscape map of the Caucasus (Beruchashvili 1979), where at a type level, depending on the degree of humidification, humid, semi-humid, semiarid, and arid regions are distinguished on

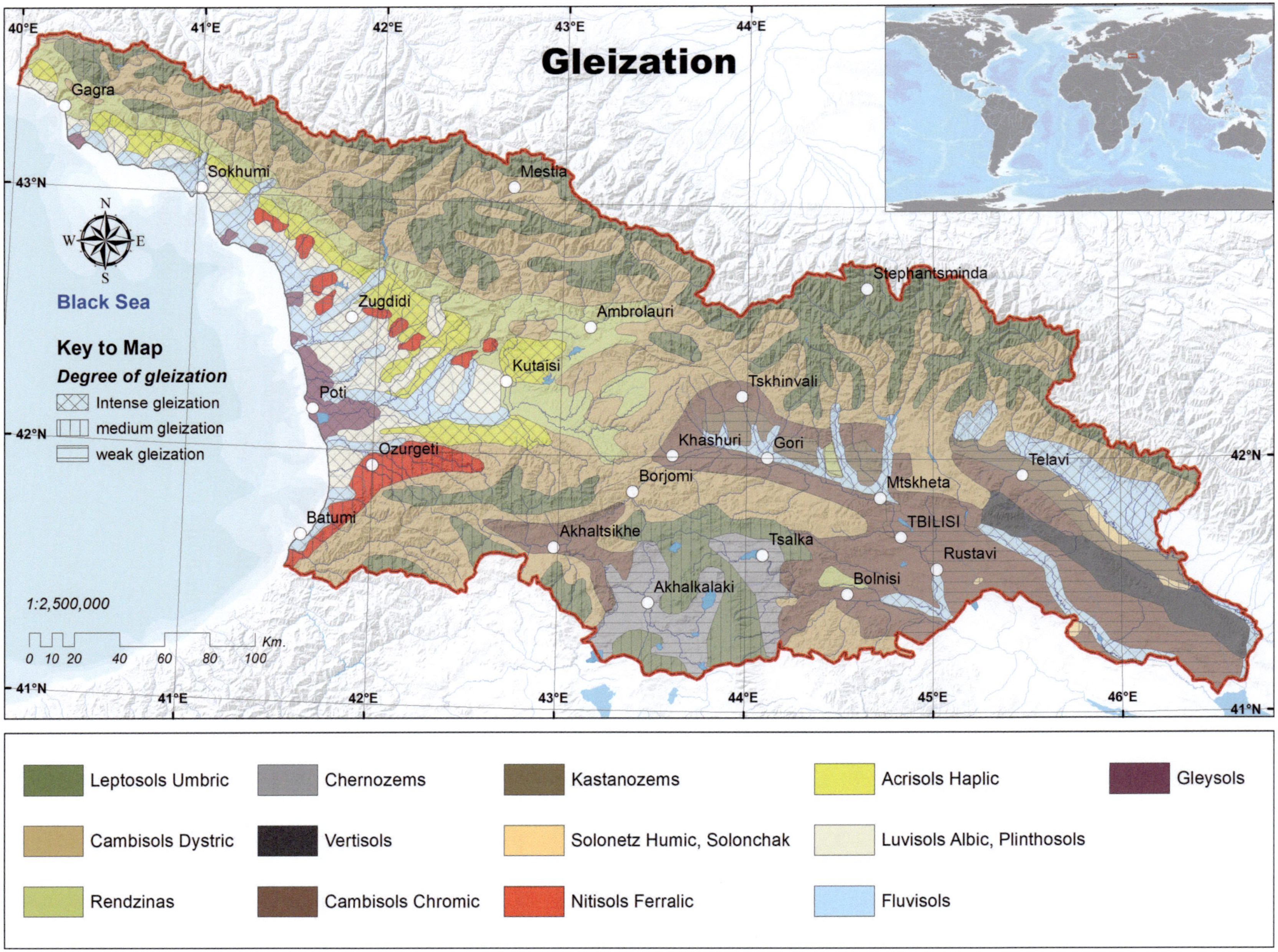

Fig. 4.8 Degrees of gleization in soils. This map is created by D. Svanadze, based on data of L. Matchavariani

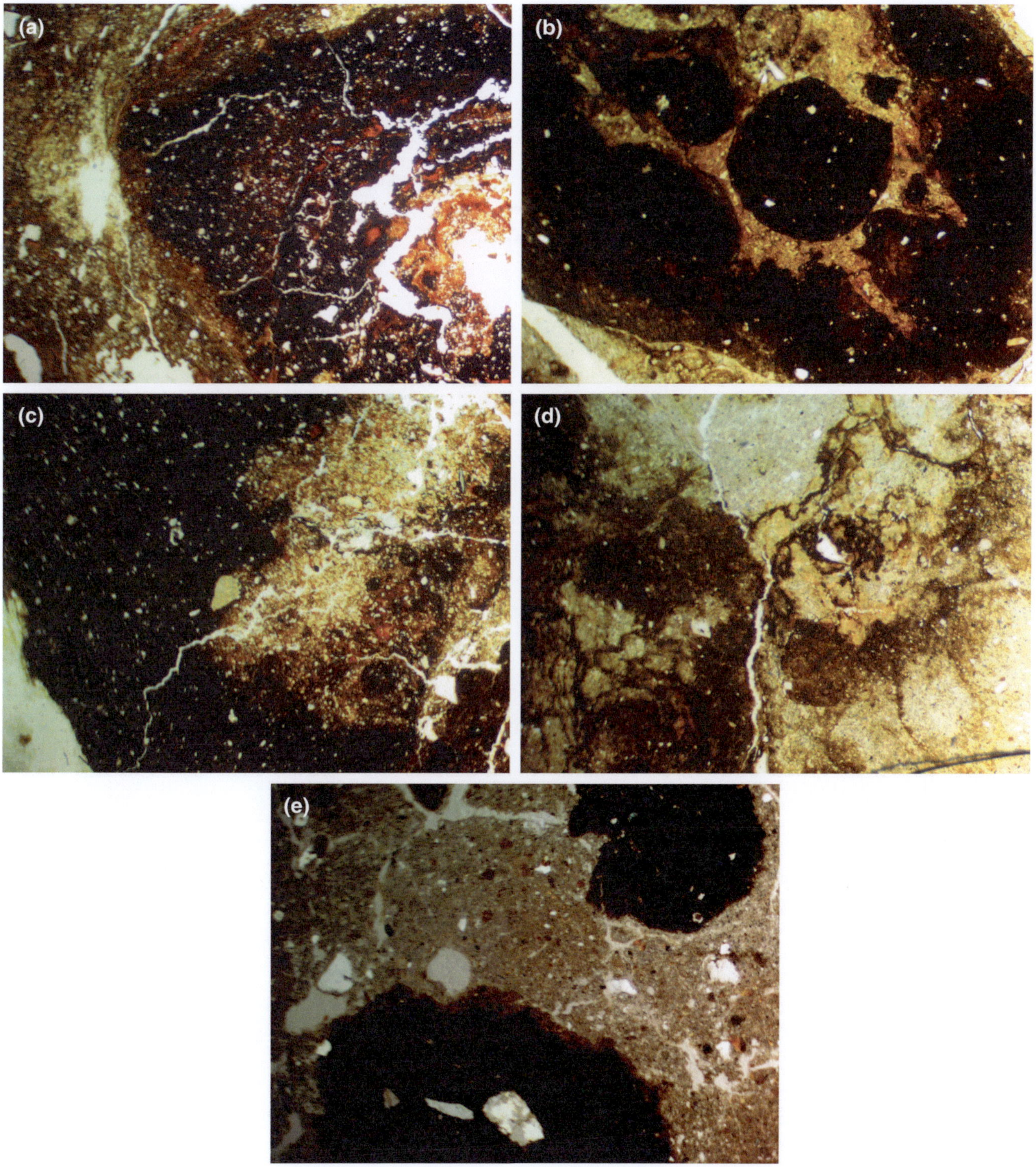

Fig. 4.9 Forms of the ferrugination process manifestation in soils at a microlevel; nic.II. Photos by L. Matchavariani

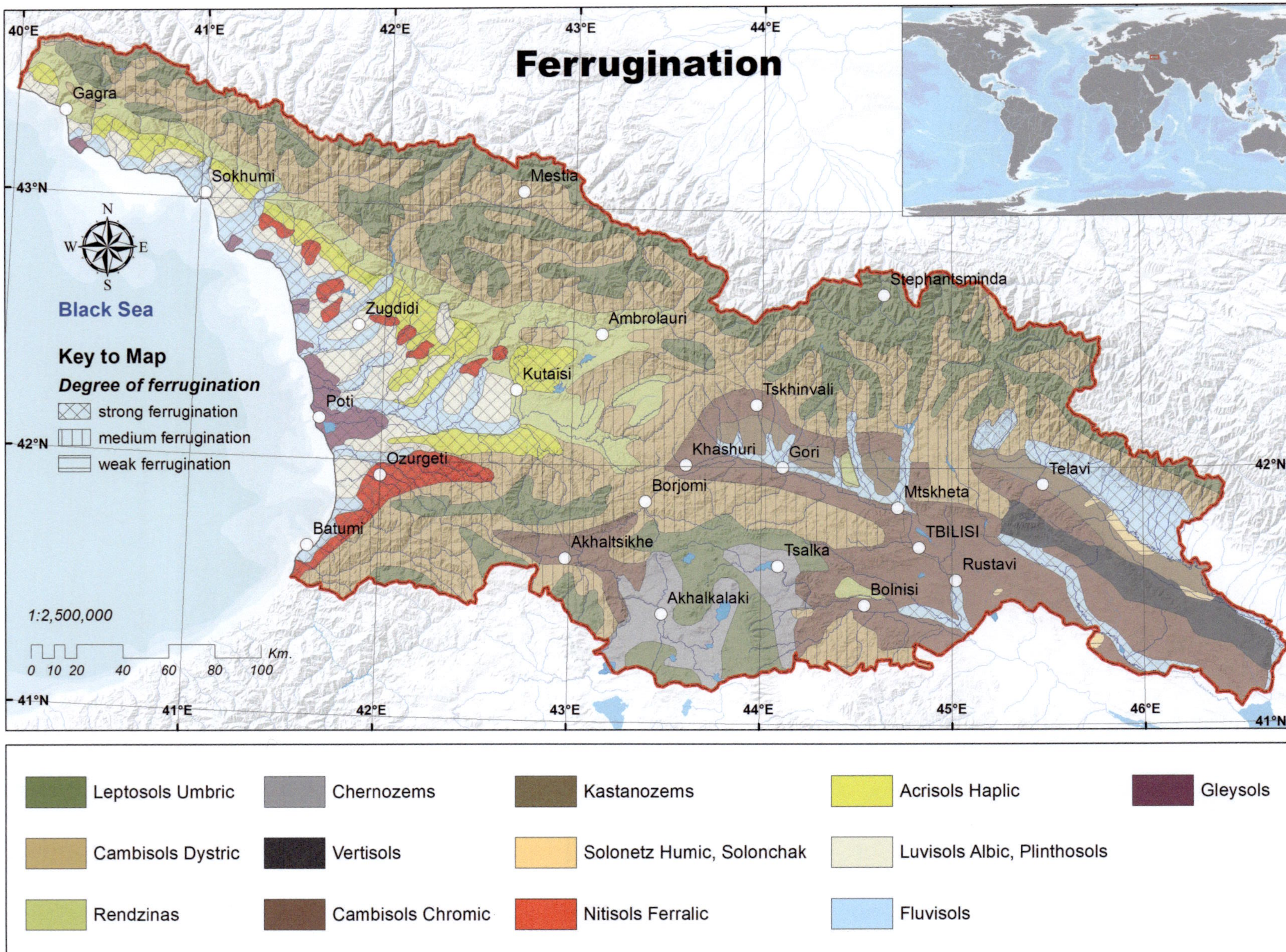

Fig. 4.10 Degrees of ferrugination in soils. This map is created by D. Svanadze, based on data of L. Matchavariani

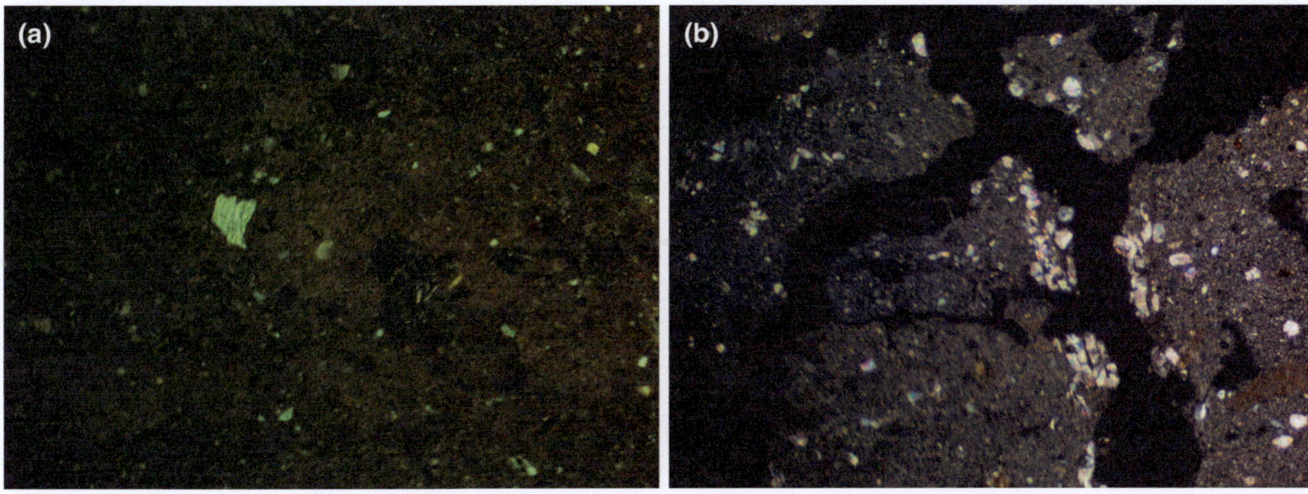

Fig. 4.11 The manifestation of the carbonization process in soils at a microlevel; nic.+ **a**—crypto-grained calcite (calcareous plasma); **b**—granular calcite crystals. Photos by L. Matchavariani

the territory of the country. The humid landscapes, which occupy the largest territory of the country and cover a large group of qualitatively different soils—Luvisols Albic, Plinthosols, Acrisols Haplic, Nitisols Ferralic, and Cambisols Dystric—combine the following ESPs: lessivage, gleization, ferrugination, and partly argilization. Arid landscapes occupying a small area of the southeastern part of the country include a small part of only Cambisol Chromic and, in fact, have their typical features. Semi-humid landscapes mainly constitute Fluvisols of East Georgia, which belong to the azonal types and are less subject to climatic factors. The same is true with intrazonal Rendzinas, in the formation of which the leading role is played by a carbonate substrate.

A more distinct is the relationship between the semiarid landscapes, incorporating Cambisols Chromic, Kastanozems, Vertisols, and Chernozems, with the following leading processes: humification, carbonization, and argillization (Fig. 4.13). Bog and floodplain soils (along river valleys), which are a part of the group of hydromorphic landscapes, have no general landscape-geographical regularities.

Thus, the ratio of the maps reflecting the spread of the major profile-forming processes in the soils of Georgia with the types of landscapes (humid, arid, semi-humid, and semiarid) has shown a very peculiar picture. A clear correlation is observed only with semiarid landscapes, where the leading ESPs are humification, carbonization, and argillization. Soil is a special component of the landscape, not always complying with certain regularities. In the process of forming of certain soils, often some factor may play a decisive role and a commonly known soil-formation process, based on the complex action of factors, may fall out of general geographic patterns. Therefore, it is not accidental that in the names of landscapes the soil appears to the least extent.

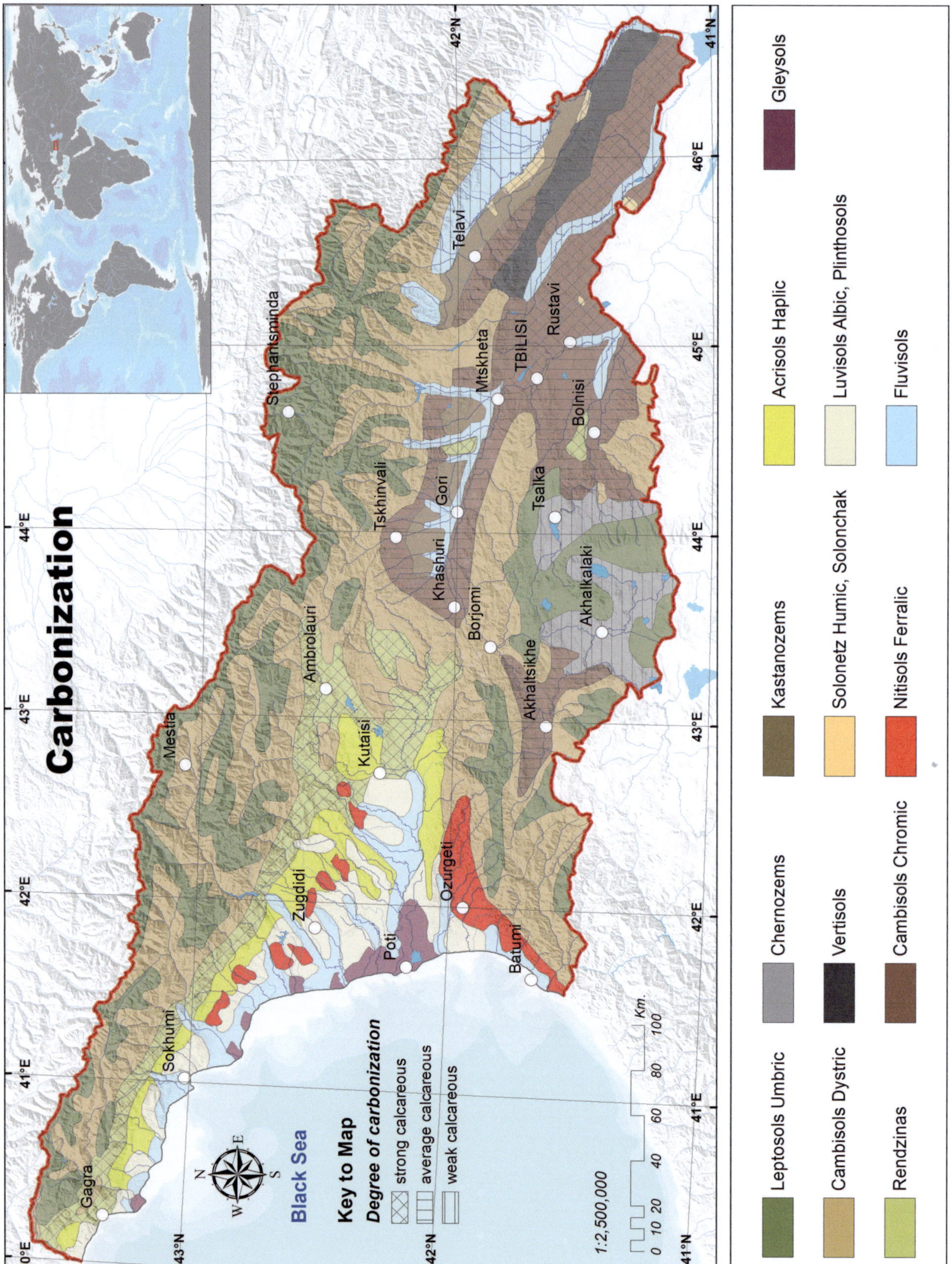

Fig. 4.12 Degrees of carbonization in soils. This map is created by D. Svanadze, based on data of L. Matchavariani

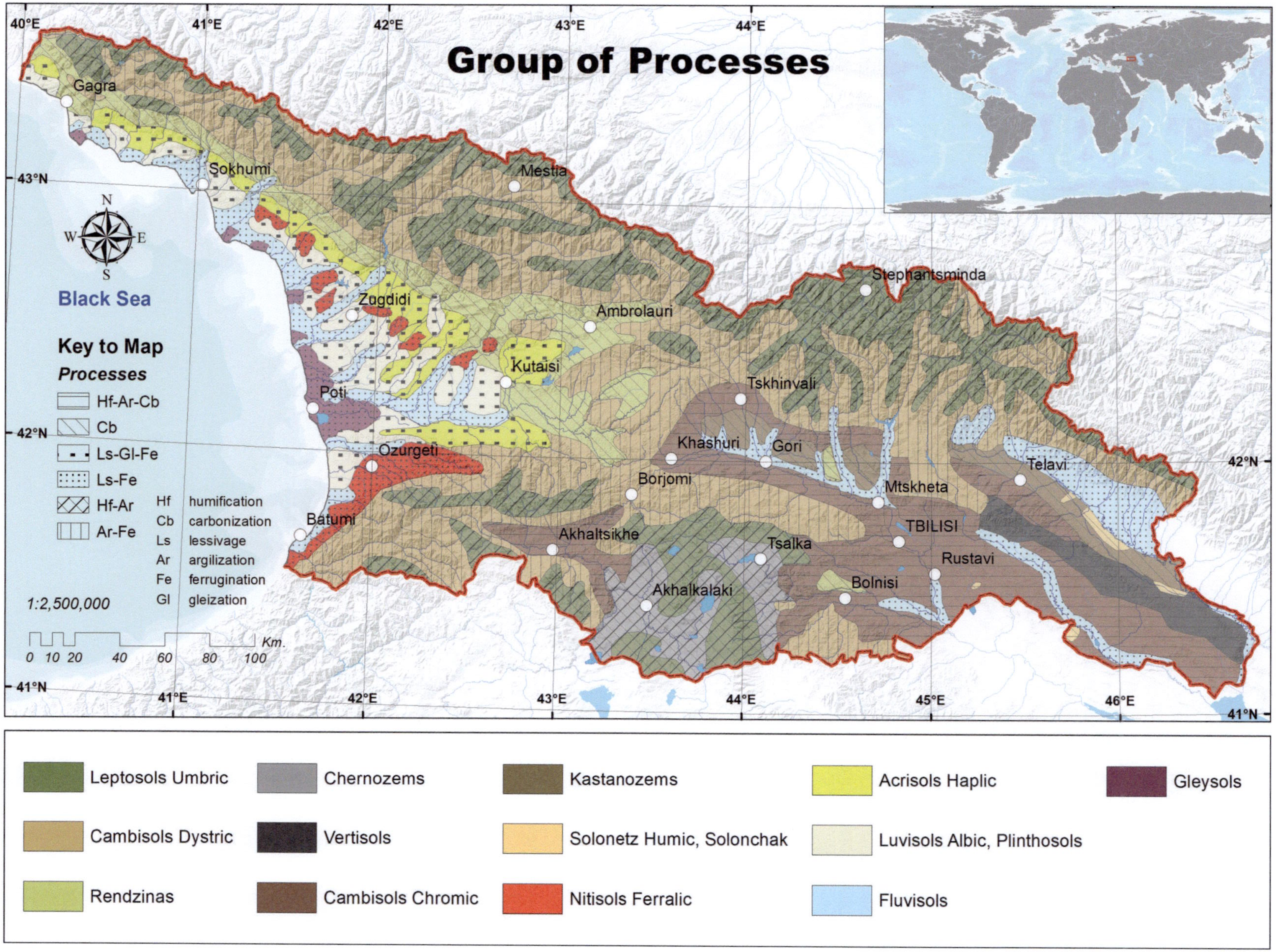

Fig. 4.13 The features of distribution of the soil processes group in soils. This map is created by D. Svanadze, based on data of L. Matchavariani

References

Ball (1973) Micromorphological analysis of soils. Wageningen

Beruchashvili N (1979) Landscape map of caucasus (1 : 1 000 000). TSU, Tbilisi

Bockheim JG, Gennadiyev AN (2000) The role of soil-forming processes in the definition of taxa in Soil Taxonomy and the World Soil Reference Base. Geoderma (Elsevier) 95(1–2):53–72

Brewer R (1964) Fabric and mineral analysis soils. Wiley, New York

Bullock P, Loveland PJ, Murphy CP (1975) A technique for selective solution of iron oxides in thin sections of soil. Soil Sci 26(3):27

Bullock P, Fedoroff N, Jongerius A, Stoops G, Tursina T (1985) Handbook for soil thin section description. Waine Research Publishing, Albrigton

Dobrovolsky GV (1977) Micromorphology of soils as a special section of soil science. Eurasian Soil Sci 3:19–23

FitzPatrick EA (1984) Micromorphology of soils. Chapman and Hall, London

Gerasimov IP (1973) Elementary soil processes as a basis for genetic diagnostics of soils. Eurasian Soil Sci 5:102–113

Gerasimov IP, Glazovskaya MA (1960) Fundamentals of soil science and soil geography. Moscow (Osnovi pochvovedeniya i geografii pochv. Moskva), 233p

Gerasimova MI, Gubin SV, Shoba SA (1992) Micromorphology of soils in the natural zones of the Soviet Union, Pushchino (Mikromorfologiya pochv prirodnikh zon SSSR. Informatsionno-spravochniye materialy), 215p

Jongerius A (1981) The role of micromorphology in agricultural research. Soil micromorphology. Rothamsted

Kubiena W (1938) Micropedology. Ames, Jowa, 187p

Kubiena W (1953) The soils of Europe. Madrid, London

Kubiena W (1970) Micromorphological features of soil geography. Rutgers University Press, New Brunswick

Matchavariani LG (2005) Morphogenetic typification of concretions in subtropical podzolic soils of Georgia. Eurasian Soil Sci 38 (11):1161–1172

Matchavariani LG (2008) Geographic paradigms of the microfabric of major soils of Georgia. Tbilisi, Universal, 309p

Müller PE (1987) Studien über die natürlichen Humusformen. Springer, Berlin

Rode AA (1948) Soil-formation process and soil evolution. Moscow, AS USSR (Pochvoobrazovatelniy process i evolucia pochv. Moskva, izd. AN SSSR)

Rozanov BG (1975) Genetic morphology of soils. Moscow, MSU (Geneticheskaya morfologia pochv. Moskva, MGU), 293p

Stoops G (2003) Guidelines for analysis and description of soil and regolith thin sections. Soil Science Society of America, Madison

Tursina TV (2002) Micromorphological methods in the analysis of debatable soil genetic problems (Mikromorfologicheskie metodi v analize diskussionnikh pochvennikh geneticheskikh problem). Pochovedenye #7:876–885

Soil Distribution and Properties

5

Lia Matchavariani and Besik Kalandadze

Abstract

The following soil classes of Georgia, corresponding of the WRB groups are considered: Leptosols Umbric, Cambisols Dystric, Rendzinas, Leptosols Molic, Chernozems, Vertisols, Cambisols Chromic, Kastanozems, Solonetz Humic, Nitisols Ferralic, Acrisols Haplic, Luvisols Albic, Gleysols, Fluvisols. The main indicators of soils distribution and their major characteristics are described there, particularly: location, soil-forming conditions, profile structure, macro- and micromorphological descriptions, morpho-chemical properties with some chemical data, agro-physical features, etc.

Keywords

Soil zoning • Taxonomic units • WRB • Soil classes • Soil distribution • Soil morphochemistry

5.1 Introduction

Soil formation and classification, as a rule, are a key field of research in the soil science (Hartemink and Bockheim 2013). There are many different national and international classifications of soils used in the world. Soil classification means grouping the soils based on their common features, properties, and fertility and implies the identification and formulation of the principles of classification; scientific treatment of the hierarchical system of taxonomic units (type, subtype, etc.); development of the soil nomenclature (system of appellations); and identification of the features used to diagnose and map the soils of all classification subsets. The basis to develop the modern classification systems is a genetic principle used to consider the features of the properties of soils as a result of the soil-formation process and to unite ecological, morphological, and evolutionary approaches. As a rule, the classifications thoroughly consider the morphological and micro-morphological properties of soil profiles, texture and properties of soils, ecological processes, qualitative content of organic substances, etc.

The main taxonomic unit of a national soil classification system is the genetic type of the soil. Lower taxonomic units are subtype, genus, species, variety, and phase.

The subtype is marked out from soil types. It is a group of soils, which is a transitional step between the soil types and is determined by the major soil-forming process.

Genus is marked out from subtypes. The qualitative genetic properties are determined under the influence of local conditions, such as the composition of soil-forming parent materials and chemism of ground waters. It can be also determined by the properties acquired during the phases preceding weathering and soil formation (relict horizons or features).

Species are marked out from soil genus and differ by the degree of the development of the soil-formation process (e.g., podzolization, gleization, argilization, etc.).

Variety is determined by the mechanical composition of the upper soil horizons and soil-forming parent materials.

Phase is determined by the genetic properties of soil-forming parent materials (e.g., alluvion).

Within every genetic type, central types with term typical or ordinary used to describe it and transitional subtypes possibly incorporating the features different from the subtype or associated with neighboring types can be distinguished. They also use additional terms to describe leading processes (e.g., Cinnamonic Calcareous); to identify morphological peculiarities (color) (e.g., light gray-brown); to locate the soils (e.g. black southern), etc.

The terms determining the properties typical to soils are used for the nomenclature of soil genus (e.g., solonetz, gley), indicators of relict properties (e.g., residual meadow, residual gley), etc.

L. Matchavariani (✉) · B. Kalandadze
Faculty of Exact and Natural Sciences, Department of Soil Geography, Ivane Javakhishvili Tbilisi State University, Tbilisi, Georgia
e-mail: lia.matchavariani@tsu.ge

B. Kalandadze
e-mail: besik.kalandadze@tsu.ge

L. Matchavariani (ed.), *The Soils of Georgia*, World Soils Book Series,
https://doi.org/10.1007/978-3-030-18509-1_5

The nomenclature of the soil species contains the terms characterizing the soil properties quantitatively and soil processes. Three categories of terms are used: texture (with little, average or high content of soil organic matter); depth of individual soil horizons or of the whole profile (of little, average or great strength, etc.) and indicating the events (slightly, averagely, intensely gley, etc.).

The names of mechanical texture are used for the soil nomenclature, while the terms describing the lithology and the genesis of the soil-forming parent materials are used for the nomenclature of phases.

The full name of a soil has the following order: type, subtype, genus, species, variety, and phase. For example, cinnamonic (type), typical (subtype), meadow (genus), average humus (species), heavy loamy (variety), and sandy loamy (phase) soil.

For the diagnostics of the soils, meaning identifying a set of features used to attribute the soils to some or other classification subcategory, usually, easily identifiable morphological properties and simple analyses are used. However, these features are not sufficient for a number of soil types, and the results of more complex analyses are used instead (content of soil organic matters and absorbed cations, results of some chemical analyses), as well as hydrothermal characteristics of soils, etc.

The national classification used in Georgia is associated with the name of Sabashvili (1948). The classification plan developed in the 1960s used the materials of soil maps and the soils on it are positioned in groups and types corresponding to the principal vertical and landscape zones. In addition to types, the plan shows the soil subtypes and genera. The description of soil groups and types start with Lowlands, continue with piedmonts, mountain-and-forest zones, and end with mountain-meadow zones. The classification also includes intrazonal soils: marsh, salt, and zonal soils—alluvial. The units of classification: soil group, type, subtype, genera, variety, as well as texture, bedrock, type of development, and degree of erosion (Urushadze 2013).

The components of the classification of the soils of Georgia developed by A. Charkseliani, R. Petriashvili, and M. Kipiani at the end of the 1980s are group, type, subtype, genera, kind (depending on the thickness of an accumulative horizon layer, content of soil organic matter, degree of development), variety, and phase.

According to Sabashvili (1948, 1965), three quite different soil regions (Western, Eastern, Southern) are distinguished in Georgia with appropriate subregions, zones, and areas (Fig. 5.1).

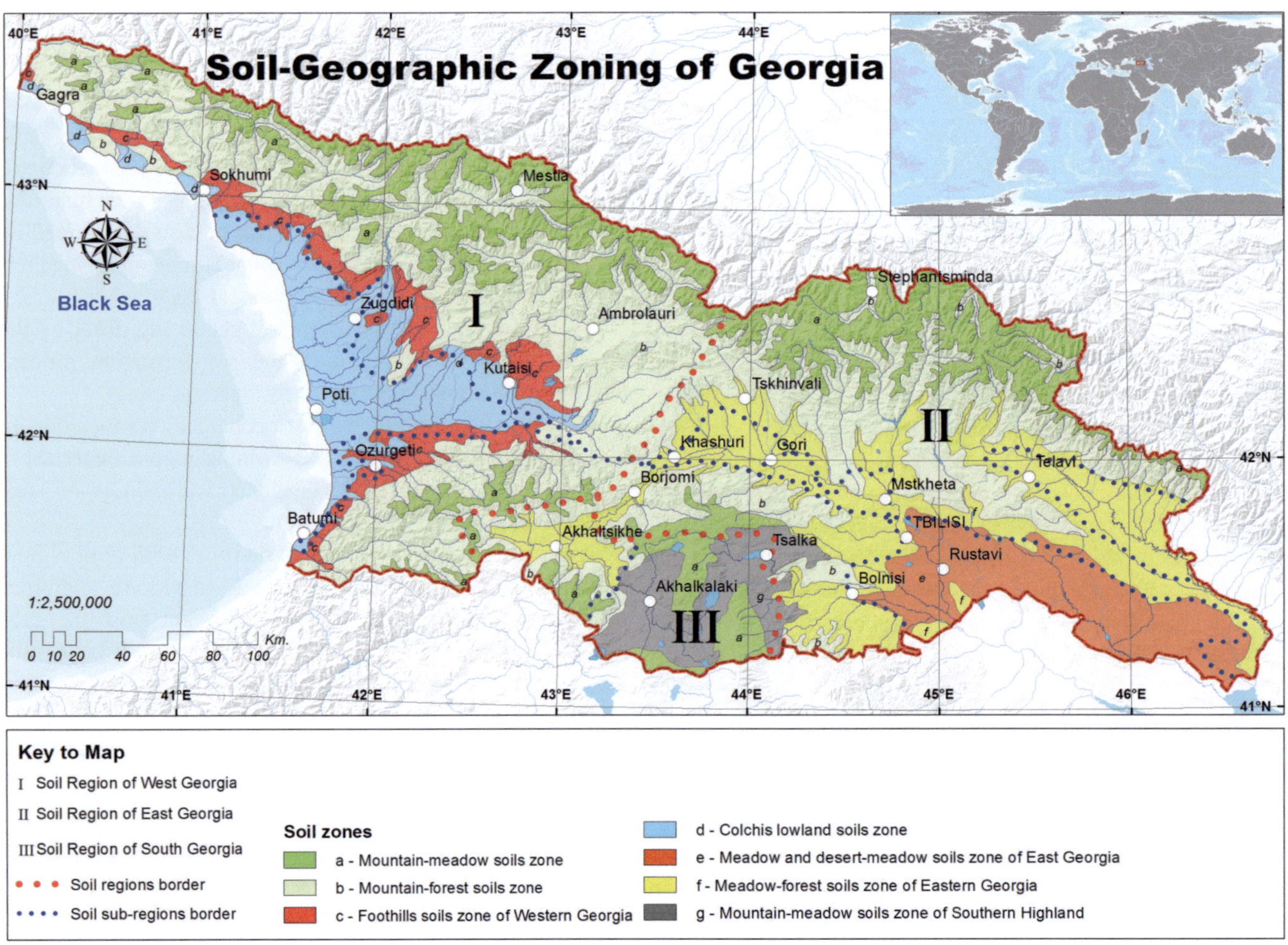

Fig. 5.1 Soil-geographic zoning. This map is created by D. Svanadze, based on data of M. Sabashvili

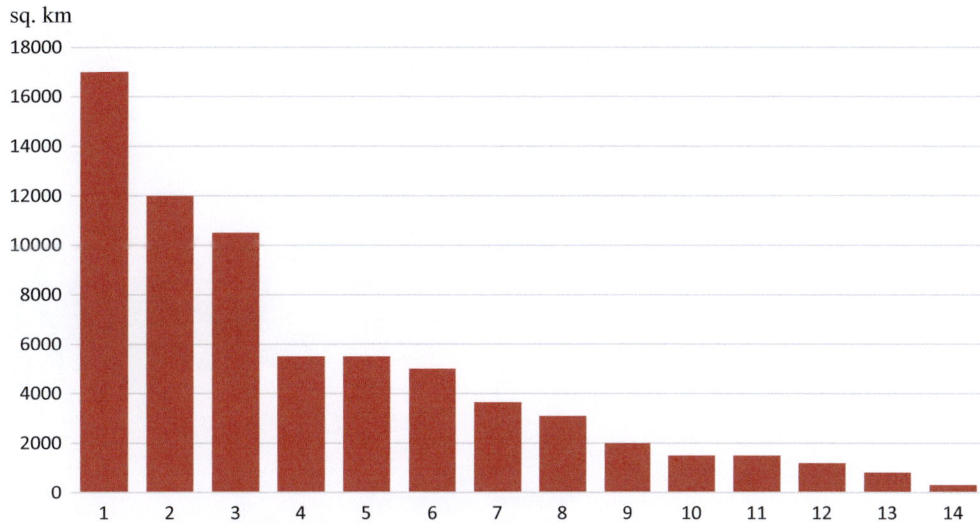

Fig. 5.2 The areas occupied by the main soil classes (according to data of T. Urushadze): 1—Cambisols Dystric; 2—Leptosols Umbric; 3—Cambisols Chromic; 4—Fluvisols; 5—Acrisols Haplic; 6—Rendzinas; 7—Luvisols Albic, Plinthosols; 8—Vertisols; 9—Chernozems; 10—Nitisols Ferralic; 11—Kastanozems; 12—Gleysols; 13—Leptosols Molic; 14—Solonetz Humic, Solonchak

In 2002, in the framework of the project "Cadastre and Land Registration" (realized with the financial support of KfW), a group of researchers, on the basis of the modern international approaches, made an inventory of Georgia's soils.[1]

Nowadays, one of the most popular classified-diagnostic systems is the World Reference Base for Soil Resources (WRB), that is, standard of soil correlation and international communication (Urushadze 2013; Urushadze and Blum 2011). This approach is based on fundamentally different principles and aims of the development of scientific relations. It is also a fundamental part of the soil resources management and rational use. WRB is not a dogmatic and legal document, and as unified common "soil language", it is developing an open system, which serves the national soil classification and correlation diagnostics (World Reference Base 2015). WRB is not intended to replace the national classifications. It is a real opportunity for individual countries—the doorway to the international scientific community and the general orientation.

Accordingly, the soil type names of the national classification are fundamentally different from the FAO and WRB soil class taxonomy. The description of soils in this chapter is given by the corresponding of the FAO-WRB groups (Fig. 5.2), combined with the national classification.

5.2 Leptosols Umbric

Leptosols Umbric, which correlate with the national classification as mountainous-meadow soils, is quite a common soil class in Georgia. It is mainly spread in the subalpine and alpine zones of Great Caucasus and southern mountains of the Lesser Caucasus, at 1800(2000)–3200(3500) m above sea level (Fig. 5.3). The hypsometric limits of its distribution vary depending on the distance from the sea, physical–geographical conditions of the mountainous massifs and economic activity of the population. The hypsometric amplitude of the distribution of the mountain-meadow soils over Great Caucasus is greater than it is in the southern mountains of the Lesser Caucasian. This type of soil adjoins to the so-called mountainous-meadow-chernozem-like soils in the subalpine and alpine zones, and "Mountain-forest-meadow" in the Subalpine zone and primitive soils in the nival zone (Fig. 5.4).

The first researcher of the "Mountainous-Meadow" soil was V. Dokuchaev, who identified the properties of this type of soil (such as turf formation, little strength of a soil profile). Detailed studies of mountain-meadow soils were accomplished by Sabashvili (1965, 1970), as well as Tarasashvili (1956), Talakhadze (1962), Talakhadze and Mindeli (1980), Iashvili (1987), Urushadze (1997), etc.

Considered soil is mainly formed on the leached hard parent materials weathering products and occupies all expositions of the upper parts of the mountains and slopes where the amount of precipitations exceeds the evaporation by 2 or 3 times what determines the washing regime of the soils.

The climatic conditions are severe, characterized by a long winter with an enduring snow cover and cool summer. During the year, the average monthly air temperature varies within a great range. The annual amount of precipitations reaches 1500 mm, with the maximum precipitations falling in May. The coefficient of humidification is high; however, in summer, despite the maximum amount of precipitations, it diminishes to one due to intense evaporation. Severe climatic conditions support the physical weathering of rocks and minerals and restrict chemical weathering. As a result, a great amount of parent material fractures has accumulated on the soil surface.

[1]This chapter uses a part of the illustrative material from the project "Cadastre and Land Registration".

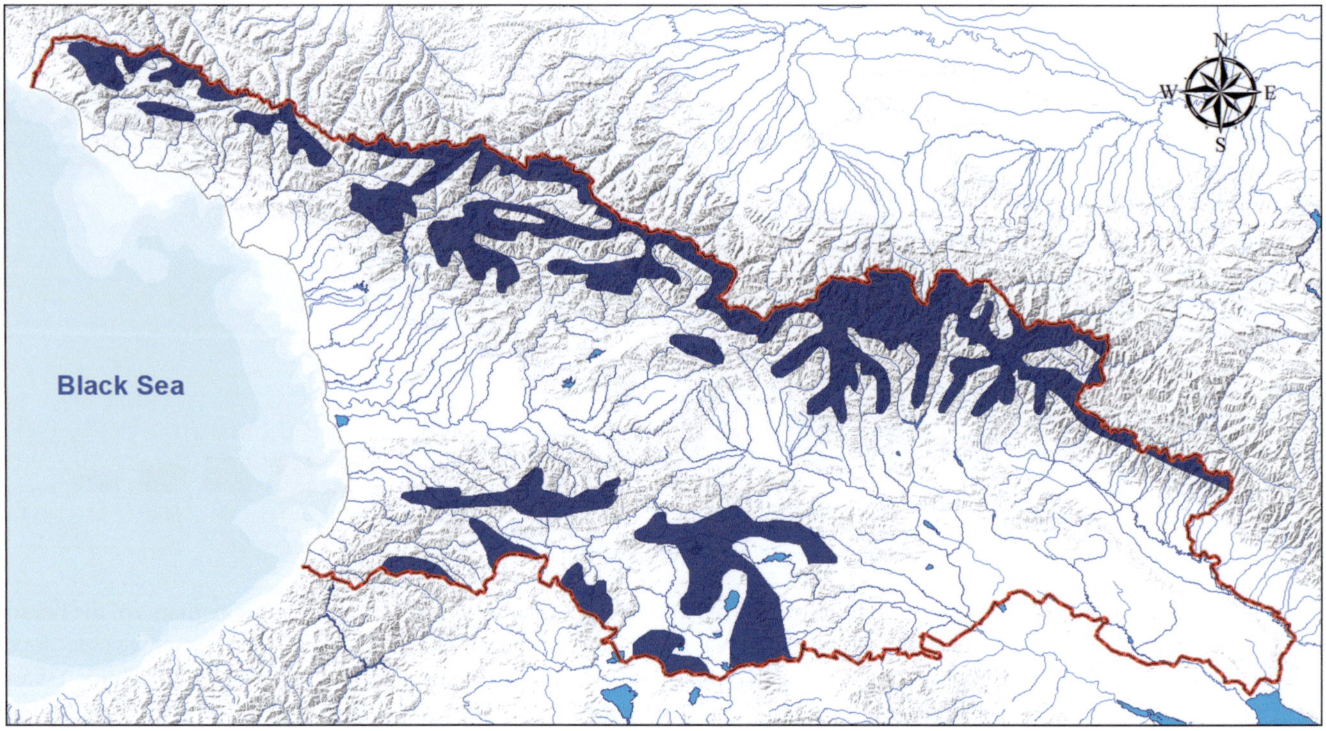

Fig. 5.3 Location of Leptosols Umbric. This map is created by D. Svanadze, based on data of L. Matchavariani

Fig. 5.4 The landscape in the area of Leptosols Umbric formation (Project "Cadastre and Land Registration", KfW)

The erosion–denudation relief dominates in the zone of the uppermost crests, where forms of the glacial origin dominate. There are also relief forms originated through the quaternary effusive volcanism. At lower altitudes, there are erosive gorges with steep slopes spread. Despite the fact that geomorphologically, the high-mountainous area is a region of a denudation and destruction type, it has smoother shapes as compared to the relief in the mountain–forest zone.

The geology of the high-mountainous area is quite complex. In West Georgia, crystal slates, quartz-mica slates, and quartz diorites are common, as well as limestones, crystal rocks, granites, and gneisses. The geology of the high-mountainous area of East Georgia is presented by shales, sandstones, and limestones. The peaks are built with effluent effusive parent materials. There are moraine sediments on the Great Caucasian, while in the mountain–meadow zone of South Georgia, there are andesites, porphyries, trachytes, and intrusive effluent rocks.

The high-mountainous vegetation is characterized by a clear zoning. The vegetation cover is mainly represented by subalpine mid-herbaceous and alpine low-herbaceous meadows, and sometimes, by bushes. The vegetation of the subalpine zone is quite diversified, including both meadow and meadow-steppe plant species and subalpine forest. There is xerophilous vegetation spread on relative dry positions.

The main diagnostic properties of the Leptosols Umbric of Georgia are a little or average strength of its profile, a non-differentiated profile and a clear accumulative horizon (Fig. 5.5). The morphological structure of the profile is Ak–A–B–BC.

The difference between the mentioned soil in the subalpine and alpine zones is negligible. Soils in the Alps are distinguished for a stronger accumulative horizon, less profile strength, and stronger profile than those in the subalpine zone.

The "Mountainous-Meadow" soils mostly have average or little strength, with turfing from surface, acid or weak acid reaction, dark-colored accumulative horizon, high (rarely average) and deep humification, fulvous or humate-fulvous type of soil organic matter, dense illuvial horizon, skeletal nature, and high content of rock fractures. They are characterized by light clay mechanical texture with unequal distribution of main fractions, with sialith weathering and high content of hydromicas and chlorites in clay minerals, with increased content of silicate of iron at great depths, low or average amount of absorbed cations, etc. The data of the gross chemical composition are presented in Table 5.1.

The data about the soil acidity (Fig. 5.6) and content of absorbed cations evidence that there is no connection between the properties of different types of soils and the soil-formation parent material what can be explained by the deluvial nature of the soil.

The Leptosols are characterized by a dark accumulative horizon turfed from the surface (Fig. 5.7). The amount of soil organic matter depends on a complex of factors—altitude, exposition, slope inclination, hydrothermal conditions, type of vertical structure of natural-territorial complexes, vegetation cover, degree of anthropogenic transformation of the area, etc. Thus, amount, reserves and distribution of SOM in the soil layers were studied in the landscapes (Nikolaishvili and Matchavariani 2010), that covers Leptosols, as well as all other main soil types of Georgia. Due to the widespread of the denudation processes, these soils characterized by a younger age of soil formation.

The Leptosols Umbric differs from the "Mountain-forest-meadow" soil (formed in the lower part of the subalpine zone) by a dark color, better and more stable structure, skeletal nature and higher content of mobile ferrum forms. The difference between the Leptosols Umbric and "Mountainous-meadow-chernozem-like" soil is that the former has a lighter color, less strong structure, more acid reaction and less absorption capacity, and higher content of fulvous-type soil organic matter.

Usually, there are hay meadows and pastures over the Leptosols Umbric soils. The necessary condition for their rational use is controlled grazing. Irregular grazing not only causes the violation of the soil cover and provokes erosive processes but also leads to the change of the vegetation.

According to our previous micropedological study (Matchavariani 2008), these soils are characterized by raw-moder and moder-raw types of SOM in the upper horizon, with dark brown color, great amount of vegetation tissues with a survived cellular structure and clear birefringence what is the evidence of humification structure, with excess excrements, spongy microstructure and inter-aggregate microstructure, nonhomogeneous microstructure, and sandy-dust-plasmic and sandy-plasmic elementary microstructure, with plasma isotropy in the surface horizons caused by masked clay particles of the SOM substance. In the lower layers, the aggregation reduces and optical orientation of a mixed-fiber structure, diversified mineralogical association and large admixtures of parent materials fragments covered with disperse calcium occur, weakened and marked; nonhomogeneous plasma and clay-dusty cutans, inleakages distributed in a micro-zonal manner.

5.3 Cambisols Dystric

Cambisols Dystric soils correlate with the national classification as Brown-Forest soil. They are widely spread in Georgia over the mountain slopes, under the forest formations, at various altitudes of mountain-and-forest zone of all soil zones (west, east, and south). As compared to the west, where these soils are spread at 800(900)–1800(2000) m above sea level, in the zone of East Georgia is common at higher altitudes, 900(1000)–1900(2100) m asl (Fig. 5.8); as for the soil zone of South Georgia, the altitude varies from 1500 to 2000 m. The area of the so-called Brown-Forest soils in Georgia is more than 18% of the total area of the country.

So-called Brown-Forest soil was first classified as an individual soil type by Raman. As for these soils of Georgia, the first scientist to study them was B. Prasolov. Basic studies of such soils belong to Tarasashvili (1956, 1965), Sabashvili (1948), Shevardnadze (1963), Urushadze (1987), etc.

Mentioned soils are mostly formed over the slopes, what, in terms of the warm and humid climate, makes for free intra-profile drainage. In the west, the Cambisols Dystric

Fig. 5.5 Profiles of Leptosols Umbric: **a**—Svaneti; **b**, **c**—Kazbegi. Photo by B. Kalandadze

Table 5.1 Gross chemical composition of Leptosols Umbric soils, % (according to data of T. Urushadze)

Horizon (cm)	Loss on ignition	SiO_2	Al_2O_3	Fe_2O_3	TiO_2	MnO	CaO	MgO	Na_2O	K_2O	SiO_2: R_2O_3	SiO_2: Al_2O_3	SiO_2: Fe_2O_3
0–12	30.0	68.6	16.7	6.3	0.9	0.1	2.6	2.1	2.0	2.1	4.3	5.4	22.5
12–25	27.7	66.2	17.1	6.7	1.0	0.1	2.2	2.1	2.1	2.1	5.3	6.6	26.3
25–40	21.0	66.2	16.7	6.6	1.0	0.1	2.7	2.1	2.1	1.9	5.0	6.3	24.9
40–80	14.3	66.3	17.2	7.3	0.7	0.1	2.1	1.9	1.9	1.8	5.1	6.5	24.6

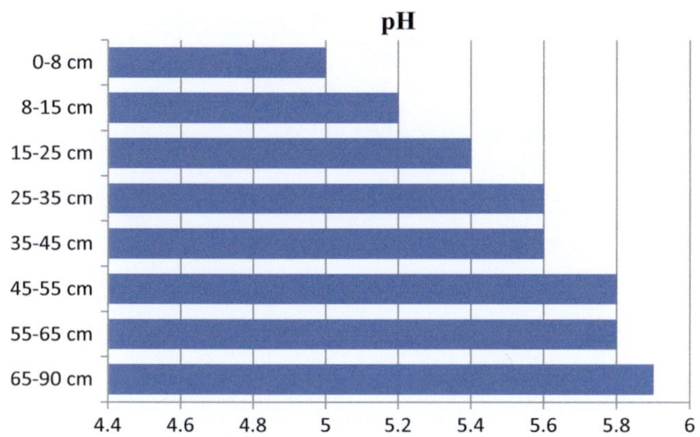

Fig. 5.6 pH distribution in profile of Leptosols Umbric (according to data of N. Iashvili)

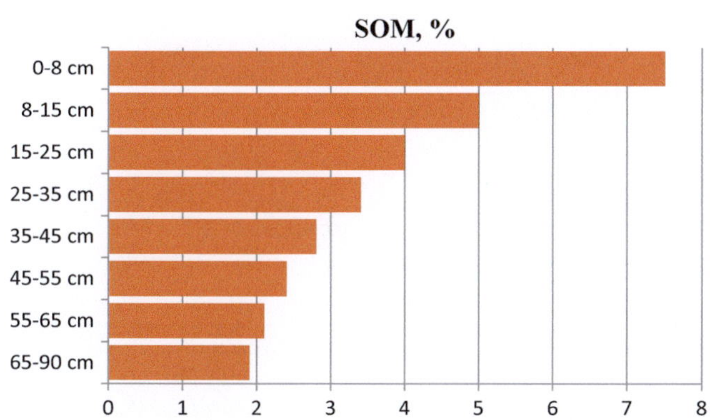

Fig. 5.7 Content of soil organic matter in Leptosols Umbric (according to data of N. Iashvili)

adjoin so-called Yellow-Brown-Forest and Mountain-Forest-Meadow soils, while in the east they adjoin "Cinnamonic" and "Mountain-Forest-Meadow" soils. Soil formation of the Cambisols Dystric soils is younger what is associated with their evolution capability in other soil types (Fig. 5.9).

Soil-forming parent materials, over which the Cambisols Dystric soils are formed, are presented as Jurassic sandy loams, shales and limestone–clay slates, while in the soil zone of South Georgia, they are presented as tertiary volcanic and sedimentary parent materials and their weathering products (porphyries, andesite-basalts, sandstones, conglomerates, etc.).

Cambisols Dystric soils are formed in a relatively warm and humid climate with an average annual temperature from +4 to +11 °C. The temperature of the warmest month of the year reaches +22 °C and that of the coldest month does not fall below +2 °C. The vegetation period lasts for up to 7 months. The atmospheric precipitation amount to 550–1700 mm a year. Humidity coefficient is more than 1 making for the wash-down water regime.

The morphological structure of the soil profile is 0–A–Bm–BC–C (Fig. 5.10). These soils are characterized by a well-established forest litter, rust color, profile skeleton (in their lower layers particularly), and acid reaction, which reduces at greater depths (Fig. 5.11). Cambisols Dystric soil is moderately or deeply containing soil organic matter (Fig. 5.12). It has a strongly pronounced organic material of a dark color. Its profile is cloddy, and partly granular in the upper profiles. With their mechanical texture, the soils are classified as loamy soils. They get heavy in the lower layers. The profile is characterized by intense weathering.

Cambisols Dystric soils are provided with nitrogen. The type of soil organic matter is fulvous. The properties of humic acids and fulvoacids are quite similar. Alumosilicates decompose easily, thus contributing to the formation of secondary clay minerals (e.g., a group of montmorillonite). One of the typical features of these soils is the accumulation of SiO_2 in upper horizons. Calcium dominates in the exchange cations. The sum of absorbed cations is average. Accumulation of Fe_2O_3 and Al_2O_3 takes place in the middle part of the soil profile. The data of the gross chemical composition are presented in Table 5.2.

There are mostly forest massifs growing over the Cambisols Dystric soils. They are usually used as arable land, hay meadows, or pastures. Due to their location over the slopes, these soils are prone to water erosion. Heavy texture and high humidity ratio protect them against erosion.

According to our previous micropedological study (Matchavariani 2008), the upper horizons of Cambisols Dystric soils are characterized by dark color, soil organic matter of a moder-mull and/or mull-moder morphological type, high micro-aggregation, masking of clay material with disperse soil organic matter, and at the depth, they are characterized by brown color, weak aggregation, fissure-like porosity, high birefringence of clay material of a scale and fiber-scale structure, argillaceous and ferrum-argillaceous

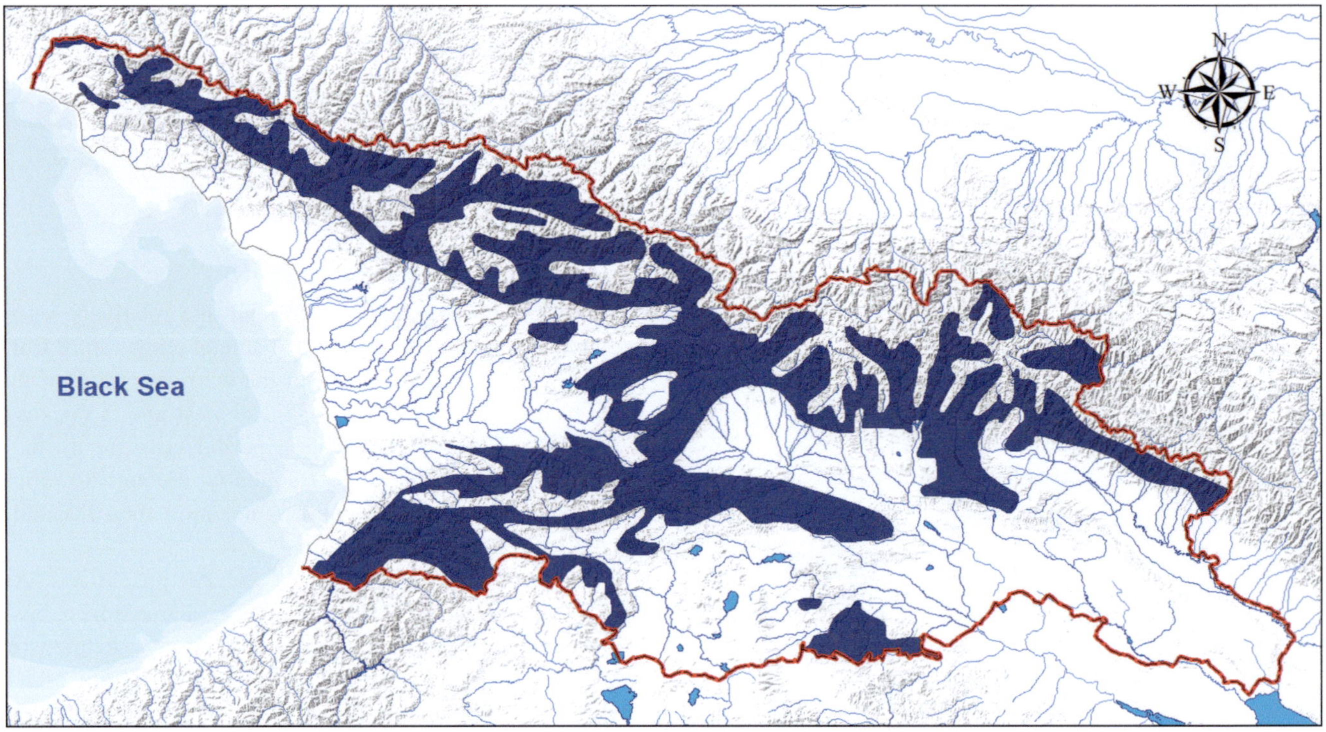

Fig. 5.8 Location of Cambisols Dystric. This map is created by D. Svanadze, based on data of L. Matchavariani

Fig. 5.9 The landscape in the area of Cambisols Dystric formation (Project "Cadastre and Land Registration", KfW)

Fig. 5.10 Profiles of Cambisols Dystric. Photo by B. Kalandadze

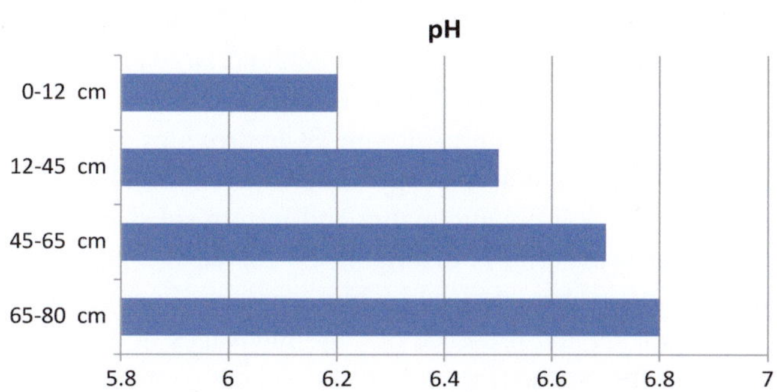

Fig. 5.11 pH distribution in profile of Cambisols Dystric (according to data of T. Urushadze)

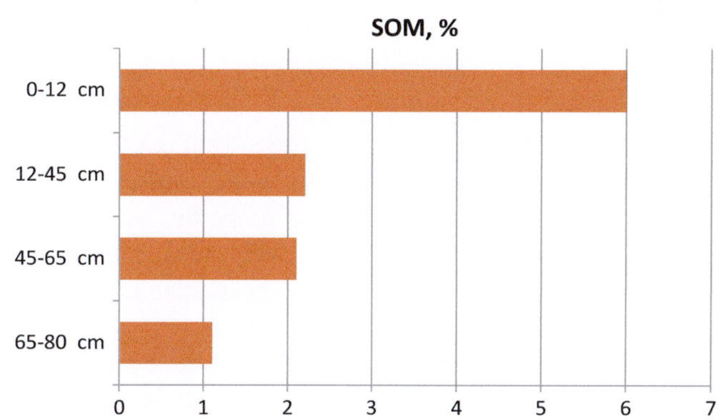

Fig. 5.12 Content of soil organic matter in Cambisols Dystric (according to data of T. Urushadze)

Table 5.2 Gross chemical composition of Cambisols Dystric, % (according to data of T. Urushadze)

Horizon (cm)	Loss on ignition	SiO_2	Al_2O_3	Fe_2O_3	CaO	MgO	SiO_2:R_2O_3	SiO_2:Al_2O_3	SiO_2:Fe_2O_3
3–10	15.00	58.5	21.7	8.6	5.3	2.4	3.6	4.6	15.7
10–25	13.77	57.2	21.7	9.4	6.8	1.9	3.5	4.4	16.5
25–48	13.83	57.1	21.2	8.4	6.5	2.5	3.5	4.4	18.4
48–69	10.81	57.2	20.5	10.3	6.0	1.8	3.5	4.6	15.3
69–80	12.41	56.6	22.4	9.3	6.7	1.5	3.5	4.4	16.5

cutans in pores and cracks, all through the profile, particularly, in its middle part and with the micro-zonal saturation of thin-disperse substance with Fe-hydroxides.

As a transitional type between the Cambisols Dystric and Acrisols Haplic soils, in the subtropical zone of West Georgia, at 400(500)–800(1000) m asl, according to national soil classification, is spread so-called Yellow-Brown-Forest soil. They occupy 1.5% of the total area of the country. Sometimes this soil considers as a subtype of "Brown-Forest" soil.

I. Gerasimov was the first to identify the "Yellow-Brown-Forest" soil in the environs of Batumi, and thus showed its transient nature from the Brown-Forest soil of a moderately warm zone to the humid subtropical soil. A doubt about the possible presence of Yellow-Brown-Forest soil in Georgia was expressed by S. Zonn as well. T. Urushadze demonstrated the need for isolating this type of soil as an individual genetic type.

The parent materials in the areas with abovementioned soil are a porphyry stratum of the middle Jurassic period and old effluent (andesite, andesite-basalt) denudation crust and their derivatives. The type of relief is erosive-denudation. The climate is subtropical humid with warm winter and warm summer. Average annual temperature is 11 °C and the sum of active temperatures varies from 3500 to 4500 °C. The duration of the vegetation period is 6 to 7 months and the average annual amount of atmospheric precipitations is great (1000–2150 mm), with over half of it falling in the warm period of the year. The annual humidity coefficient is more than one. The vegetation is presented by chestnut forests with the fragments of Caucasian hornbeam, oak, oriental maple, and other plantations. A peculiar sign of these forests is wide areas of evergreen understory.

The genesis of "Yellow-Brown-Forest" soil is the result of the joint action of Cambisols Dystric and Acrisols Haplic soil-formation processes, and consequently this type of soil has much in common with both soils. As a result, such a combination of processes forms new properties determining the individual nature of this type. Besides the vegetation, the hydrothermal conditions also play a particular role in the formation of this kind of soil.

Morphologically, this soil is characterized by a clearly expressed illuvial horizon of yellow-brown color with a strong soil organic matter and cloddy structure. The main diagnostic properties are allitic weathering and ferrum concentration. Profile has the following structure: A–AB–B1–B2–C1–C2.

As the analytical data suggest, soil has acid reaction, particularly in its accumulative horizon. The soil acidity shows a decreasing trend (an increasing pH value) as the depth increases. The content of soil organic matter is high; however, distribution of SOM is not subject to the regularities typical to the forest soil. As the depth increases, the content of SOM in the profile reduces gradually and insignificantly reaching great depth. Nitrogen distribution across the profile shows similar regularities.

Soil organic matter is of a fulvous type. The soil is unsaturated with bases. The amount of absorbed hydrogen is quite great. The "Yellow-Brown-Forest" soil is poor in calcium and manganese. The amount of these elements in the soil depends on the eluvial processes on the one hand and on the lithological and the petrographic structure of the soil-forming parent materials on the other hand.

With its texture, the "Yellow-Brown-Forest" soil belongs to the category of heavy loams. Movement of a micron fraction across the profile is not typical to this type of soil. The mineral portion of the soil is characterized by eluvial processes. The clay minerals are presented by chlorine-montmorillonite and contain great amounts of kaolin and average amount of chlorines. The amount of mica is relatively little.

The most part of the soil is covered with forest, and small part of it is used to grow perennial crops, vine, fruit, etc. Unlike the "Brown-Forest" soil, which is formed in cooler conditions, the mentioned soil is of a yellowish and sometimes, of a reddish color and has no forest litter, is characterized by stronger ferralitic weathering, higher content of soil organic matter, less absorption capacity, more content of different forms of iron, and more acidic reaction. The accumulation of nonsilicate ferrum in the illuvial horizon can be explained by an intense wash down. Unlike the so-called Yellow and Red soils, which are formed in warmer conditions, the "Yellow-Brown-Forest" soil has light yellowish or reddish color, strong accumulative horizon, better structure, and less weathering.

According to our previous micropedological study (Matchavariani 2008), the "Yellow-Brown-Forest" soil is characterized by an intense coloration of aggregated accumulative horizon with organic mass, favorable microstructure, mull-moder type of soil organic matter, and large amounts of the vegetation remain at different stages of decomposition. Plasma is intensely saturated with Fe-hydroxides across the whole profile what is seen as spots and concretions. Due to the masking with soil organic matter and ferrum substances, the fine-disperse material is distinguished for weak optical orientation and is characterized by high content of parent materials fragments and presence of clay inleakages in the lower portion of the profile (Fig. 5.13).

5.4 Rendzinas

Rendzinas (Leptosols Rendzic), which correlate with the national classification as an intrazonal type—raw-humus-calcareous soil, are widely spread in West Georgia (Abkhazeti, Samegrelo, Racha-Lechkhumi, Zemo Imereti), as well as East Georgia (Mtiuleti, Samachablo, Kakheti, Kartli). Their area coincides with the areas with limestones and marls. In addition to the mountain–forest zone, these soils are spread in the humid and dry subtropics and high-mountainous regions (Fig. 5.14). They occupy 4.5% of the total territory of the country.

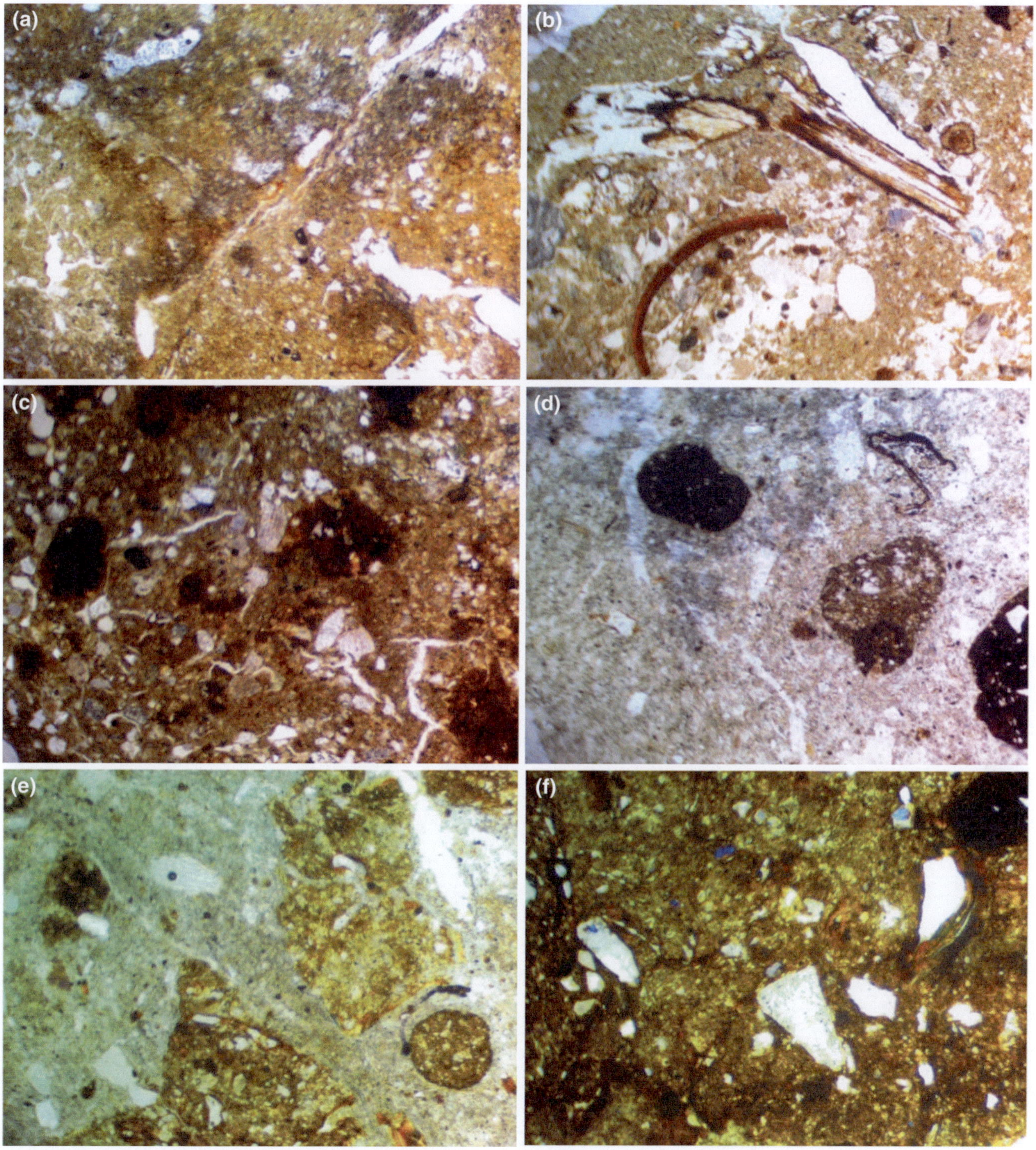

Fig. 5.13 Microstrucutre of "Yellow-brown-forest" soil, nic.II; Pit-3, horizons: **a** 0–14 cm; **b** 14–28 cm; **c** 28–55 cm; **d** 55–80 cm; **e**–**f** 80–100 cm. Photos by L. Matchavariani

The "Raw-humus-calcareous" soil of Georgia was studied by Zakharov (1924), Talakhadze (1964), Sabashvili (1965), Chkheidze (1977), etc. Sabashvili was the first to explore the chemical content of this type of soil and to develop the issues of its classification; Talakhadze, together with the ordinary "Raw-humus-calcareous" soils, identified Rendzic Terra Rossa.

Rendzinas is mainly formed in the forest zone, over the parent materials enriched with $CaCO_3$ (gypsum, marble, dolomite, marl), and is characterized by a flushing or periodically flushing regime of moisture (Fig. 5.15). There are two main types of relief in the area with carbonate rocks: glacial and karst. The glacial relief is developed with old glaciers and it runs as a continuous strip in the high-mountainous region of

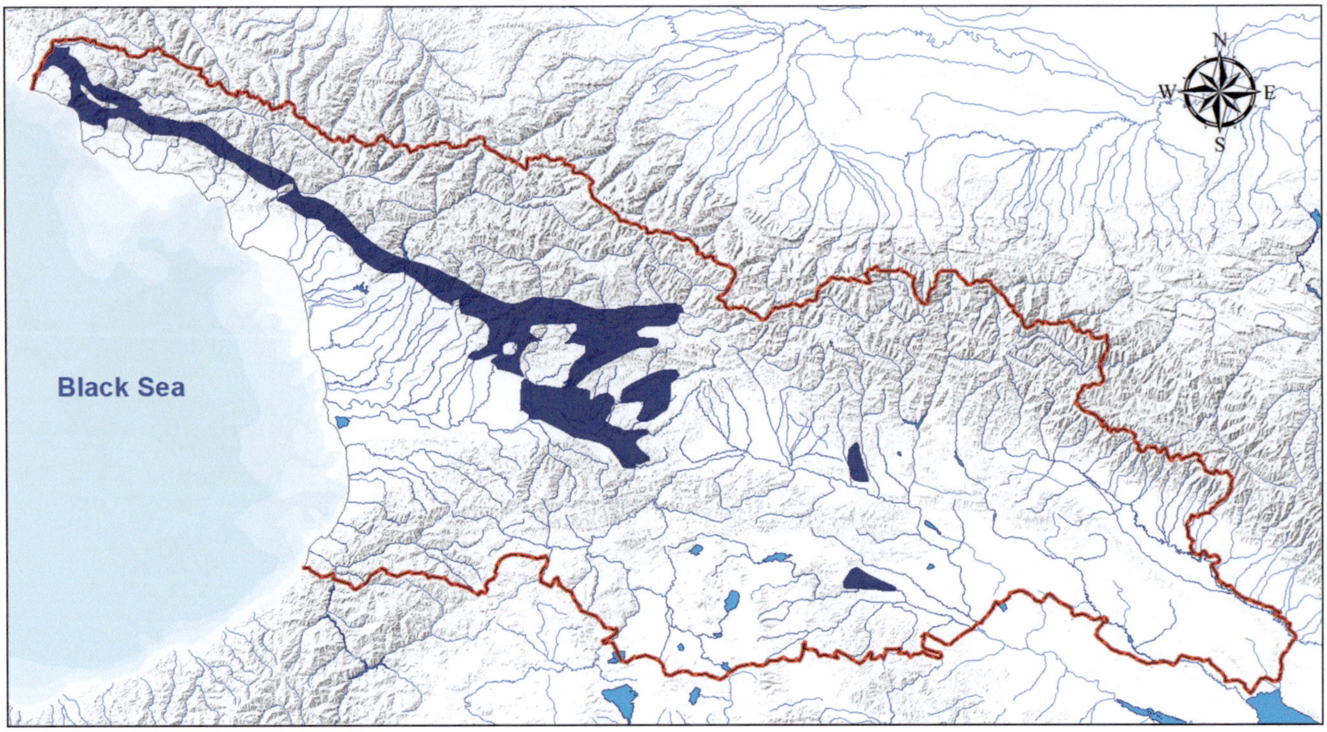

Fig. 5.14 Location of Rendzinas. This map is created by D. Svanadze, based on data of L. Matchavariani

Fig. 5.15 The landscape in the area of Rendzinas formation (Project "Cadastre and Land Registration", KfW)

West Georgia. Karst relief is widely spread in the middle zone and its development is associated with the sediments of a Cretaceous system. The relief in the area of these soils is of an erosive type and is presented as denudation, denudation-accumulation, and denudation-landslide forms.

The climate in the mountain–forest zone of Georgia, where the Leptosols Rendzic soil is widely spread, is moderately warm. The temperature of the coldest month is −1, −4 °C and that of the warmest month is +18, +20 °C. The sum of active temperatures is 2000–3500 °C.

Fig. 5.16 Profiles of Rendzinas soils. Photo by B. Kalandadze

The annual amount of atmospheric precipitations reaches 1400–1600 mm.

The vegetation in the region is presented by a hardwood forest (oak-and-hornbeam forests) with wide areas of grass. The cultivated areas are used to grow vineyard, orchards, bay trees, and other perennial plants.

The mentioned soils are characterized by a slightly differentiated profile (Fig. 5.16), which usually has the

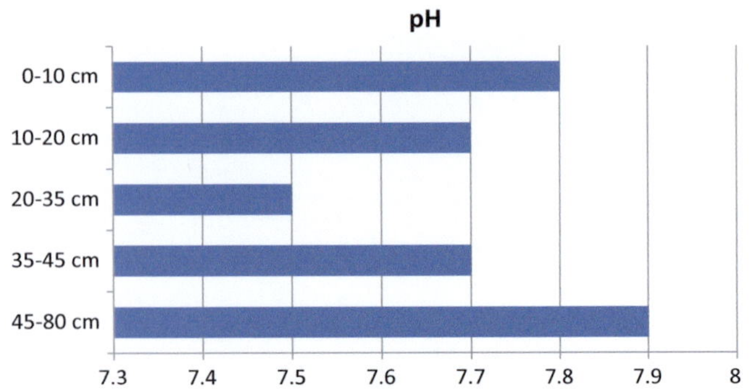

Fig. 5.17 pH distribution in profile of Rendzinas (according to data of T. Chkheidze)

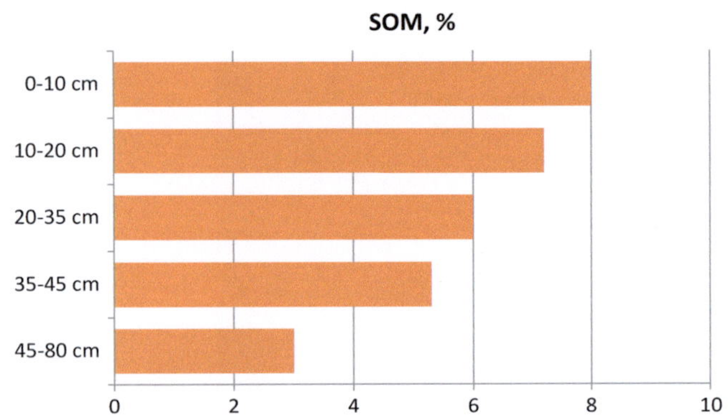

Fig. 5.18 Content of soil organic matter in Rendzinas (according to data of T. Chkheidze)

following structure: A–AB–CD or A–AB–BC or A–AC. The soil is distinguished for a clear accumulative horizon and granular or fine-cloddy-granular structure. Soils developed over the limestones are more skeletal than those developed over the marls. However, the latter type has a stronger profile than the former one.

As the analytical data suggest, Rendzinas is characterized by a neutral or weak alkaline reaction (Fig. 5.17). The soil organic matter is a humic type with an average or little content. In addition, the soil developed over the marls is distinguished for a less content of soil organic matter. As a rule, the soil is deeply humificated (Fig. 5.18). The amount of carbonates varies within great limits. The content of nitrogen is average or low and the amount of calcium constitutes 92% of the absorbing complex. The soils developed over the limestones are characterized by clay mechanical content, while those formed over the marls have a loamy texture. Rendzinas have a predominant content of silicate ferrum. The content of nonsilicate or amorphous ferrum is within the horizon transient to the maximum. The data of the gross chemical composition are presented in Table 5.3.

Rendzinas differ from the Cambisols Dystric soil by a dark color, alkaline reaction, weak argilization, and carbonate content.

The Leptosols Rendzic soil incorporates typical, leached and red (Terra Rossa) subtypes. The carbonates in the typical soils are spread on the surface or in the accumulative horizon and develop in the area of the Cambisols Dystric over such parent materials, which contain large amounts of calcium carbonates. The carbonates in the leached soil are found in the illuvial horizon and develop over relatively stronger elluvion-delluvion layer of the carbonate parent materials. Red-colored Terra Rossa develops over the dense limestones and marls and has a carbonate nature, red color, and weak acid or neutral reaction.

According to our previous micropedological study (Matchavariani 2008), the diagnostic properties of Rendzinas are black color of the upper portion of the profile, moder type of soil organic matter, carbonate nature of the whole profile, even saturation of the clay material with organic hydroxides, masking of optically oriented clay from carbonates, presence of microgranular calcite in the upper horizons, and reddish-chestnut color of the lower part of the profile as a result of the participation of R_2O_3 oxides and organic acids in the soil solutions.

5.5 Leptosols Molic

According to Georgian national soil classification, mountainous-meadow-chernozem-like soils, which correlate with WRB as Leptosols Molic, are spread in the Subalpine and Alpine zones of South Georgia (Fig. 5.19), at an altitude of over 1800(2000) m above sea level. This type of soil covers 1.6% of the total area of the country and borders primitive mountain-meadow soils of the nival, subalpine, and alpine zones and mountainous-forest-meadow soils of the subalpine zone (Fig. 5.20).

The abovementioned soil was the subject of study of I. Liverovskyi and V. Friedland. They associated the formation of this soil in the Caucasus with the rocks rich in carbonates, limestones, and carbonate slates. When studying the soils of the Caucasus, Zonn (1974, 1987) established that this soil is formed on the carbonate-free parent materials in the dry regions of high mountains. They are extracted on the erupted lava and tuffs on the South Caucasus Plateau. As the most recent studies suggest (Urushadze et al. 2010), most of these soils spread in Georgia belong to so-called Andosols, or soils formed in volcanic tephra, tuffs, pumice, and other effusive volcanic material, and partially, on other silicate sediments in terms of hilly or mountainous relief, under various thermal conditions and vegetation communities. Swift weathering of a porous substrate causes the accumulation of sustainable, organic mineral compounds, and origination of slightly crystallized minerals.

The principal diagnostic morphological features of the Leptosols Molic are clearly seen intense accumulative

Table 5.3 Gross chemical composition of Rendzinas, % (according to data of T. Chkheidze)

Horizon (cm)	Loss on ignition	SiO_2	Al_2O_3	Fe_2O_3	TiO_2	CaO	MgO	$SiO_2:R_2O_3$	$SiO_2:Al_2O_3$	$SiO_2:Fe_2O_3$
0–10	26.90	56.4	11.8	7.4	0.60	19.7	1.76	5.85	8.18	20.63
20–40	26.62	60.7	12.7	6.7	0.62	15.0	1.39	5.99	8.09	24.10
60–80	25.84	42.6	4.6	4.0	0.36	45.1	1.08	10.13	15.75	28.36
110–150	32.36	44.4	8.2	4.9	0.44	37.6	1.68	7.39	9.23	24.63

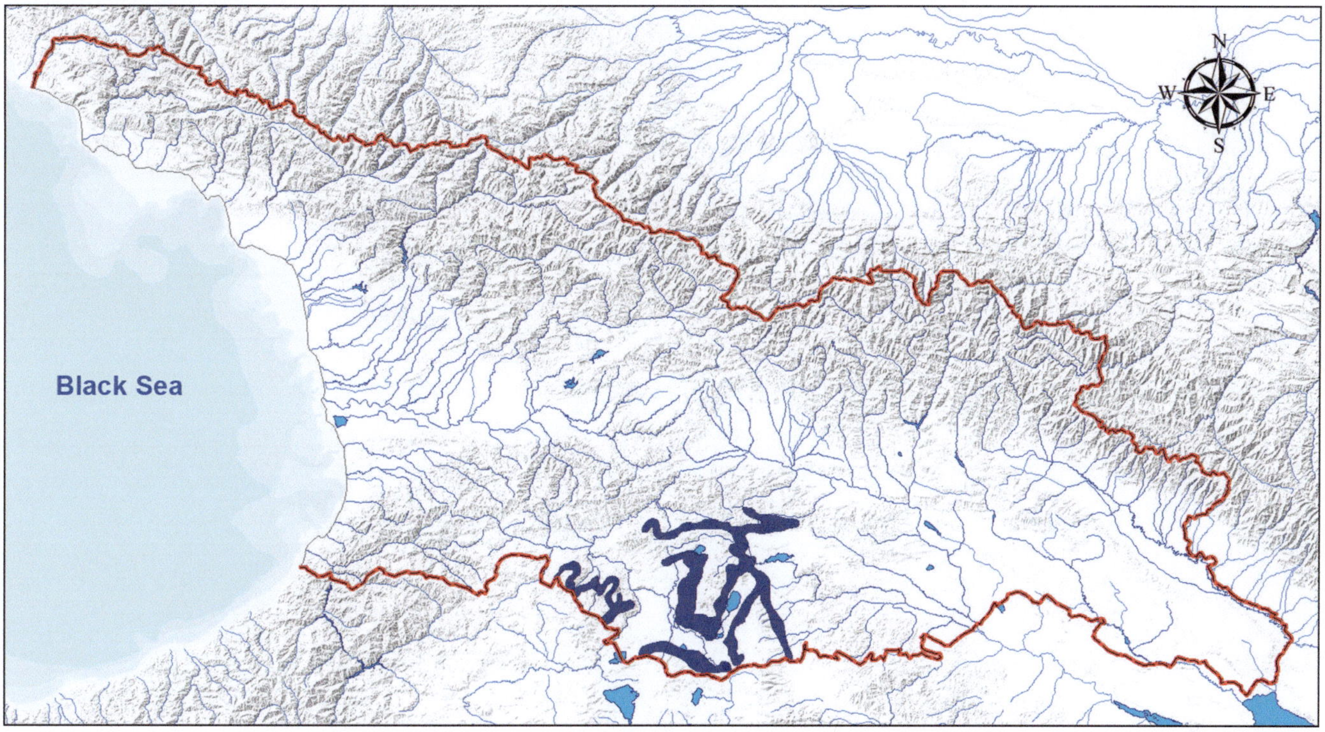

Fig. 5.19 Location of Leptosols Molic. This map is created by D. Svanadze, based on data of L. Matchavariani

Fig. 5.20 The landscape in the area of Leptosols Molic formation (Project "Cadastre and Land Registration", KfW)

Fig. 5.21 Profiles of Leptosols Molic soils. Photo by B. Kalandadze

horizon, little or average strength, and non-differentiated profile (Fig. 5.21) with the following structure: A_1'–A_1''–BC or A_1'–A_1''–B–BC.

The Leptosols Molic develops under alpine and subalpine stepped meadows and meadow steppes in high-mountainous regions and is used as pasture and hayfields consequently. The relief is a volcanic plateau, with its central part occupied by two meridian ridges of volcanic cones. The bedrocks are mainly presented by base volcanic rocks, andesite-basalts, and basalts.

The climate is cold with cool short summer and long severe winter. The temperature of the coldest month (January) is −7.8 °C and that of the warmest month (August) is +13.6 °C. Average annual temperature is +3.2 °C. The duration of the vegetation period is up to 4 months. The duration of the period without frosts is up to 2 months. Annual amount of precipitations is 600 mm. Maximum precipitations fall from April to June. Average annual relative air humidity is 78%; humidity coefficient is 1–3. The water regime of the soils is washing down and is periodically washing down in the moderately humid regions with a drough period.

As the analytical data suggest, the mentioned soil is characterized by weak acid reaction (Fig. 5.22), high content of humate type of soil organic matter (Fig. 5.23), deep humification, high absorption capacity, weak unsaturation, clay or loamy texture, higher content of sludge fraction and

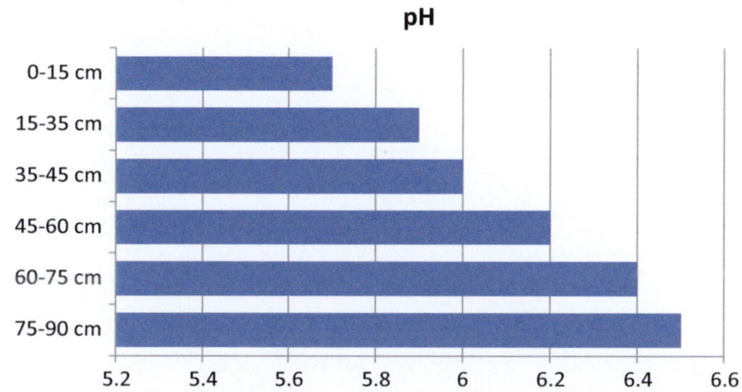

Fig. 5.22 pH distribution in profile of Leptosols Molic soil (according to data of T. Urushadze)

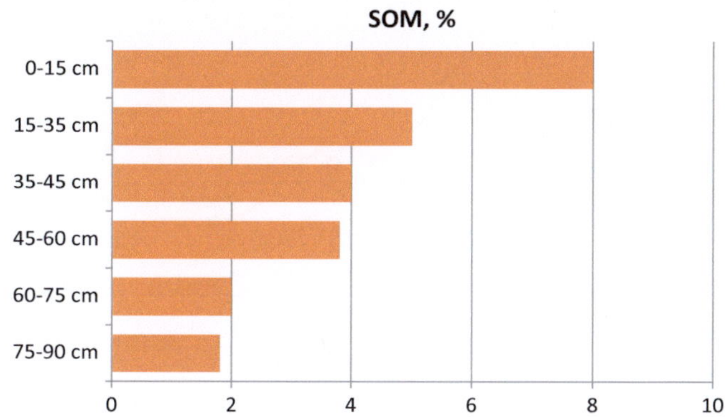

Fig. 5.23 Content of soil organic matter in Leptosols Molic soil (according to data of T. Urushadze)

physical clay at greater depths or in the middle of the profile, and bulk of hydromica in clay minerals. The data of the gross chemical composition are presented in Table 5.4.

Leptosols Molic differs from the neighboring Leptosols Umbric by a dark color, solid structure, weaker acid reaction, high absorption capacity, high content of soil organic matter, deep humification, and presence of humatc type of soil organic matter, and it differs from the Chernozem by the absence of carbonates, more intense porosity, and less clear differentiation.

5.6 Chernozems

Chernozems of Georgia occupied 1.4% of the total territory of the country nature and are spread at 1200–1300 m asl (Fig. 5.24). In a national soil classification, they called as mountain Chernozems unlike the Vertisols (black soils) of plains. The mountain Chernozems are common over the volcanic plateau of the southern mountainous area of Georgia having a mountain plain (Fig. 5.25). These soils belong to the group of mountain-meadow steppe soils located between the mountain-forest and mountain-meadow soils of the subboreal zone.

The first researcher of the mountain Chernozems of Georgia was V. Dokuchaev. These soils were also studied by Zakharov (1924), Talakhadze (1962, 1964), etc.

The volcanic, mountainous region is built with andesite, andesite-basalt, and basalt rocks. They are covered with lacustrine sediments in the depressions. The presence of moraine sediments in the region evidences that the southern mountains were subject to the influence of the Glacial Age.

The origination of the Chernozems is associated with secondary meadow formation—the processes of retreat of the subalpine forests and evolution of the lakes. Typical and leached soils can be identified as subtypes of mountain Chernozems.

The area of mountain Chernozems spreads in the cold climate zone with an average annual temperature of +6 °C. The temperature of the warmest month is up to 17 °C and that of the coldest month is −7.5 °C. The vegetation period lasts up to 5 months. The annual amount of atmospheric precipitations is up to 800 mm, much of which falls as snow.

Following the climatic conditions, the processes of weathering and soil forming take an intense course, and consequently the soil is of a heavy texture, it is rich in clay minerals (montmori-llonite, illites, kaolin), and the thickness of its profile does not exceed 1 meter.

In a morphological respect, Chernozems are characterized by quite a strong black accumulative horizon, with a cloddy-nutty or prismatic structure and profile argilization (Fig. 5.26). The structure of the soil profile is A_1'–A_1''–AB–BC.

As the analytical data suggest, the mountain Chernozems are characterized by clay or heavy loamy texture. The silt fraction is usually equally distributed in the upper horizons and decreases gradually at greater depths. The soil has a weak acid, neutral or weak alkaline reaction (Fig. 5.27). It is enriched with bases and Fe-oxides; calcium is found in the greatest amounts in the exchange cations. The content of soil organic matter is high and soil organic matter penetrates deep in the profile (Fig. 5.28). The data of the gross chemical composition are presented in Table 5.5.

Table 5.4 Gross chemical composition of Leptosols Molic, % (according to data of T. Urushadze)

Horizon (cm)	Loss on ignition	SiO_2	Al_2O_3	Fe_2O_3	TiO_2	MnO	CaO	MgO	Na_2O	K_2O	SiO_2: R_2O_3	SiO_2: Al_2O_3	SiO_2: Fe_2O_3
0–10	24.0	65.8	18.4	6.9	0.7	0.1	2.2	1.9	7.1	1.9	4 9	6.1	25.5
10–30	17.8	67.6	57.0	5.7	0.8	0.1	2.4	1.8	2.4	2.0	5.6	6.8	31.3
30–60	17.9	67.5	17.0	5.6	0.6	0.1	2.4	2.0	2.4	2.1	5.6	6.8	32.1

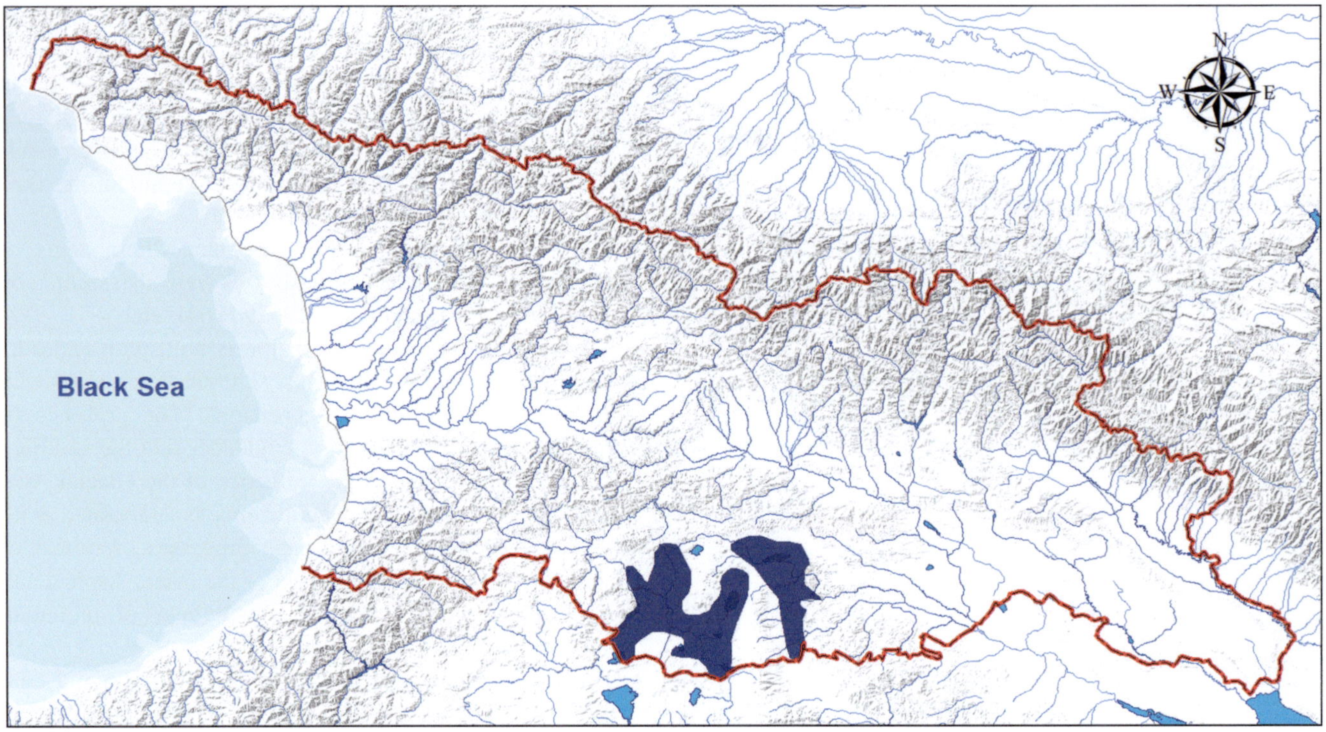

Fig. 5.24 Location of Chernozems. This map is created by D. Svanadze, based on data of L. Matchavariani

Fig. 5.25 The landscape in the area of Chernozems formation (Project "Cadastre and Land Registration", KfW)

Despite the fact that the Chernozems are considered a high-productive soil, their use is limited in agriculture due to the severe climatic conditions.

According to our previous micropedological study (Matchavariani 2008), the principal properties of the Chernozems' microstructure were identified: loose and sponge-like microstructure, plasma with an organic-clay content, mull-type organic matter strongly bound with clay, bulk of vegetation remains and excrements in the decomposition phase, presence of coprolite macro- and

Fig. 5.26 Profiles of Chernozems. Photo by B. Kalandadze

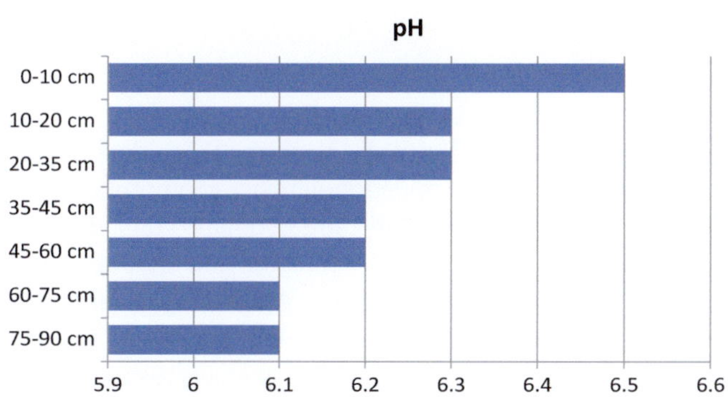

Fig. 5.27 pH distribution in profile of Chernozems (according to data of T. Urushadze)

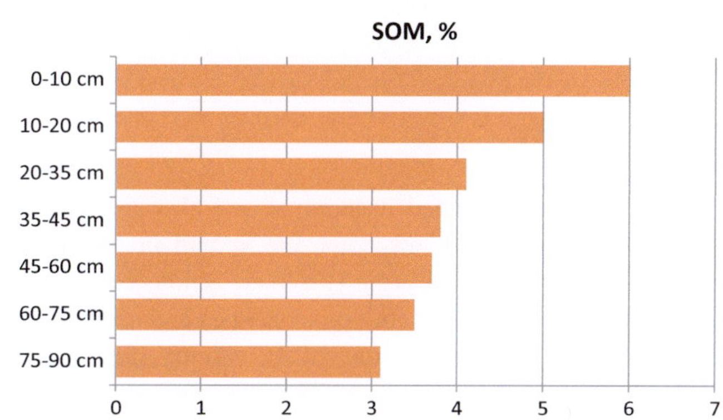

Fig. 5.28 Content of soil organic matter in Chernozems (according to data of T. Urushadze)

micro-aggregates, signs of intense pedogenic treatment from mesofauna, sandy-dusty-plasmatic elementary microstructure, light color at greater depths, carbonate-clay plasma, presence of ferrum and organic-Fe micro-concretions, presence of new carbonate formations as dispersed calcite grains, and presence of needle-like calcite in the pores.

Table 5.5 Gross chemical composition of Chernozems, % (according to data of T. Urushadze)

Horizon (cm)	Loss on ignition	SiO_2	Al_2O_3	Fe_2O_3	TiO_2	CaO	MgO	$SiO_2:R_2O_3$	$SiO_2:Al_2O_3$	$SiO_2:Fe_2O_3$
0–15	15.0	60.4	18.9	8.7	0.4	3.7	3.1	5.3	4.2	18.5
15–35	14.7	60.4	18.9	8.9	0.4	3.3	3.2	5.3	3.7	18.3
35–65	14.0	61.2	19.3	8.9	0.4	3.2	3.1	5.4	4.1	18.4
65–90	14.0	58.9	20.9	8.8	0.3	3.2	3.2	4.9	3.8	17.8

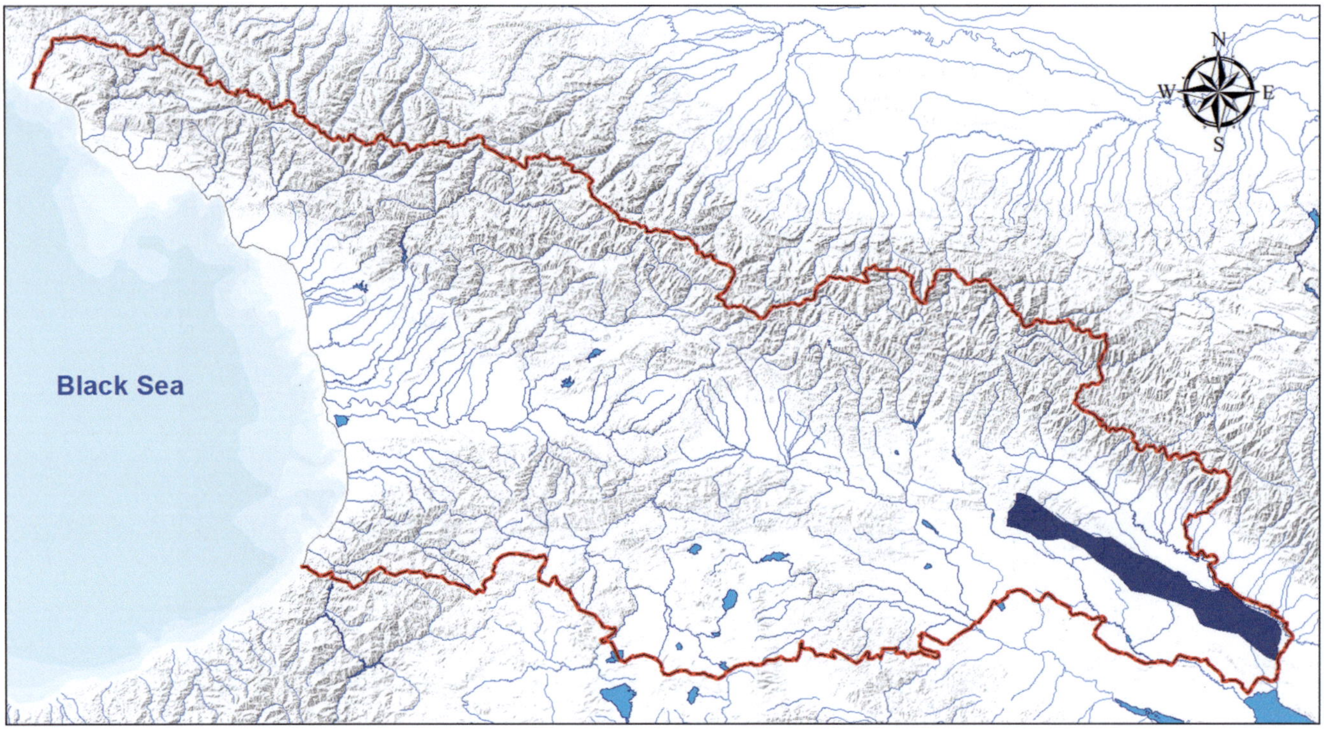

Fig. 5.29 Location of Vertisols. This map is created by D. Svanadze, based on data of L. Matchavariani

5.7 Vertisols

Vertisols, which are called in national soil classification as plain Chernosems, or Black soils, are spread in the intermontane zone of East Georgia (Kakheti, Shida Kartli, and Kvemo Kartli), in the dry subtropical steppes and occupy the area of 3.9% of the total territory of the country (Fig. 5.29).

The so-called Black soil of Georgia was studied by Zakharov (1924), Sabashvili (1948, 1965), Talakhadze (1962, 1964), Mardaleishvili (1973), Mardaleishvili and Pipia (1988), Mardaleishvili et al. (2005), Pipia (2008), etc.

As per Sabashvili (1948), the high content of gypsum in the loess-like sediments evidences that they are of a lacustrine origin. In opinion of Talakhadze (1962), the formation of one part of this soil is associated with the evolution of the alluvial plains, while the formation of another part is associated with the evolution of the lakes and relief forms of a depression type. It was him to develop the classification of this type of soil.

The zone of the Vertisols is formed by denudation-accumulation and accumulation-genetic geomorphological types. The relief forms of the intermontane lowland zone in East Georgia are relatively younger and belong to the Upper Tertiary and Quaternary Ages (Fig. 5.30). The deluvial–proluvial sediments are widely spread on this territory. The zone where this is spread contains also a sloping terrace-like plain, with its hypsometric levels varying between 650 and 750 m above sea level. The accumulation relief types are widely spread in the area where the Black soil is spread.

The Vertisols develop in terms of dry subtropical climate, with winter with almost no snow and hot, dry summer, where the average annual temperature is +10, +12 °C; the temperature of the warmest month (June) is +23 °C and that of the coldest month (January) is −0.3, −4 °C. The sum of

Fig. 5.30 The landscape in the area of Vertisols formation (Project "Cadastre and Land Registration", KfW)

Fig. 5.31 Profiles of Vertisols. Photo by B. Kalandadze

active temperatures reaches 4000 °C; the duration of the vegetation period is 6 to 7 months. The atmospheric precipitations usually fall as rain, with the average annual amount of 400–600 mm. The precipitation minimum is fixed in winter months, and the maximum is fixed in the May or June. During the year, the evaporation exceeds the amount of the atmospheric precipitations (consequently, the humidity coefficient is 0.3–0.9). Average annual relative air humidity is 64–70%. During the year, the soil temperature does not fall below 0 °C, and as a result, the soil biogenicity

Table 5.6 Gross chemical composition of Vertisols, % (according to data of A. Nanaa)

Horizon (cm)	Loss on ignition	SiO_2	Al_2O_3	Fe_2O_3	CaO	MgO	Na_2O	K_2O	SO_3	$SiO_2:R_2O_3$	$SiO_2:Fe_2O_3$
0–20	18.8	63.8	15.9	5.9	4.2	3.3	1.4	3.7	1.0	5.5	29.1
20–40	17.0	63.3	16.0	6.7	3.3	3.2	1.5	4.0	1.1	5.3	25.1
40–60	16.1	60.4	16.0	6.8	4.9	2.9	1.6	3.4	3.5	5.0	23.6
60–100	17.1	62.2	15.3	7.3	5.0	2.8	1.6	3.2	1.9	5.3	22.7
100–125	17.1	61.4	15.5	7.4	3.7	3.4	1.7	3.1	2.6	5.2	22.1
125–150	16.8	60.4	15.4	7.8	3.7	3.7	2.4	3.2	2.1	5.1	20.7

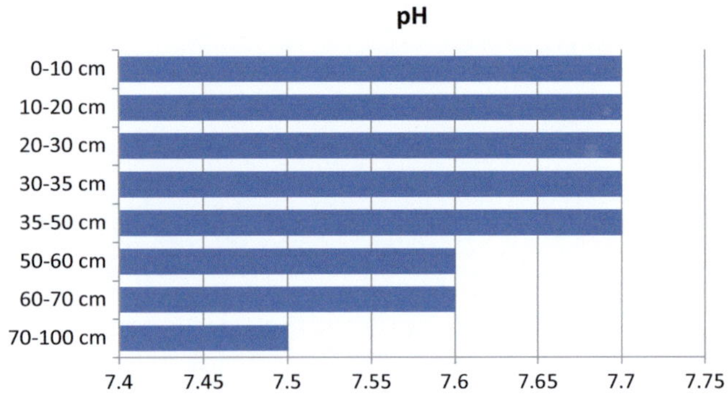

Fig. 5.32 pH distribution in profile of Vertisols (according to data of A. Nanaa)

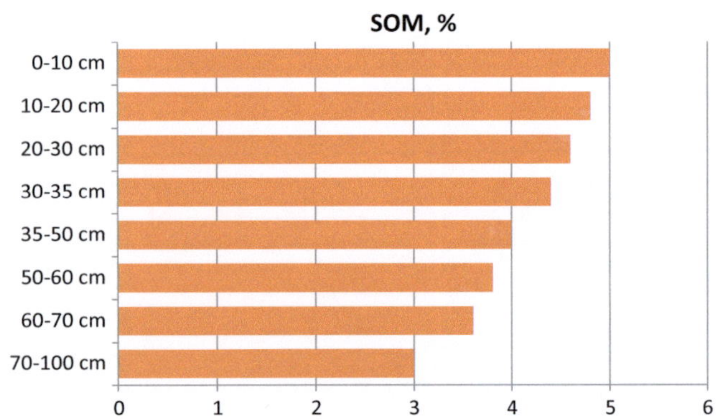

Fig. 5.33 Content of soil organic matter in Vertisols (according to data of A. Nanaa)

is quite high during the year and the soil-forming processes take place all year long with different intensities.

The Vertisols are spread in dry subtropical steppes, and as Ketskhoveli (1935) states, it is classified as two groups: of the primary and of the secondary origin. The steppe vegetation is made up of thornbush, beard grass, feather grass, and grasslands.

In a morphological respect, the Vertisols of Georgia divide into a typical, carbonate, leached, and meadow-gleied subtypes. The morphological structure of the profile is A_1'–A_1''–AB–B–BC–C. The principal diagnostic indicators of these soils are the black color of the upper layers, argilization, and carbonization (Fig. 5.31). Vertisols are distinguished by a strong accumulative horizon and increasing density at greater depths, characterized by a clear accumulative horizon, cloddy-granular and nutty-prismatic structure at greater depths, heavy texture, signs of compactness, carbonate-illuvial horizon, and white spots of carbonates.

The data of the gross chemical composition of Vertisols are presented in Table 5.6. As the analytical data suggest, the Vertisols are characterized by the weak alkaline reaction, with the calcium carbonates spread right from the surface with a gradually increasing content at greater depths (Fig. 5.32). As for the soil organic matter, its content shows an opposite trend and decreases as the depth increases. The SOM is of a humatic type, with little content of mobile humic acids evidencing the high stability of soil organic matter (Fig. 5.33). The silt fraction contains great amounts of R_2O_3; besides, the content of SiO reduces at greater depths, while that of Fe_2O_3 increases gradually. A dominant Fe-form is a silicate one fixed in the middle part of the profile, while amorphous forms are accumulated in the upper horizons.

The texture of Vertisols is classified as light and average clays. The content of physical clay reaches 60–80%. They are characterized by a high content of sludge fraction.

With its mineralogical content, the light fraction is presented by quartz, feldspars, and fractures with clay and earth silicon; the coarse fraction is presented by magnetite-ilmenite, augite, zircon, biotite, and anhydride. Clay minerals are presented by smectites and illites, while chlorite, kaolin, feldspars, and quartz are spread as admixtures. This type of soil is characterized by a greater content of a silicate ferrum as compared to that of a nonsilicate ferrum. Amorphous ferrum is fixed in little amounts and is accumulated in the upper part.

Unlike the mountain Chernozems, the texture of the Vertisols is heavier, with more intense argilization and compactness; the content of Fe-forms is high and the soil is sometimes characterized by the accumulation of easily soluble salts. They distinguish between the typical, carbonate, leached, and meadow-gleized soil subtypes.

According to our previous micropedological study (Matchavariani 2008), the Vertisols of Georgia are

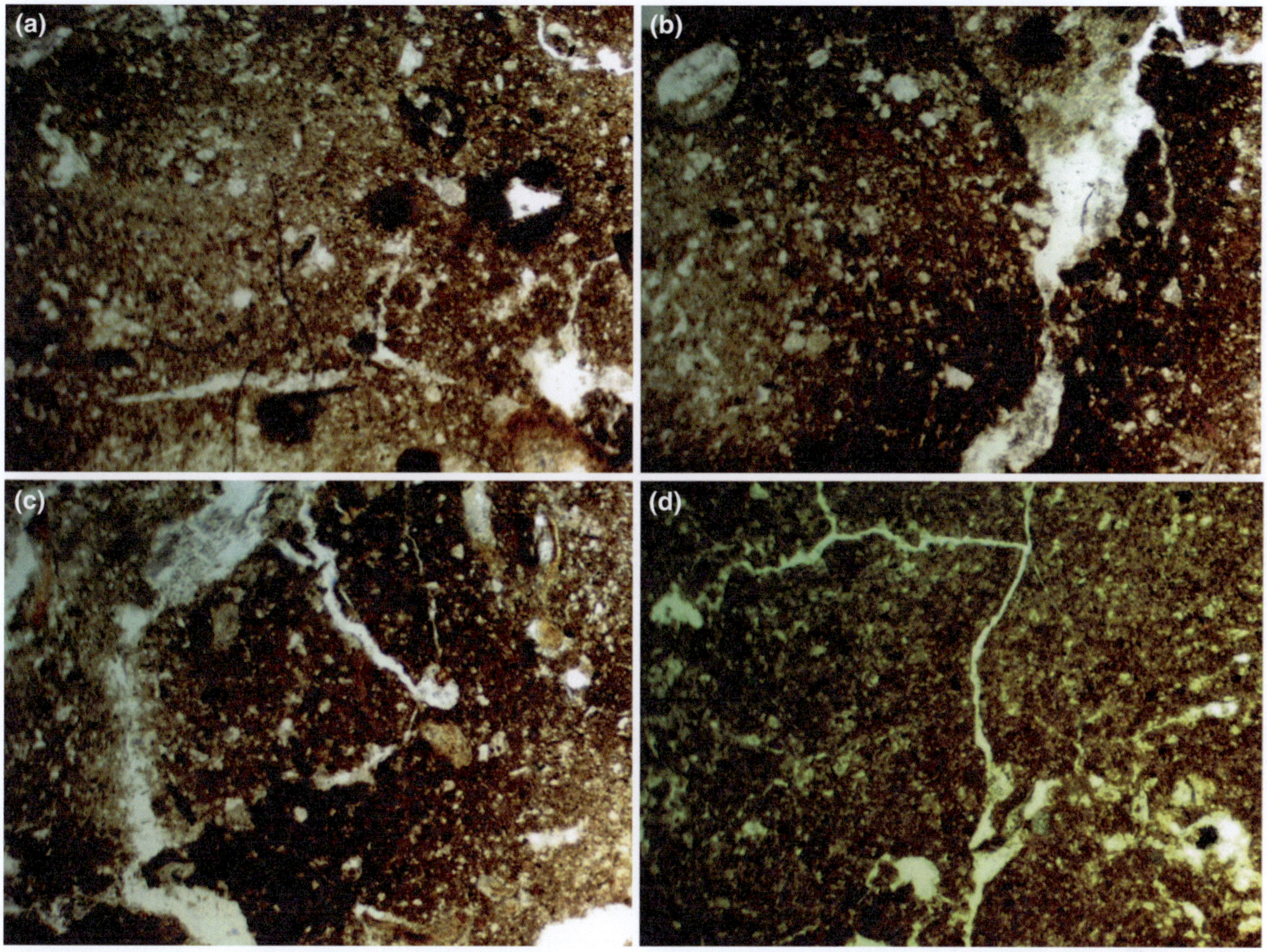

Fig. 5.34 Microstructure of Vertisols, nic.II; Pit-66, horizons: **a–b** 0–6 cm; **c** 15–20 cm; **d** 20–25 cm. Photos by L. Matchavariani

characterized by loose, fragmental, and compact microstructure, with a dark, chestnut-brownish mass, channel-like pores of irregular shapes, fine biogenic aggregation, organic-clay or carbonate-organic-clay plasma structure, mull-type of organic matter, remains of coprolites, fine microzones of organic matter and intensely decayed vegetation remains, carbonate nature of the main mass seen as fine-grain calcites, with the weak optical orientation, dusty-plasmatic elementary microstructure, significant densification, and intensified carbonate content at greater depths (Fig. 5.34).

5.8 Cambisols Chromic

Cambisols Chromic, called as Cinnamonic soils, are spread in the subtropical forest-steppe zone of East Georgia, at the altitude of 500(700)–900(1300) m above sea level (Fig. 5.35). Its lower border adjoins Kastanozems and Vertisols and its upper limit adjoins the Cambisols Dystric soils. They cover 4.8% of the total territory of the country.

The first who reflected the "Cinnamonic" soil on the Trans-Caucasus soil map were S. Zakharov (1924) and M. Sabashvili (1939, 1948). They made a valuable contribution to the study of this type of soil. I. Gerasimov theoretically confirmed the necessity to classify this soil as a type of its own genesis. The properties and peculiarities of abovementioned soils were studied by Anjaparidze (1979), Nakaidze (1977), Urushadze (1987, 1997), and others.

Cambisols Chromic are formed in a dry subtropical climate with warm winter, almost without snow and hot, dry summer with an average annual temperature of +9 to 12.5 °C. The duration of the vegetation period is up to 7 months. The sum of active temperatures is 2800–3800 °C. The amount of average annual precipitations is 300–800 mm, with two maximums at the end of spring and at the beginning of autumn. The humidity coefficient is 0.5–0.8, i.e., evaporation exceeds the amount of fallen precipitations.

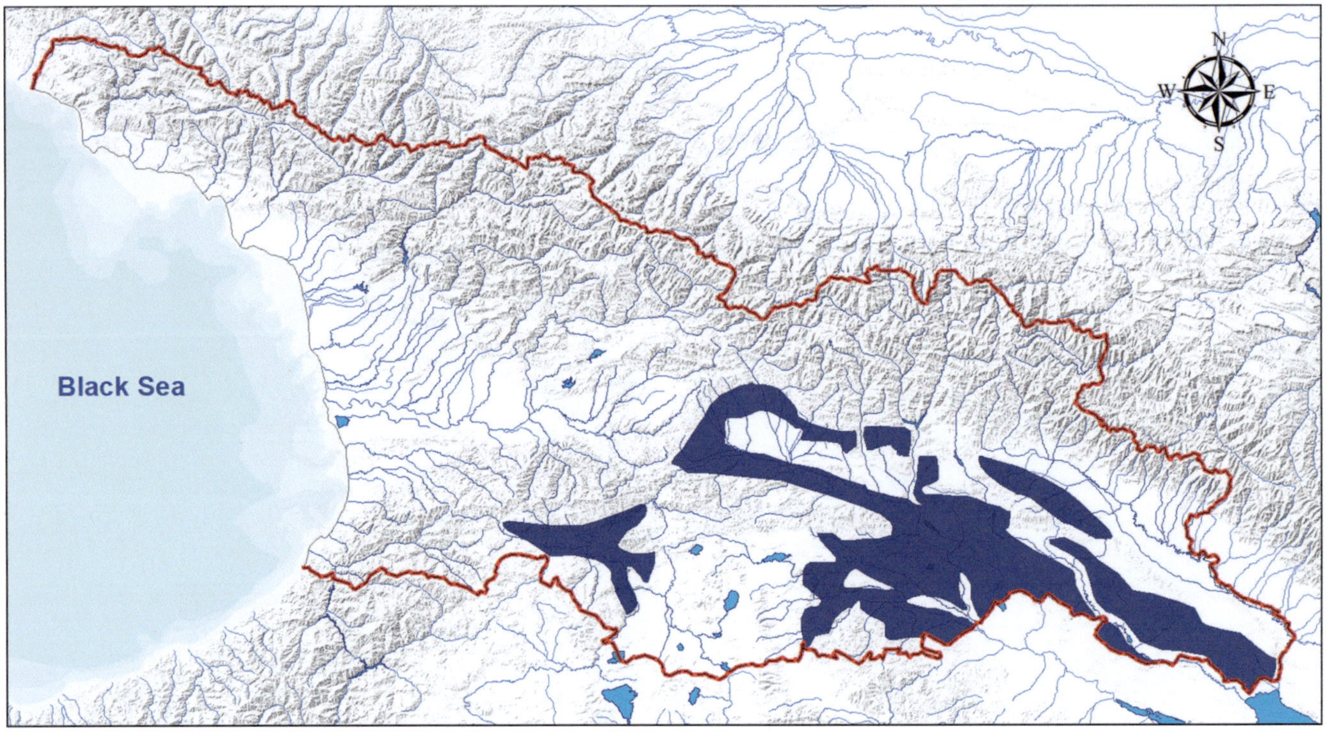

Fig. 5.35 Location of Cambisols Chromic. This map is created by D. Svanadze, based on data of L. Matchavariani

The geology of the northwestern and northeastern areas of the region is mainly presented by sandy clay and volcanogenic formations of the Paleogene Age and conglomerates, sandstones, and limestones of the Neogene Age. Inclined slopes and trains are presented by alluvion. The geology of the southern and southwestern areas of the region is presented by volcanogenic rocks of the Neogene Age: porphyry tuffs, lava flows, Upper-Cretaceous limestones, etc. The climate, which is peculiar as it is rich in parent material cations, promotes the formation of the weathering crust rich in carbonates.

The formation of the largest areas of the relief is mostly associated with erosive processes. At some locations, the relief is presented as landslide forms. The slopes are crossed by a number of wide gullies at many places. In the lower zone of some slopes, there are whole formations of flattened platforms observed.

The vegetation is presented by sparse arid and oak forests (Fig. 5.36). Sparse arid or light forests belong to the savannas of the subtropical climate. The grass cover is mainly presented by yellow bluestem. All plant species in the area are light demanding and drought resistant with a strong root system.

The age of soil formation in Cambisols Chromic is great. It is divided into light, carbonate, typical, alkaline, and Rendzic-Brown subtypes. Most of the areas of Cambisols Chromic are cultivated and the existing landscapes are almost totally anthropogenic.

Morphologically, the profile of the so-called Cinnamonic soil has a clear differentiation with the following common structure: $A–B_{Ca}–BC(BC_{Ca})–C_{Ca}$. The main diagnostic properties are clearly observed accumulative horizon, dark grayish-brown color, cloddy structure, heavy texture (clay and heavy loam), metamorphous horizon, and profile carbonization (Fig. 5.37). As for the gross chemical composition, the data are presented in Table 5.7.

As the analytical data suggest, the Cambisols Chromic soil is characterized by a light alkaline or neutral reaction (Fig. 5.38). Alkalinity increases at greater depths. The content of soil organic matter is average, but the soil is deeply humificated. Carbonate subtype contains carbonates right from the surface, while typical soils contain them from horizon AB and alkaline soils contain them from horizon C. Calcium carbonates form carbonate-illuvial horizon at some depth. The content of silicate iron exceeds that of nonsilicate iron. Besides, free (amorphous and crystallized) iron is accumulated in the upper part of the profile. The hydrothermal regime of Cambisols Chromic soil supports deep weathering of the primary minerals. Among sludge fraction minerals, montmorillonite and hydromica are found in the largest amounts. The hydro-physical properties of the mentioned soil are quite favorable. These soils have their

Fig. 5.36 The landscape in the area of Cambisols Chromic formation (Project "Cadastre and Land Registration", KfW)

Fig. 5.37 Profiles of Cambisols Chromic. Photo by B. Kalandadze

Table 5.7 Gross chemical composition of Cambisols Chromic, % (according to data of A. Nanaa)

Horizon (cm)	Loss on ignition	SiO_2	Al_2O_3	Fe_2O_3	CaO	MgO	Na_2O	K_2O	SiO_2: R_2O_3	SiO_2: Al_2O_3	SiO_2: Fe_2O_3
0–14	14.6	61.1	20.3	8.4	2.7	1.7	2.81	1.87	4.1	5.1	19.5
14–28	14.4	61.7	20.1	8.9	2.4	1.5	2.33	1.87	4.1	5.2	18.6
28–43	13.2	61.4	20.0	8.0	2.7	2.2	2.53	1.96	4.2	5.2	20.4
43–75	11.6	61.3	19.6	8.7	2.8	1.8	2.72	1.92	4.1	5.3	18.8
75–100	12.9	60.6	19.5	8.8	3.2	1.9	2.78	1.94	4 1	5.3	18.4

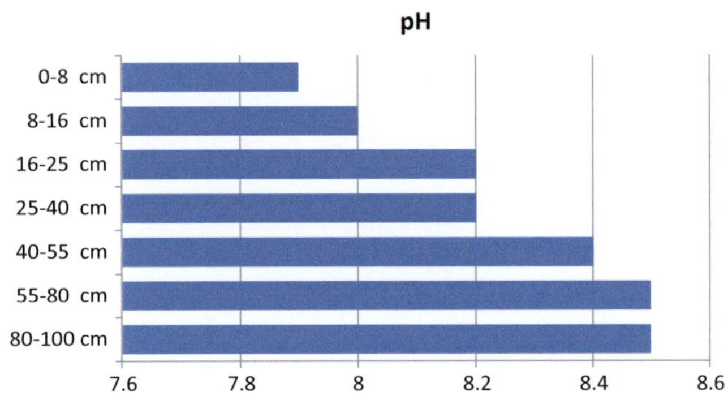

Fig. 5.38 pH distribution in profile of Cambisols Chromic (according to data of A. Nanaa)

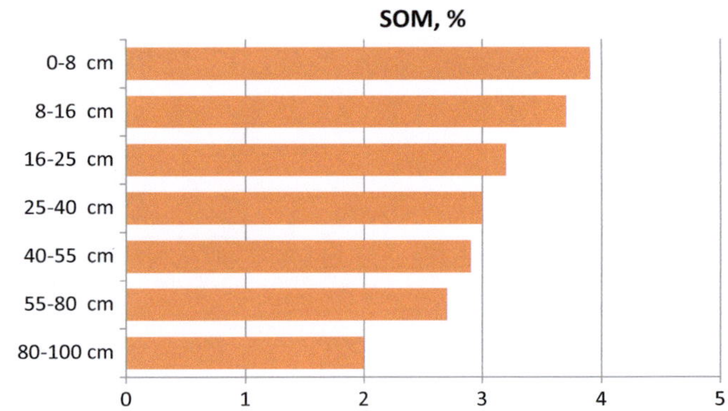

Fig. 5.39 Content of soil organic matter in Cambisols Chromic (according to data of A. Nanaa)

greatest density in Horizon B. General porosity is 40–52% and the least moisture content is 30–45%.

Unlike the "Meadow-cinnamonic" subtype, Cambisols Chromic is characterized by a clearer argilization and great amount of new carbonate formations. Unlike the Kastanozems (which has less moisture and is formed in terms of higher thermal provision), they have a lighter color, higher content of soil organic matter (Fig. 5.39), strong accumulative horizon, presence of carbonates at different depths in some subtypes, and less content of different forms of iron. Unlike the Vertisols, formed in similar terms of humidification, the Cambisols Chromic has a horizon with less soil organic matter, brown color, grain and prismatic structure, compacted and metamorphous horizon, with a less sharp transition from the accumulative horizon to the lower layers, less porosity, and water conductivity.

Unlike the Cambisols Dystric, formed under colder and more humid conditions, the Cambisols Chromic has a brown color, illuvial-carbonate horizon and intense argilization of the central part of the soil profile, less content of soil organic matter in the upper horizon, alkaline and neutral reaction, etc.

The Cambisols Dystric has quite high productivity and together with Vertisols and Chernozems is the most productive soil on the territory of the country. With its agricultural properties, it is one of the most favorable soils to grow high-productive vine and fruit. This soil is also used to grow cereals, vegetable, and other crops.

According to our previous micropedological study (Matchavariani 2008), the diagnostic properties of Cambisols Chromic are dark chestnut homogeneous structural anisotropic plasma; loose and sometimes spongy microstructure with the participation of complex aggregates of an irregular form; dusty-plasmatic elementary microstructure; presence of mull-type soil organic matter evidenced by the saturation of plasma with dark disperse organic substance and presence of numerous fine spots; organic-clay content of the main mass and weak optical orientation; presence of dispersed fine-grain calcites in plasma, bulk plasma on skeleton, almost total carbonization of plasma at great depths (carbonate content of the carbonate subtypes from the surface), etc. (Fig. 5.40).

In the area of the Cambisols Dystric, soils in the lower parts of relief are formed so-called meadow-cinnamonic soils. They are common in Kvemo Kartli and Zemo Kartli, Kakheti (on the right bank of the Alazani River) and Meskheti; present in the subtropical forest-and-steppe zone of Georgia, with a higher ground and surface humidity and occupied 2% of the total territory of the country. Sometimes this soil considers as subtype of "Cinnamonic" soils.

Fridland was the first to classify these soils in Georgia as Meadow-Cinnamonic soils as a soil type of the plains and foothills in East Georgia. Sabashvili (1948, 1965) referred to them as an "old alluvial meadow". Later (in 1965), he called them as Meadow-Cinnamonic soils. M. Sabashvili identified

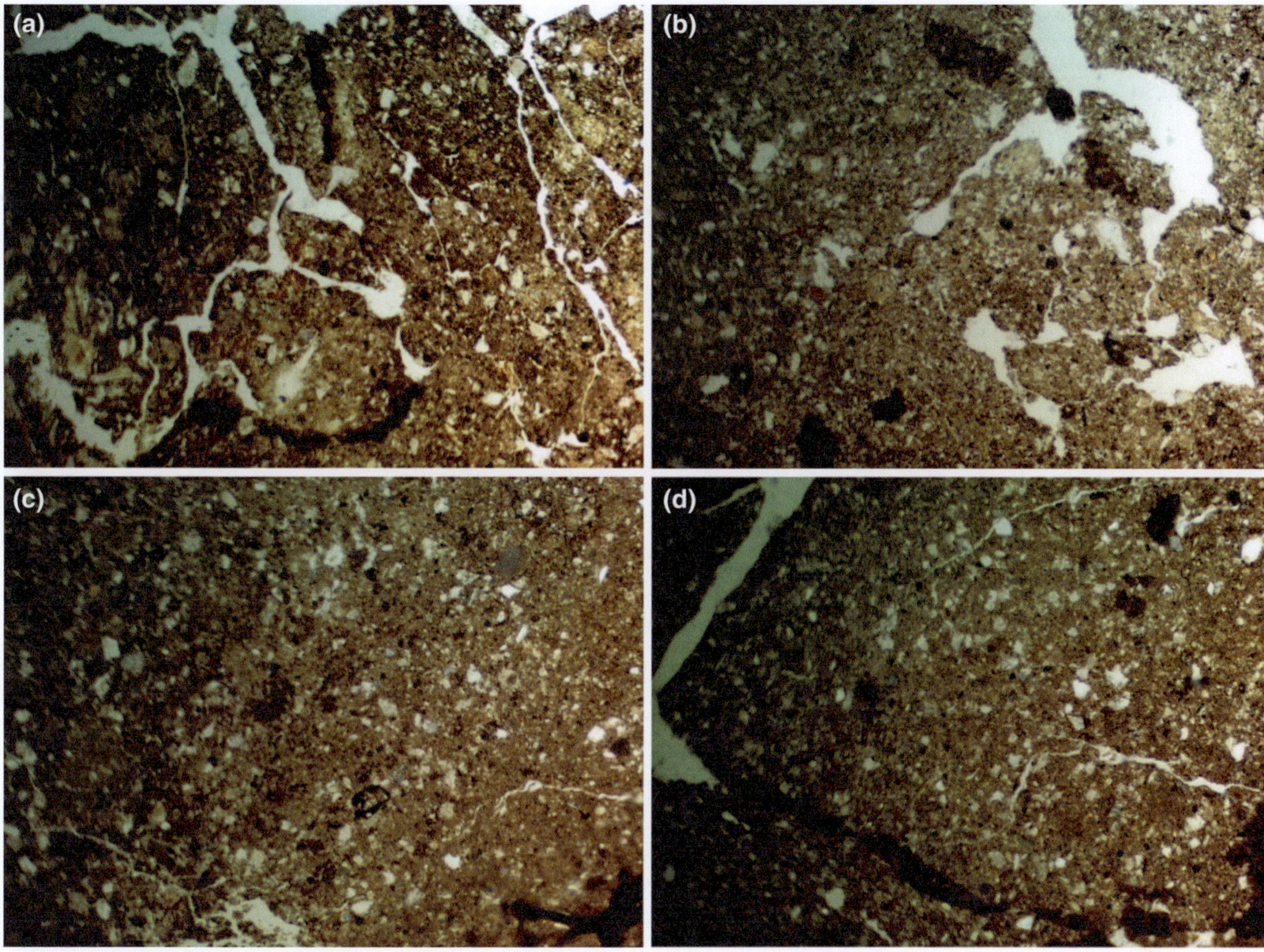

Fig. 5.40 Microstructure of Cambisols Chromic soils, nic.II; Pit-17, horizons: **a** 0–10 cm; **b** 22–27 cm; **c** 57–68 cm; **d** 80–128 cm. Photos by L. Matchavariani

two stages in the genesis of these soils: (1) the stages of the development of alluvial floodplain and meadow soils toward the "Cinnamonic" soils are clearly seen and (2) under the climate and human impact, the forest vegetation tends to change by the steppe vegetation. Talakhadze (1964) considered this type of soil as the step following the "Cinnamonic" soil evolution, as a partial elevation of the groundwater level supported the process of meadow formation and directed the development of the "Cinnamonic" soil towards the "Meadow-cinnamonic". These soils were also studied by Nakaidze (1977), R. Kirvalidze, K. Mindeli, E. Lataria, etc.

The soil-forming parent materials spread in the area of so-called meadow-cinnamonic soils are presented by strong alluvial and deluvial–proluvial deposits of a heavy texture and stone admixtures. The climate is moderately warm. Average annual temperature is 10–11 °C; the sum of active temperatures is 2800–3800 °C; the duration of the vegetation period is 6 to 7 months; the amount of precipitations varies between 460 and 520 mm; and the humidity coefficient is 0.5–0.9. The natural vegetation is presented by floodplain forest (oak forests). The major part of the area is used as arable lands, orchards, or vineyards and is mostly irrigated (Fig. 5.41).

The difference between the "Cinnamonic" and "Meadow-cinnamonic" soils is that the latter has a darker color, the signs of gleyzation expressed as bluish or rusty-colored spots, by less argilization and presence of new carbonate formations.

Morphologically, the mentioned soils have a dark brown color and slightly differentiated and relatively stronger profile than the brown soils; signs of gleyzation almost through the whole profile, heavy texture, carbonization of the whole profile, and slightly expressed carbonate-illuvial horizon (Fig. 5.42). The morphological structure of the profile is A–AII–AB–$B_{Ca(g)}$–$BC_{Ca(g)}$.

Based on the profile analysis of "Meadow-Cinnamonic" soils, they are characterized by alkaline or weak alkaline reaction, with little sum of absorbed bases and little content of soil organic matter in the arable horizon, but by deep

Fig. 5.41 The landscape in the area of "Meadow-cinnamonic" soil formation (Project "Cadastre and Land Registration", KfW)

Fig. 5.42 Profiles of "Meadow-cinnamonic" soil. Photo by B. Kalandadze

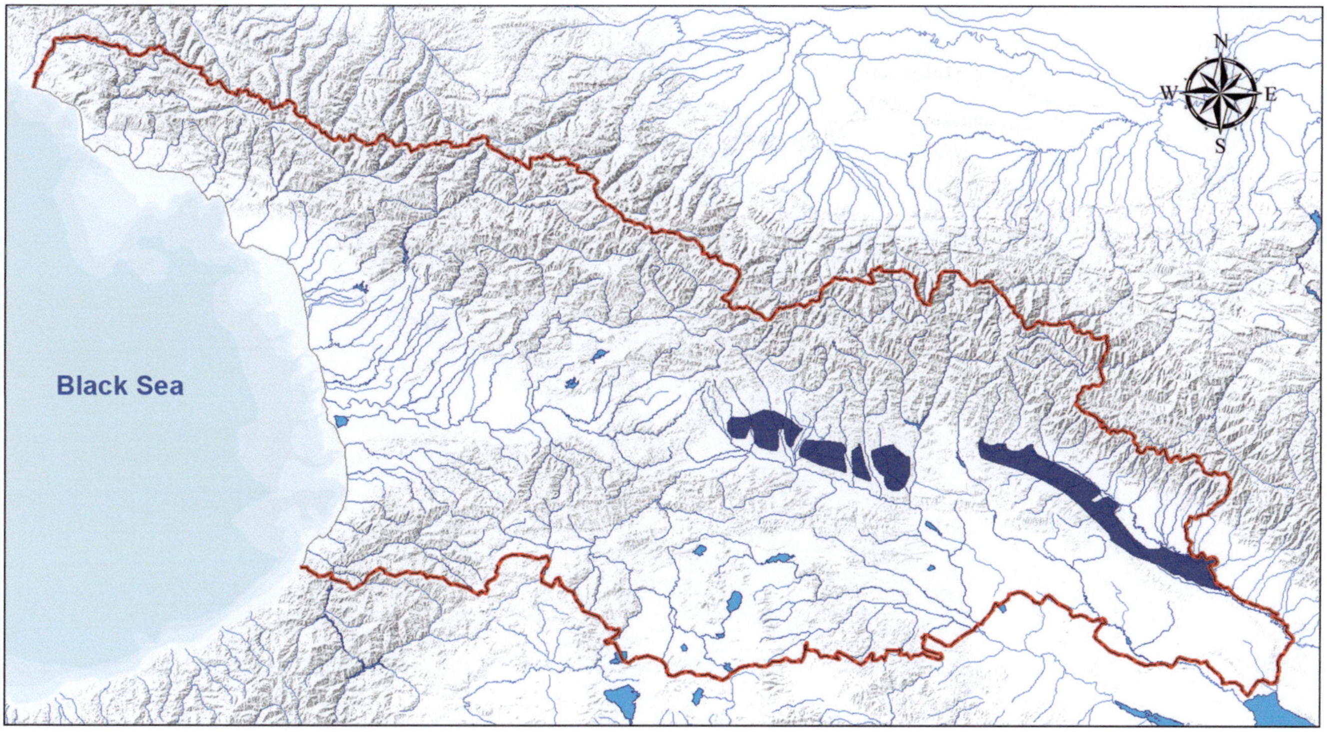

Fig. 5.43 Location of Kastanozems. This map is created by D. Svanadze, based on the data of L. Matchavariani

humification. Carbonates are observed right from the surface, with their amount significantly increasing in the bedrock. With their texture, soils belong to the category of light and average clay. A dominant part in the mineralogical content of silt is hydromicas. Argilization is clearly observed in the middle portion of the profile. Silicate exceeds nonsilicate one with its maximum in the mid-profile.

According to our previous micropedological study (Matchavariani 2008), "Meadow-Cinnamonic" soil characterized by: slight aggregation of upper horizons, dark chestnut homogeneous structural mass; loose and sometimes spongy microstructure with the participation of complex aggregates of an irregular form; fragmental porosity, with the participation of dusty-plasmatic elementary microstructure; presence of inter-aggregate porosity; saturation of plasma with dark disperse soil organic matter substance and presence of numerous fine organic spots; organic-clay content of the main mass and weak optical orientation; presence of individual organic-iron spots; slight optical orientation of fine-disperse substance; high content of calcite microcrystal; presence of dispersed fine-grain calcites in plasma, bulk plasma on skeleton, almost total carbonization of plasma at great depths (carbonate content of the carbonate subtypes from the surface); dense illuvial horizon, etc.

5.9 Kastanozems

Kastanozems, which correlate with the national classification as Gray-Cinnamonic soils, are common in the dry steppes of the subtropical zone of southeast Georgia (Fig. 5.43). They border Vertisols, Cinnamonic, Meadow-Cinnamonic soils. The area occupied by the Kastanozems in Georgia is 9.1% of the total territory of the country.

First, this type of soil was studied by Zakharov (1924) under the name of Kastanozems. D. Gedevanishvili was the first soil scientist to use term Gray Cinnamonic with the modern meaning. The initiative to classify the Gray-Cinnamonic soils as an individual zonal type of soil belongs to A. Rozanov. These soils were thoroughly studied by Sabashvili (1948), Nakaidze (1977), etc.

Kastanozems of Georgia have an old age. They are formed in terms of moderately dry subtropical climate with the average annual temperature of 12–13 °C. The sum of active temperatures is 4000–4500 °C; the duration of the vegetation period is more than 7 months; the average annual amount of precipitations is 300–500 mm with its maximum in spring and autumn; the humidification coefficient is 0.4–0.6; the relief is presented as plains, piedmonts, and low

Fig. 5.44 The landscape in the area of Kastanozems formation (Project "Cadastre and Land Registration", KfW)

mountains; the soil-forming parent materials are presented by proluvial, alluvial, and eluvial–deluvial deposits of different granular, mineralogical, and chemical contents (sometimes, saline); the vegetation is a dry-steppe one. Most of the territory is used as arable or sowing lands to grow agricultural crops (wheat, barley, maize, and sunflower). A relatively small area is occupied by perennial plants (orchards and vineyards). A large area is used as winter pastures (Fig. 5.44).

The properties of Kastanozems soil are associated with modern bioclimatic conditions. The process of soil formation mainly takes place in terms of severe moisture deficit. Consequently, the vegetation remains, and newly formed soil organic matter are subject to intense mineralization. The peculiarities of the dry subtropical climate (high temperature combined with short humidification periods) result in inter-soil weathering by accumulation of clays, Fe-hydroxides and carbonates. In humid conditions, the soil solutions (with dominant calcium and manganese hydrocarbon content) have a descending dislocation, while in dry periods, they have an ascending dislocation. One of the main properties of the Kastanozems is the spaciousness of a carbonate-illuvial horizon.

Morphologically, Kastanozems soil has a brown or grayish color; it is slightly differentiated, mudded, with carbonate and little accumulative horizons, presence of soil organic matter and carbonate horizons, well-expressed argilization in the middle part of the profile and presence of carbonates right from the surface (Fig. 5.45). The morphological profile of the soil usually has the following structure: A_{Ca}–Bm_{Ca}–BC_{Ca}, or A1′–A1″–AB–$B1_{Ca}$–C1–C2.

As the analytical data suggest, Kastanozems have weak alkaline or alkaline reaction and little content of soil organic matter—type of soil organic matter: fulvous humate (Figs. 5.46 and 5.47). The upper and middle parts of the profile have heavy clay texture, which is lighter in the lower part. The argilization of the middle part of the profile is one of the diagnostic properties of this soil. The main oxides are distributed evenly in the profile. The content of carbonates varies from 4 to 23%. Generally, the carbonates are fixed from the surface, and in this way, these soils differ from the soils found in dry subboreal steppe zone. The absorption complex is saturated with bases. Calcium dominates in the exchange cations. The content of calcium decreases at the expense of the increased content of exchange manganese at greater depths. Exchange sodium is a part of the absorption complex of Kastanozems. Montmorillonite and hydromicas dominate in the silt fraction, while the content of kaolin, quartz, and other minerals is low. The content of silicate iron is more than that of a nonsilicate Fe in the Kastanozems. The maximum content of ferrum is observed in the upper part of the soil profile. The data of the gross chemical composition are presented in Table 5.8.

Unlike the Cambisols Chromic, which are formed in terms of higher humidification and less thermal provision, the Kastanozems are characterized by a darker color, less content of soil organic matter, the carbonate nature of the whole profile, and higher alkalinity. They differ from the Vertisols by a less strength of an accumulative horizon, lighter texture, and presence of an illuvial-carbonate horizon.

According to our previous micropedological study (Matchavariani 2008), Kastanozems is characterized by a

Fig. 5.45 Profiles of Kastanozems. Photo by B. Kalandadze

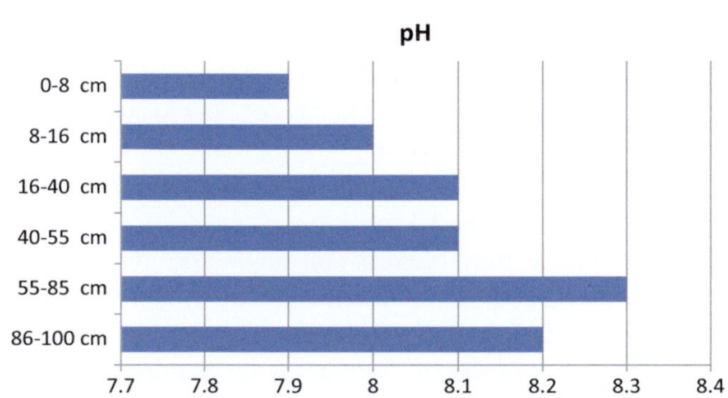

Fig. 5.46 pH distribution in profile of Kastanozems (according to data of A. Nanaa)

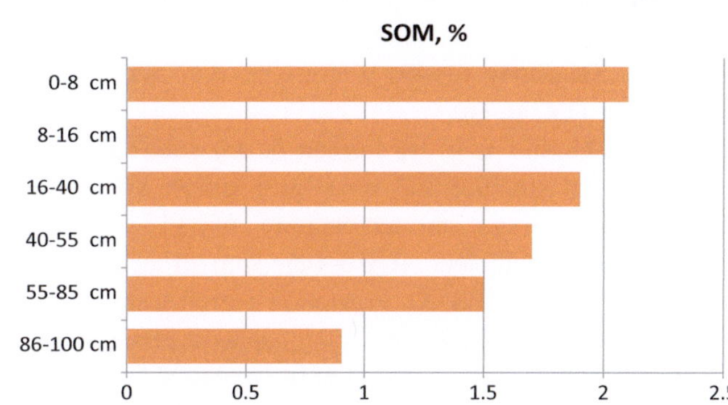

Fig. 5.47 Content of soil organic matter in Kastanozems (according to data of A. Nanaa)

cloudy structure, inter-aggregate, and fracture porosity at some locations, moder-mull or mull-moder morphological type of SOM, elementary sandy-plasmatic microstructure, soil organic matter coloration, mineralization of most organic remains, organic-carbonate-clay content of plasma, point structure of optical orientation and presence of dispersed grains of fine-crystal calcite in the arable horizon; by compact microstructure and increased clay content, carbonate-clay plasma, presence of numerous nodules of fine-grain calcite, point and mixed-fiber structure in the carbonate horizon; and by less argilization of fine-disperse mass and lower optical orientation of carbonate-clay plasma and dusty-plasmatic elementary microstructure in the lower horizons.

In the area of Kastanozems, in the lower parts of relief, in terms of higher humidification, there are spread so-called Meadow-Gray-Cinnamonic soils. They are common in the

Table 5.8 Gross chemical composition of Kastanozems, % (according to data of A. Nanaa)

Horizon (cm)	Loss on ignition	SiO_2	Al_2O_3	Fe_2O_3	CaO	MgO	Na_2O	K_2O	SiO_2: R_2O_3	SiO_2: Al_2O_3	SiO_2: Fe_2O_3
0–20	14.4	62.0	17.0	8.2	2.8	3.6	2.7	2.9	4.7	6.2	20.1
20–40	13.9	61.1	17.2	8.1	3.4	3.3	3.1	3.1	4.7	6.0	20.3
40–55	14.7	60.5	16.8	7.9	4.0	3.7	3.4	2.8	4.7	6.1	20.5
55–90	13.9	58.5	16.9	7.7	6.9	3.6	3.5	2.2	4.6	5.9	20.2
90–110	15.3	60.4	16.5	8.0	4.4	3.8	3.1	2.9	4.8	6.2	20.2
110–140	11.4	61.1	17.4	7.9	3.5	3.1	4.1	2.1	4.6	6.0	20.7

southeast part of the country (on the right bank of the Alazani River), in Kvemo Kartli and as fragments in Shida Kartli. This soil is formed in terms of dry subtropical climate, over the plain relief with dominant negative forms. The soil formation is affected by an anthropogenic factor (irrigation). The morphological structure of the profile is A_{Ca}–$B1_{Ca}$–$B2_{Ca}$,t,g–BC_{Ca},g–Cg (Fig. 5.48).

5.10 Solonetz Humic

The area occupied by Solonetz Humic soil group is making 1.6% of the total territory of the country. This soil is spread as saline soil, salt soil, and solonetz in eastern Georgia on the accumulation plains, sloping plains, and slopes of the erosive watershed plateau of deserts Kakheti, as well as plains fragments in Shida Kartli (Fig. 5.49). Salt soils have easily soluble salts right from their surface, while saline soils have such salts at various depths and Solonetz contains absorbed sodium in their illuvial horizon. The group of soils was studied by Sabashvili (1948, 1965), Chkhikvishvili (1970), Akhvlediani (1973), etc.

Solonetz Humic soils are formed in dry subtropical climate, with hot summer and warm and almost non-snowy winter with average air temperature of +12, +13 °C; humidification coefficient of 0.3–0.5, sum of active temperatures of 4000–4500 °C, vegetation period lasting up to 7 months, and annual amount of atmospheric precipitations of 350–500 mm with its maximum in May and June.

The soil-forming parent materials where Solonetz Humic are spread are mainly presented by alluvial, proluvial–deluvial, and saline sediments with saline clays. The type of soil-forming parent materials results in high salinity of ground waters to a certain extent.

The relief is presented by intermontane depressions, alluvial plains, and elements of former lakes. Solonetz soil mostly develops on young depression relief, while salt soil develops on the elements of an old, elevated relief. Salination is more intense on a flat plain relief than on the inclined slopes where ground waters are located deep and the process of periodic salination takes place (Fig. 5.50).

Saline and salt soils are characterized by heavy texture (mostly, clay). Ca dominates in the absorbed cations; however, Na and Mg are also found in great amount. The content of soil organic matter is low and it sharply decreases as the depth increases. These soils contain different amounts of easily soluble salts with their amount increasing at greater depths. Clay minerals are presented by montmorillonite and hydromicas. The main oxides are distributed evenly in the profile. Saline soils are characterized by low productivity; however, owing to proper melioration measures, their productivity is possible to improve significantly.

The profile of typical salt soils is slightly differentiated and is characterized by a high salt content (Fig. 5.51). The morphological structure of the profile is A–BC–C. Salt soil is characterized by strong alkaline reaction and air–water regime of a poor structure. These soils have unfavorable physical (water–air) properties. Among saline soils, one can identify automorphous (with maximum salt content on the surface and deeper) and hydromorphous soils formed near the ground waters under the periodic wash-down regime. With the hydrological conditions, Solonetz soils are classified as hydromorphous formed in case of near location of the mineralized groundwaters (1.5–3 m) and automorphous soils, which are spread where the mineralized ground waters are located deep (up to 10 m).

The profile of the Solonetz soil has a specific structure with a differentiated eluvial–illuvial structure, heavy texture, alkaline reaction (Fig. 5.52), column and prismatic structure, dense Solonetz horizon Bt^{Na+} (what is its diagnostic index), increased amount of absorbed sodium, and poor water conductivity in the lower horizons. Clay minerals are also presented by montmorillonite and hydromicas. Content of soil organic matter varies between great limits and it is higher in weak Solonetz soils than in intense or average Solonetz soils (Fig. 5.53). The content of easily soluble salts also varies. Saltiness, as a main genetic feature, is determined by the content of absorbed Na and reaches 30% in the intense and average Solonetz soils. Content of absorbed magnesium is another feature. Solonetz has very poor water conductivity. The data of the gross chemical composition are presented in Table 5.9.

Fig. 5.48 Profiles of "Meadow-gray-cinnamonic" soil. Photo by B. Kalandadze

According to our previous micropedological study (Matchavariani 2008), Solonetz characterized by a non-aggregated mass and high content of easily soluble salts and dispersed crystals, clay-carbonate or clay-salty anisotropic plasma (with the participation of carbonate micro-concretions in the lower horizons), relict forms of a raw-type soil organic matter of a dark color, which are the remains of the meadow-forming stages, with the signs of disperse soil organic matter displacement, accumulation of sinter forms, densification at greater depths, signs of gleyzation, optical orientation of a current-like and scale structure, presence of snowflake Fe-spots at the level of microzones, etc.

5.11 Nitisols Ferralic

Nitisols Ferralic soils, which correlate with the national classification as Red soils, are spread in the southwest part (Adjara, Guria) and partly in the western (Samegrelo) and northwestern part (Abkhazeti) of Georgia (Fig. 5.54). The total area on the territory of Georgia occupied by these soils is about 2% of the total territory of the country. Nitisols Ferralic soils are spread at 100–300 m above sea level, in a humid subtropical zone, on hilly reliefs. They border Yellow-Black, Yellow, Subtropical Podzolic, and Gley Podzols.

The first researchers to study the "Red" soils were A. Krasnov and V. Dokuchaev, who equalized it with laterites. In K. Glinka's view, it is a relict soil with a gley-formation process taking place in it at present. It was him to classify the so-called Red soils, Laterites and Zheltozems (Yellow soils) into different groups in the first world map of soils. The fundamental studies of "Red" soils were accomplished under the leadership of B. Polinov making it clear that in the humid subtropics of West Georgia, the acid forms of soil formation develop on the red-color weathering crust and modern soil is younger than the weathering crust. These soils were also studied by Sabashvili (1948, 1965). The works by the latter consider the regularities of the geographical distribution of these soils. Daraselia (1939, 1949, 1974) studied the physical properties, hydrological regime, and humidity of them.

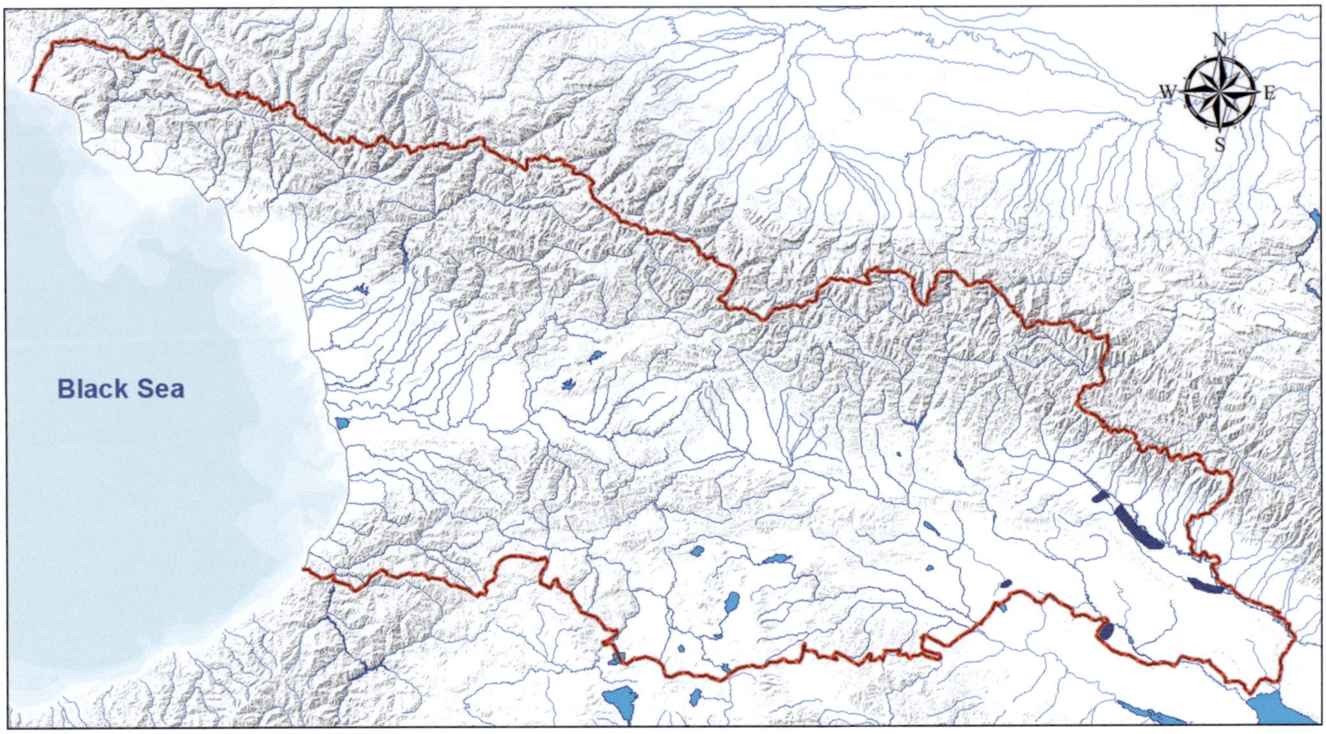

Fig. 5.49 Location of Solonetz. This map is created by D. Svanadze, based on the data of L. Matchavariani

Fig. 5.50 The landscape in the area of Solonetz formation (Project "Cadastre and Land Registration", KfW)

Monographs about the "Red" soils of Georgia belong to Romashkevitch (1966, 1974a, b) and Palavandishvili (1987).

The Nitisols Ferralic soils are formed on the base-effluent parent materials (andesites mainly) and red-colored weathering products. The color of the parent materials results from the presence of the closely associated Fe-hydroxides on the surface of clay particles. The soils formed on them well maintain the red color and all properties of weathering products, and the name of this type of soil comes from its color.

The climate in the zone with Nitisols Ferralic is humid subtropical; the average annual temperature is quite high (+14, +15 °C); the vegetation period lasts for 8 months and the annual amount of atmospheric precipitations is 1200–2500 mm, with the minimum falling in spring. The natural vegetation is presented by the mixed fragments of subtropical forest and evergreen undergrowth (Fig. 5.55).

The morphological structure of Nitisols Ferralic is A–AB–B–BC–C. They are characterized by a strong profile of a red coloration (Fig. 5.56). Soil reaction is acid (pH = 4–4.5)

Fig. 5.51 Profiles of Solonetz and Solonchak (Project "Cadastre and Land Registration", KfW)

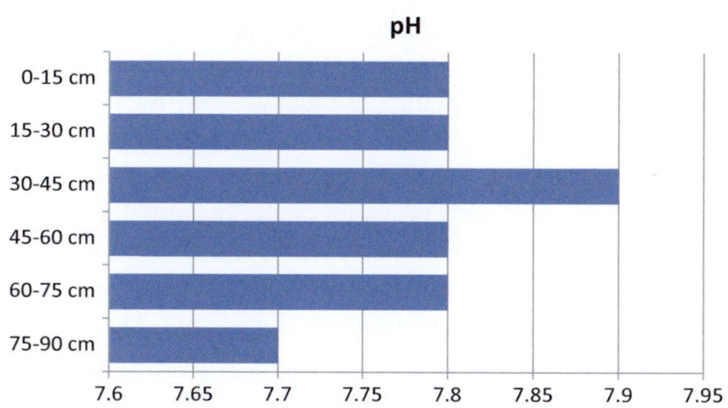

Fig. 5.52 pH distribution in profile of Solonetz (according to data of T. Urushadze)

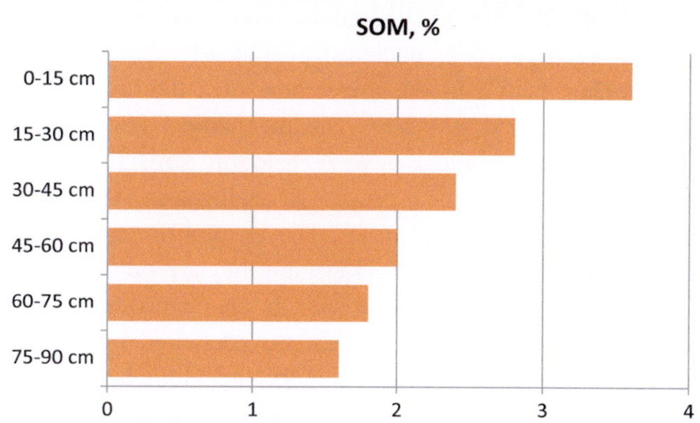

Fig. 5.53 Content of soil organic matter in Solonetz (according to data of T. Urushadze)

Table 5.9 Gross chemical composition of Solonetz, % (according to data of T. Urushadze)

Horizon (cm)	SiO_2	Al_2O_3	Fe_2O_3	TiO_2	CaO	MgO	Na_2O	K_2O	$SiO_2:R_2O_3$	$SiO_2:Al_2O_3$	$SiO_2:Fe_2O_3$
0–6	70.9	12.7	6.1	0.6	2.2	1.4	3.7	1.4	7.3	9.5	31.1
6–12	68.6	15.2	6.3	0.5	2.2	1.4	2.9	1.8	6.1	7.7	19.3
12–28	64.6	16.7	7.7	0.5	1.7	1.0	3.6	1.9	5.1	6.6	22.4
28–40	66.3	16.1	7.3	0.4	2.3	2.0	2.7	2.0	5.4	7.0	24.6
58–70	66.7	17.2	7.8	0.6	1.4	2.23	1.2	1.2	5.1	6.6	22.7
110–115	67.2	17.1	7.8	0.4	0.6	2.21	1.8	1.3	5.2	6.7	22.9
180–192	67.0	16.4	7.9	0.5	0.7	1.53	2.4	1.6	5.3	5.9	22.8

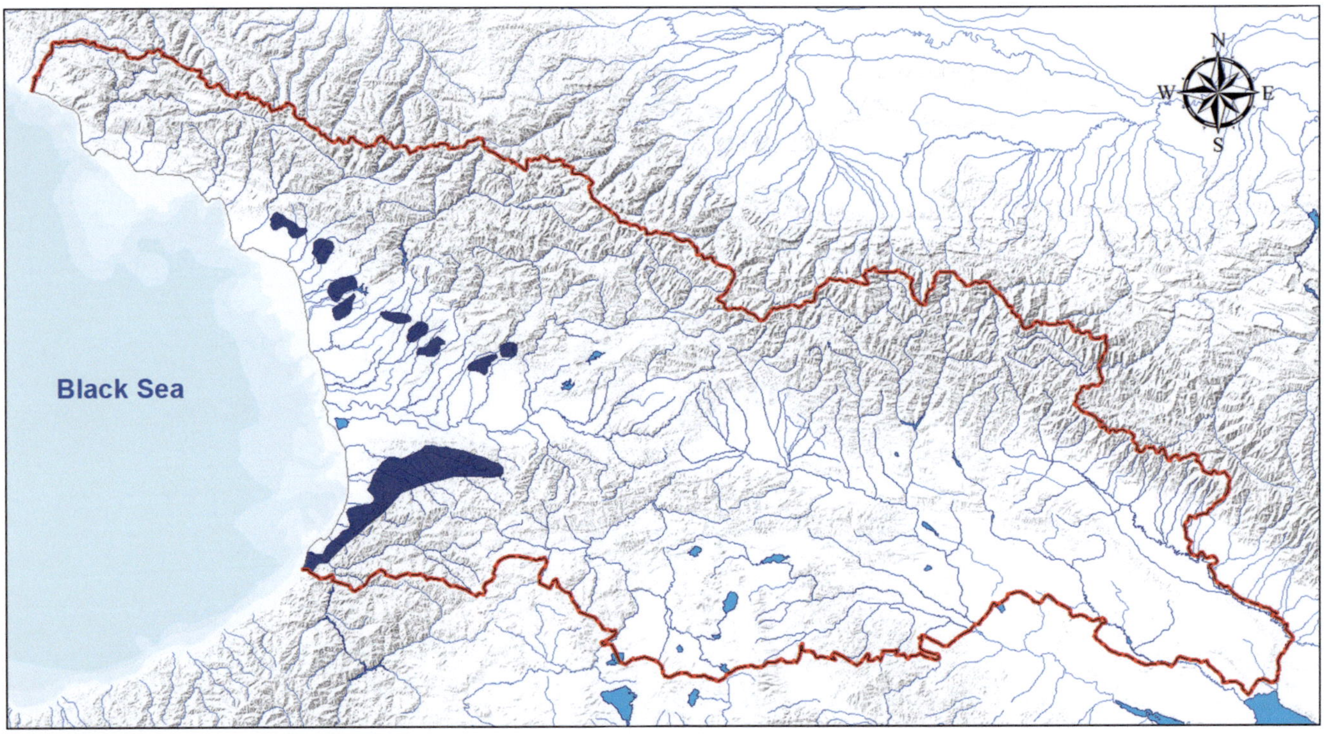

Fig. 5.54 Location of Nitisols Ferralic. This map is created by D. Svanadze, based on the data of L. Matchavariani

and changes insignificantly through the profile (Fig. 5.57). The soil organic matter is of a fulvous type; its content is from average to high (6%, decreasing gradually as the depth grows) (Fig. 5.58). In the upper horizons, the structure is cloddy, nutty, and granular. The mechanical composition (texture) of the profile is heavy—a loamy soil, average or heavy clay, with a typical process of argilization. The soil is poor in earth silicon and bases. The absorption capacity of Nitisols Ferralic is low to average and they are rich in R_2O_3. The mineral portion of these soils is characterized by ferralitic weathering. Clay minerals are presented by kaolin, halloysite, goethite, and gibbsite. In the Nitisols Ferralic, silicate ferrum dominates over the nonsilicate ferrum. Individual forms of ferrum are distributed more or less evenly with their profile. The data of the gross chemical composition are presented in Table 5.10.

Unlike the Acrisols Haplic, which develop in the same bioclimatic conditions, on the parent materials rich in earth silicon, the Nitisols Ferralic are distinguished for a red coloration, less coarse structure, and greater weathering. They are used to grow subtropical crops and tea plantations.

According to our previous micropedological study (Matchavariani 2008), Nitisols Ferralic are characterized by a compact microstructure and crumbling porosity and dust-plasmic elementary microstructure (Fig. 5.59). Their upper horizons are reddish-brown clay mass, with a moder-mull and mull-moder morphotype of soil organic matter, with the presence of Fe-segregation, plasma with intense optical

Fig. 5.55 The landscape in the area of Nitisols Ferralic formation (Project "Cadastre and Land Registration", KfW)

orientation with a fine-scale and fiber structure (Fig. 5.60). The transient horizon has a dusty-loamy structure of a yellow color. Plasma has an Fe-loamy structure with a weak optical orientation. Illuvial horizon is yellowish, disperse, dense, dusty-plasmic, and intensely ferruginated. The Fe-concretions in the profile are dissected and are often presented as spot microzones saturated by Fe-hydroxides. The skeletal nature is intensified as the depth increases, with the large rock fragments present. The local clay movement is evidenced by clay and ferrous-clay cutans. The horizon transient to parent material is of a light gray color with yellowish strips, sandy-dusty-plasmic microstructure, and slight optical orientation. Intensely ferruginated strong clay cutans are fixed.

5.12 Acrisols Haplic

Acrisols Haplic soils, which correlate with the national classification as Yellow soils, are widely spread on old-marine terraces, dissected and adjoining foothills in the humid subtropical zone of Georgia, at 300–600 m asl (Fig. 5.61). The total area of these soils in Georgia is making 4.5% of the territory of the country. In Guria, Imereti, and Abkhazeti, these soils are spread adjacent to the Nitisols Ferralic, adjacent to the Luvisols Albic and Gleysols in Samegrelo and Cambisols Dystric in Abkhazeti, Imereti, and Guria.

Some data about the so-called Yellow soils (Zheltozems) are found in the works by P. Kosovitch, I. Vitin, and S. Zakharov. The name "zheltozem" was introduced following the similarity with the lower horizons of the "Red" soils developed on a yellow-colored weathering crust in West Georgia. Sabashvili (1965) explored these soils thoroughly. He was the first to fix the dependence of this kind of soils on the nature of the soil-forming parent materials. Earlier studies identified the "Yellow" soils as a subtype of "Red" soils. However, at present, they are considered as a separate genetic type.

Soil-formation parent materials, over which these soils are formed, are presented on the acid and averagely acid (mainly slates) weathering products. This kind of soil usually develops on loose clay parent materials. The soil-forming parent materials belong to sialith clays; however, sometimes there are ferrallitized ones, too. The area of Acrisols Haplic, like that of Nitisols Ferralic, is determined by the scales of spreading of parent materials.

Acrisols Haplic soils are formed in a humid subtropical climate, with an average annual temperature of +13, +15 °C; the temperature of the coldest month (January) is +3, +7 °C; the temperature of the warmest period of the year (July and August) is +19, +25 °C. The vegetation period lasts for 8 months. The annual amount of precipitations is 1100–2500 mm, but the precipitations are distributed unevenly in different months, with the maximum in April, May, and June.

The natural vegetation is presented by a mixed subtropical forest (oak, elm zelkova, chestnut, wing nut, beech, maple) (Fig. 5.62). At present, the vegetation cover on the

Fig. 5.56 Profile of Nitisols Ferralic. Photo by B. Kalandadze

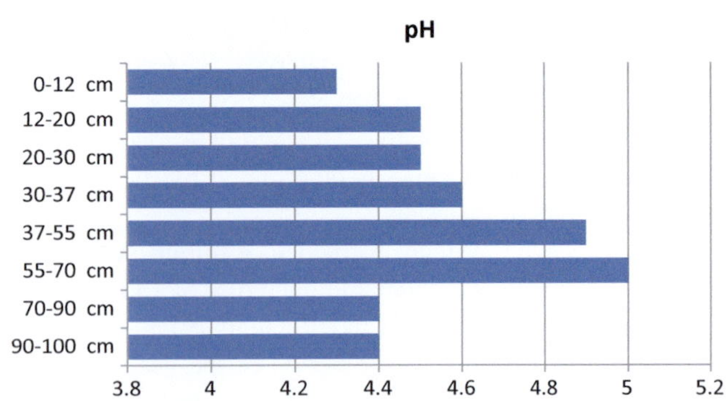

Fig. 5.57 pH distribution in profile of Nitisols Ferralic (according to data of T. Ramishvili)

most of the territory is destroyed and agricultural fields and plantations are cultivated instead.

The morphological structure of Acrisols Haplic is A0–A–AB–B–BC. These soils are characterized by a yellow color,

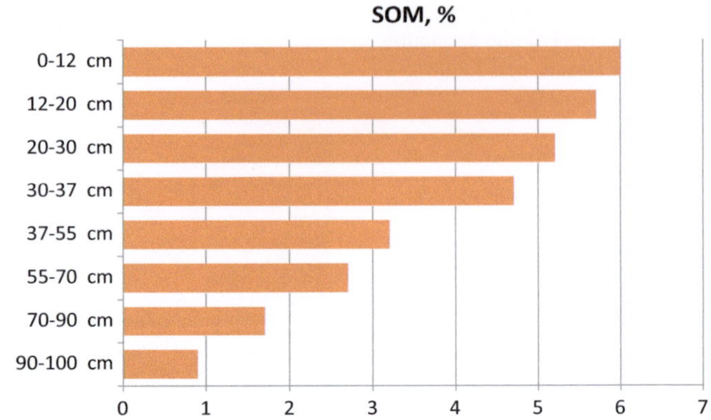

Fig. 5.58 Content of soil organic matter in profile of Nitisols Ferralic (according to data of T. Ramishvili)

argilization, and strong profile (Fig. 5.63). The soil reaction is averagely acidic (pH = 5–6) and changes slightly through the profile (Fig. 5.64). The content of soil organic matter is 4–5% and decreases drastically as the depth increases (Fig. 5.65). The soil organic matter has fulvous composition. The absorbed complex is not saturated with bases, but the degree of unsaturation varies within the great limits (4–7 to 60–70%). The mechanic composition (texture) of the Acrisols Haplic is heavy and slightly changes at greater depths; the content of physical clays (<0.001 mm) is 40–45%. The amount of amorphous iron is little and that of nonsilicate ferrum is quite high. The main oxides are distributed unevenly. SiO_2:R_2O_3 ratio in the sludge/silt fraction evidences both, ferralitic and sialithic weathering. The data of the gross chemical composition are presented in Table 5.11.

Unlike the Nitisols Ferralic soil, the Acrisols Haplic, which develop in the same bioclimatic conditions, but on the parent materials poor in earth silicon, have a yellow color, weaker weathering, and less solid and coarser structure.

The distribution scales and properties of Acrisols Haplic depend on the nature of parent materials. In the soil-forming process, ferrum moves and intense hydration of its compounds result in the yellow color of the profile. Intense hydration of the mentioned soils has resulted from the retention of a great amount of water by clays and shales.

Acrisols Haplic soils are poor in nutrition elements and rapidly lose their productivity when exploited. Their physical properties are not favorable, either. They have poor water conductivity, aeration, and weak structure. Therefore, they can be most efficiently used under permanent plantations (tea, citruses, tung tree). Rich harvest can be gained only by applying large amounts of organic and mineral fertilizers.

According to our previous micropedological study (Matchavariani 2008), the Acrisols Haplic soils are characterized by nonhomogeneous dense clay mass, inter-aggregate porosity, pores with roundish or figurative shapes (Fig. 5.66), moder type of soil organic matter in the upper horizon and

Table 5.10 Gross chemical composition of Nitisols Ferralic, % (according to data of T. Ramishvili)

Horizon (cm)	SiO_2	Al_2O_3	Fe_2O_3	TiO_2	CaO	MgO	SiO_2:R_2O_3
0–20	52.70	25.33	16.76	1.63	1.28	1.44	2.48
58–86	51.87	25.84	16.66	1.61	0.84	1.52	2.46
86–136	57.24	25.54	16.79	1.52	0.92	0.85	2.97
136–180	44.92	29.97	19.15	1.57	2.04	2.69	1.83

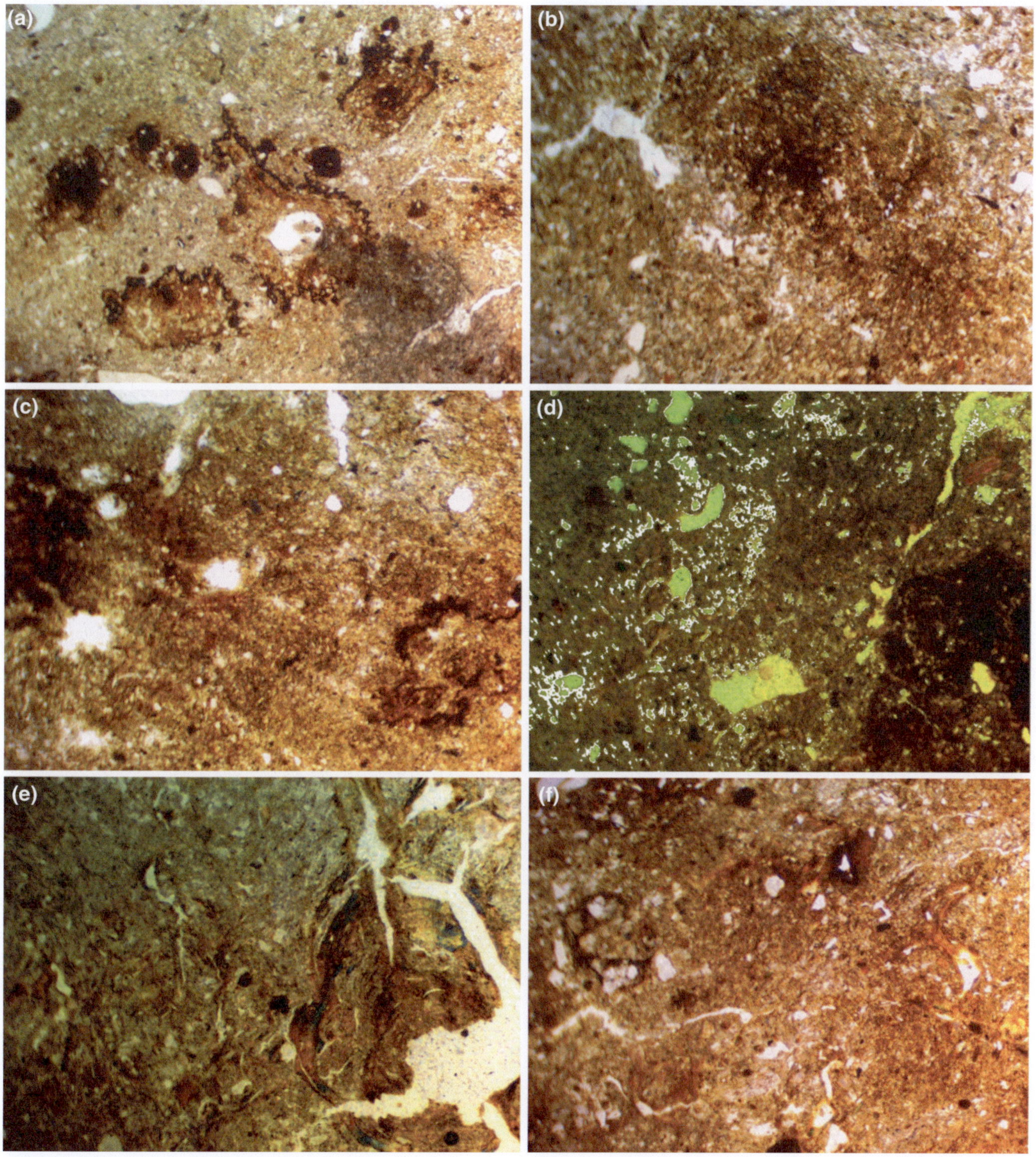

Fig. 5.59 Microstructure of Nitisols Ferralic, nic.II; Horizons: **a** 0–20 cm; **b** 20–58 cm; **c** 58–86 cm; **d** 86–130 cm; **e–f** 136–180 cm. Photos by L. Matchavariani

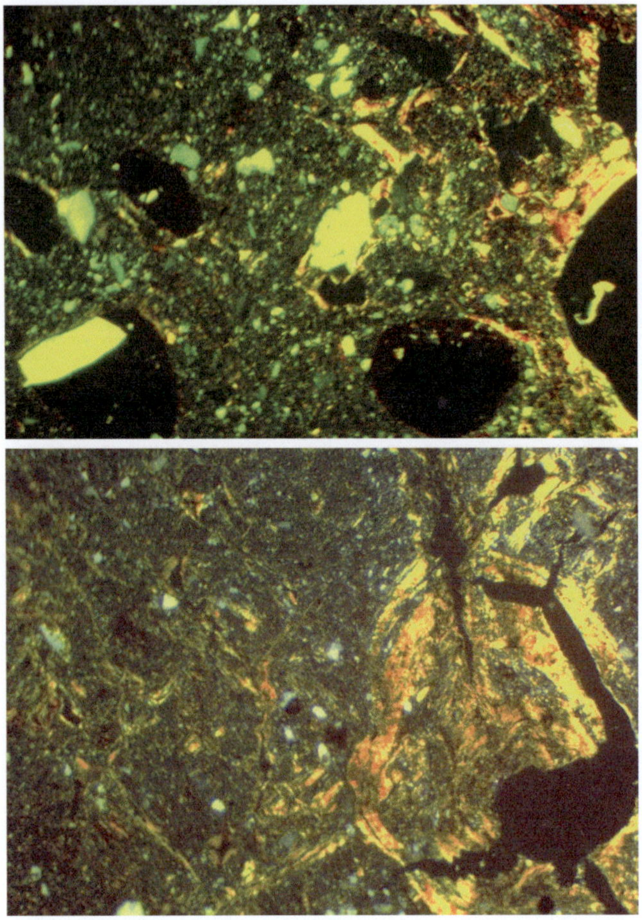

Fig. 5.60 Microzones with optical orientation plasma in Nitisols Ferralic, 136–180 cm, nic.+. Photos by L. Matchavariani

plasmic saturation from disperse soil organic matter (Fig. 5.67), dusty-plasmic structure at great depths, with Fe-clay and clay plasma at some locations, intense ferrugination, uneven distribution of Fe-formations, rich mineralogy of skeleton (Fig. 5.68), and intense optical orientation of clay in lower structures.

5.13 Luvisols Albic

Luvisols Albic soils, which correlate with the national classification as Subtropical Podzolic or Yellow Podzolic soils, are widely spread in the humid subtropical zone of West Georgia, at 30–200 m above sea level, in the northern and eastern peripheries of Kolkheti Plain, but are less widely spread in the western part of the plain, over the marine terraces (Abkhazeti); they are spread as fragments in the southwestern part of Kolkheti Plain (Fig. 5.69). Large massifs of this type of soils are spread on the old terraces of the Kodori, Enguri, Khobi, Rioni, Kvirila, and other rivers. This type of soil covers 2% of the whole territory of the country and borders both, Acrisols Haplic (Yellow soils) and Leptosols Rendzic (Raw-Humus-Calcareous soils) and Gleysols.

Luvisols Albic is quite a peculiar soil and is quite argumentative in a genetic respect. Due to the morphology of its profile (with a clearly seen upper light, whitish horizon), the first researchers, starting from V. Dokuchaev, attributed it to the Podzol soils. D. Gedevanishvili was the first to call this type of soil Subtropical Podzolic, later the term used by scientists: I. Vitin, S. Zakharov, V. Kovda, B. Polinov, M. Sabashvili, M. Daraselia, and many others. M. Sabashvili extended the area of the process of podzolization to "Red" and "Yellow" soils. However, B. Polinov noted: "Even if admitting the genetic connection of these soils to the Podzols, attributing them to podzolized soils is as much inadmissible as e.g. attributing Grey soils to Chernozems". V. Kovda named several features evidencing the soil podzolization in the study region. K. Bogatiriov was the first to try to isolate this type of soil from the group of podzolized soils and pointed to another way of formation of the light horizon. He linked the lighting of the upper horizon to the elluviation resulted by excess surface humidity and ferrum segregation what was considered a sign of podzolization earlier. In his article "What are Subtropical Podzols of Abkhazia?", Gerasimov (1966), criticized and strongly doubted about the podzolized nature of these soils. He called the possibility of founding Podzols in the Black Sea coastal area of the Caucasus an absurd geographical paradox. He proposed title "Subtropical Pseudo-Podzol", i.e., strongly lessivaged, superficially gleized, illuvial-ferruginated soil, formed under the influence of seasonal surface excess moisture. Zonn (1974, 1987), Zonn and Shonia (1971), Romashkevitch (1974a, b, 1979), and others evaluate this type of soil in the same respect.

Later, aiming at explaining the complex genesis and contradictory opinions, Matchavariani (1987, 2002, 2005, 2008) thoroughly studied so-called Subtropical Podzol soil and concretion formations. As a result of the study it was established that the mechanism of formation of a texturally differentiated profile is associated with an originally nonhomogeneous lithological background—the heterochronic nature of the sediments building the profile, which take part in modern pedogenesis. As for the process of lessivage, it takes place on such a nonhomogeneous lithological background having no profile-forming function. Despite the fact that humid subtropical climate must make for the intense movement of plasma across the profile, the process of lessivage takes place locally and has an intra-horizon (not an intra-profile) nature, as the presence of an underlying heavy clay horizon and complicated mode of filtration hampers the migration of fine-disperse substance across the concretion nodules and local lessivage. As for the process of podzolization seen in the name of this type of soil, virtually, it is not diagnosed because as the results of the thorough studies

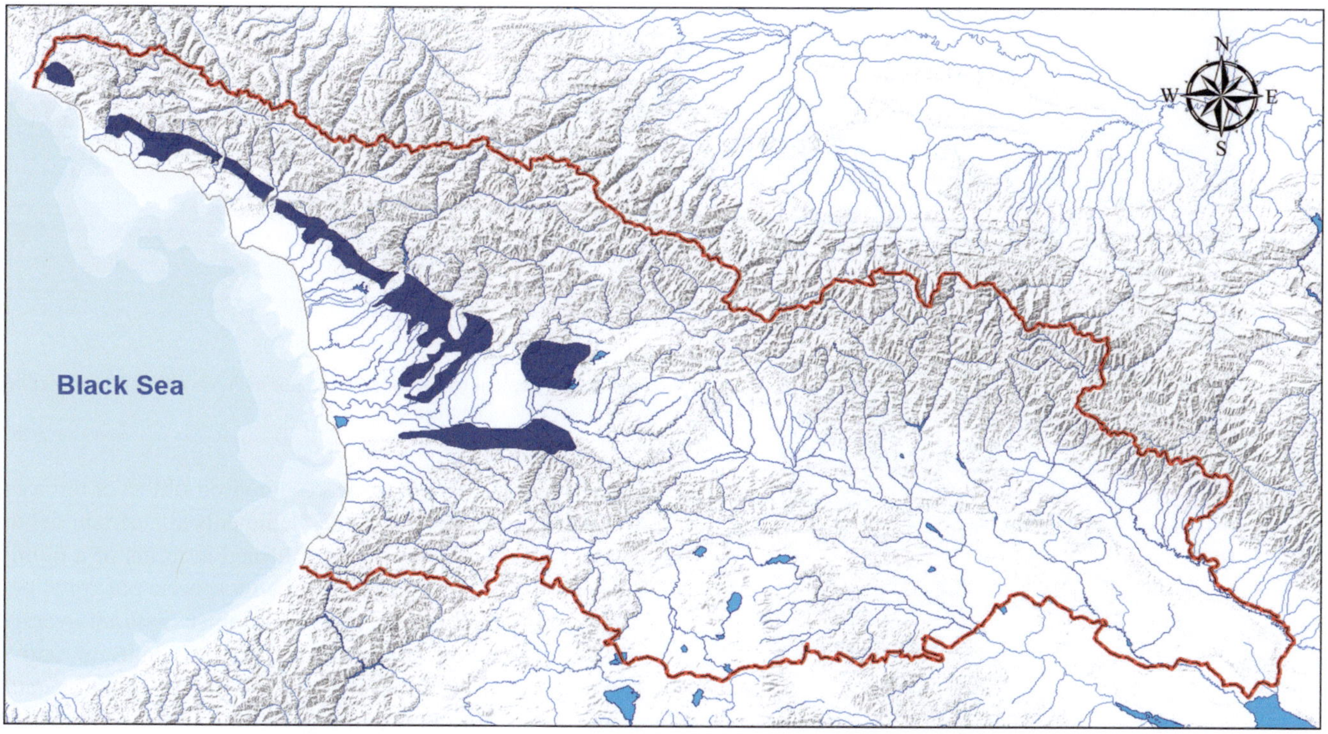

Fig. 5.61 Location of Acrisols Haplic. This map is created by D. Svanadze, based on the data of L. Matchavariani

Fig. 5.62 The landscape in the area of Acrisols Haplic formation (Project "Cadastre and Land Registration", KfW)

accomplished by L. Matchavariani suggest, no process of decomposition of primary minerals in the upper horizon is fixed at macro-, meso-, micro- or submicro-levels and no signs of movement of chemically modified talus material across the vertical profile are seen.

The Luvisols Albic soil is formed on old-marine and river terraces. The general inclination of the terraces is from the peripheral part of the Colchic Lowland toward the Black Sea (Fig. 5.70). A high hypsometric zone of the terraces is relatively dissected and drained, while the lower part is less

Fig. 5.63 Profile of Acrisols Haplic. Photo by B. Kalandadze

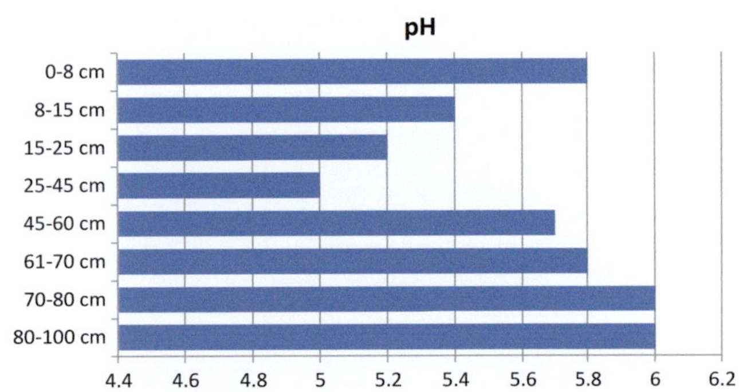

Fig. 5.64 pH distribution in profile of Acrisols Haplic (according to data of T. Ramishvili)

water permeable. Soil-forming parent materials are loose and are, usually, heterogenous. The low terraces in the north-western part of Colchic Lowland are presented by loamy sediments, which cover shingle and clay sediments at some locations. The central and northwestern piedmonts of Colchic Lowland are presented as high terraces. There are tertiary parti-colored clays, zebra-like clays, and shingle

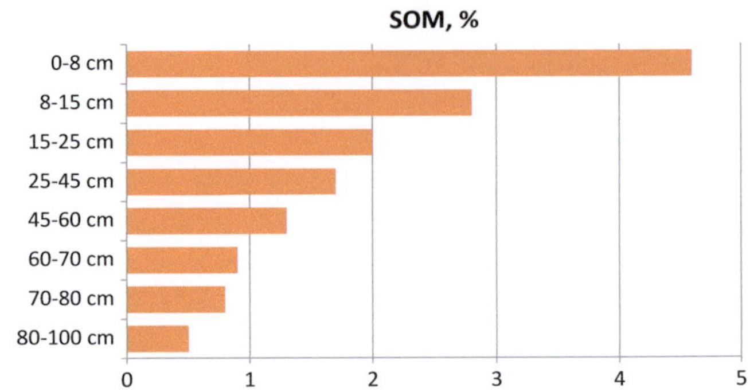

Fig. 5.65 Content of soil organic matter in Acrisols Haplic (according to data of T. Ramishvili)

spread here. Heavy clays are common on old river terraces, which are substituted by lighter sediments at a certain depth.

The Luvisols Albic soils are formed in terms of a humid subtropical climate with abundant atmospheric precipitations (1500 mm), warm winter and hot summer (with an average annual temperature of +14, +19 °C); the sum of active temperatures is 4000–4500 °C; the duration of the vegetation period is 8 months and the duration of frost-free period is 250–290 days. The periods with abundant precipitations are often changed by droughty periods. The relative humidity in summer and autumn reaches 90% and it is minimal (67–79%) in spring and autumn.

The Polydominant Colchic forest grew in this zone, but at present, the natural vegetation is disturbed due to the felling of trees. The areas of former forest massifs are now used to grow agricultural crops: tea, citruses, tobacco, and maize. Colchic forest has survived as fragments.

Morphologically Luvisols Albic of Georgia is characterized by a clearly differentiated profile with a following morphological structure: A1A2n–BSf–Btg–BGt or A1A2n–BSf–Btg–BGt–[B/Cgh]. The main diagnostic properties of this type of soil are clearly expressed light eluvial horizon, which is depleted with a sludge fraction and oxides (Fig. 5.71). Texture differentiation, bulk of Fe-concretions in the upper horizons and ferruginated Ortshtein horizon in the middle part of the profile often as cemented layer "petroplinthic" or "plinthic" are peculiar common features for this type of soils. The Ortshtein layer is usually fixed at the spots of lithological transitions between the upper horizon with the light texture and the lower, heavy clay layer. The intensified hydromorphism in the profile is followed by less concretion nodules and reduced density of an Ortshtein layer. When the process of soil formation occurs in nonhomogeneous sediments, which have a heavy water-proof underlying layer, an Ortshtein horizon of different strengths and structures is fixed in the profile. The Luvisols Albic is characterized by poor natural productivity and unfavorable physical properties. The presence of an Ortshtein horizon is the principal

Table 5.11 Gross chemical composition of Acrisols Haplic, % (according to data of T. Ramishvili)

Horizon (cm)	SiO_2	Fe_2O_3	Al_2O_3	CaO	MgO	$SiO_2:R_2O_3$
0–11	74.2	5.2	17.1	0.65	0.78	6.52
12–22	75.6	4.4	15.9	0.64	0.64	6.63
26–36	72.9	5.2	19.1	0.73	0.73	5.76
43–49	66.2	6.8	23.2	0.85	0.85	4.23
64–70	65.8	6.5	25.1	0.96	0.96	3.88

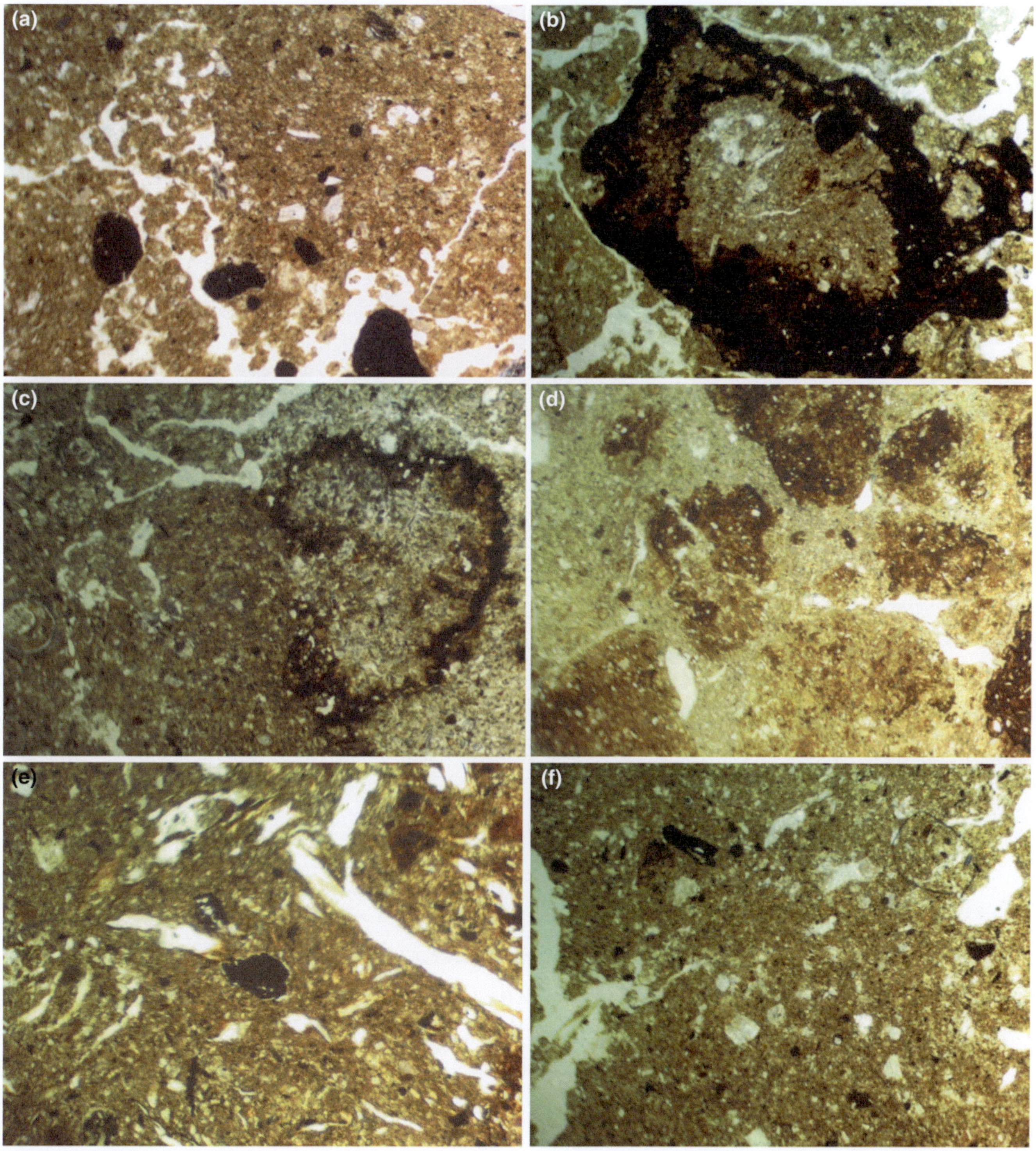

Fig. 5.66 Microstructure of Acrisols Haplic, nic.II; Horizons: **a–b** 0–18 cm; **c** 18–33 cm; **d** 33–48 cm; **e** 48–75 cm; **f** 75–100 cm. Photos by L. Matchavariani

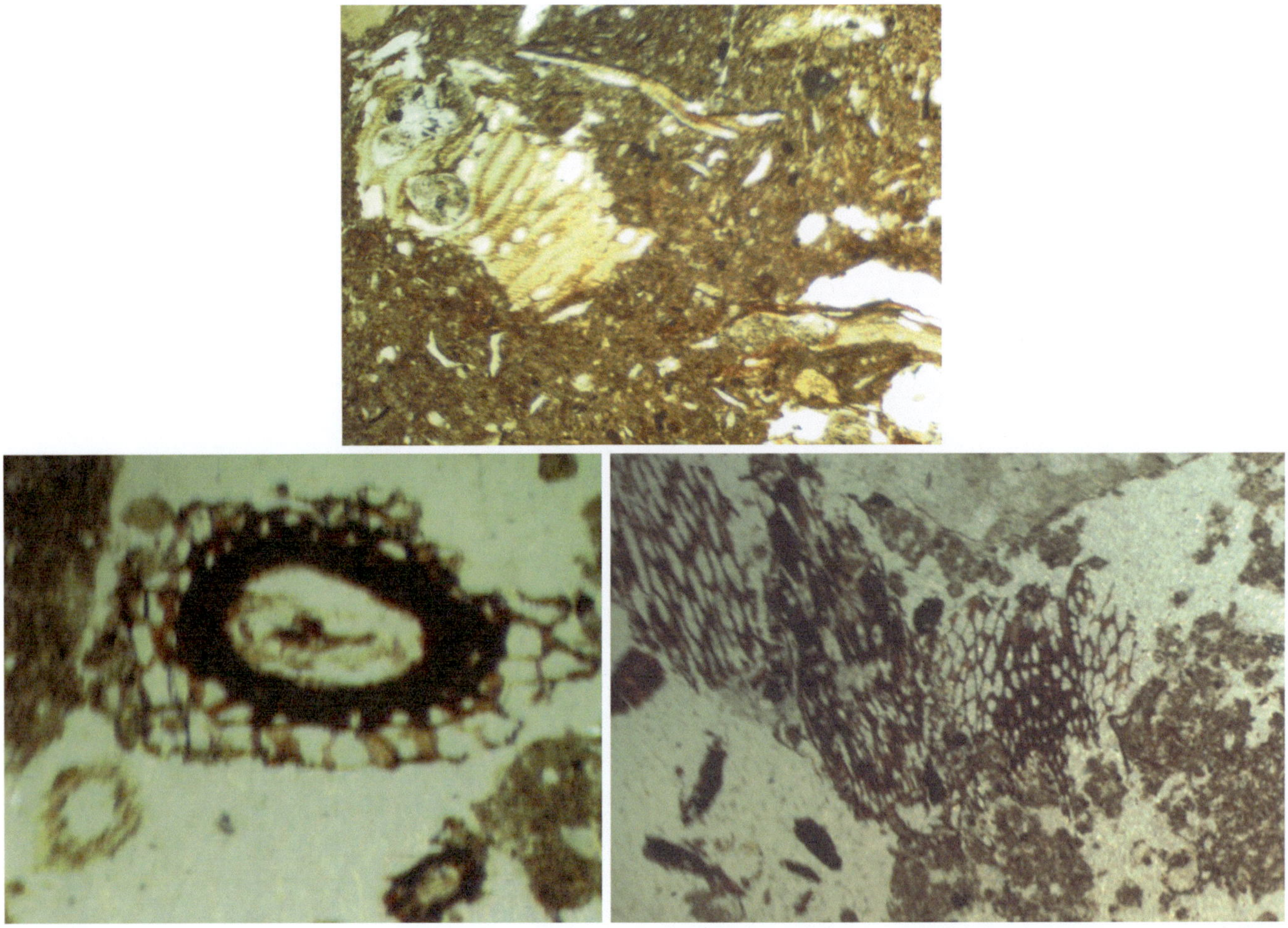

Fig. 5.67 Plant residues in the profile of Acrisols Haplic, nic.II. Photos by L. Matchavariani

Fig. 5.68 Fragments of the parent rocks in the soil profiles of Acrisols Haplic, nic.+. Photos by L. Matchavariani

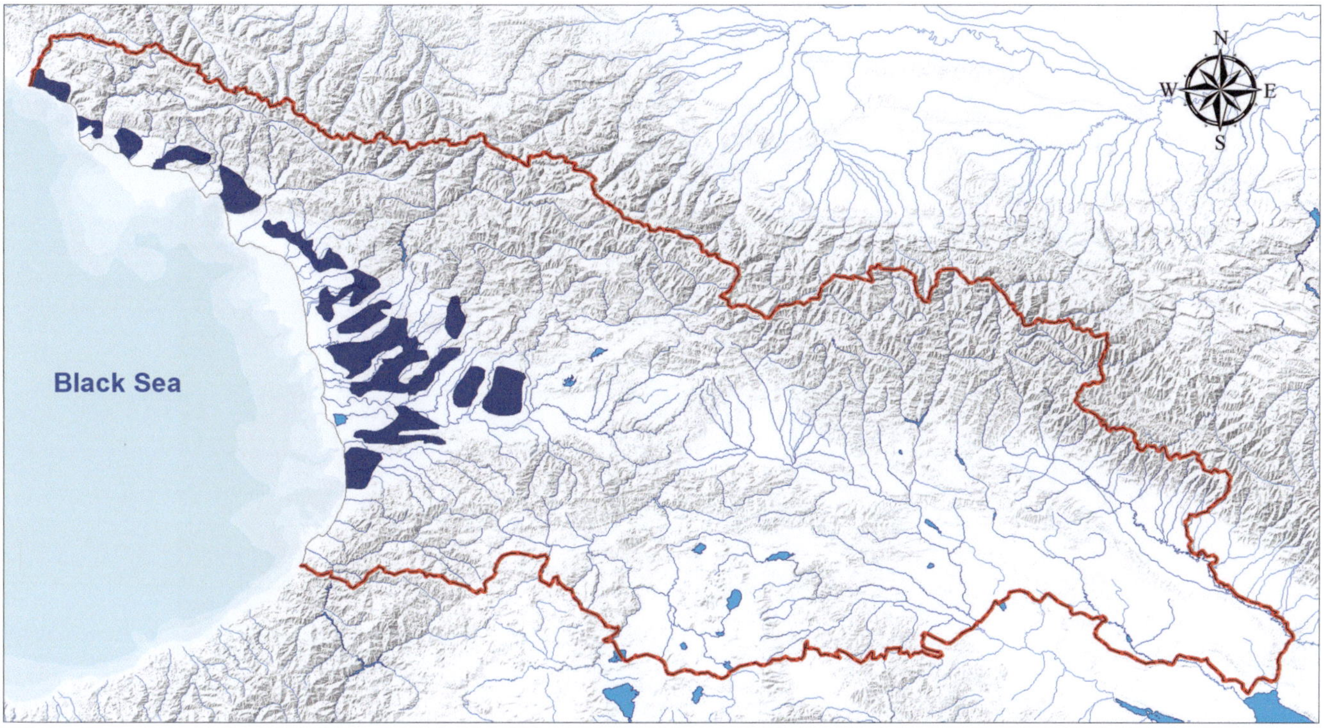

Fig. 5.69 Location of Luvisols Albic. This map is created by D. Svanadze, based on the data of L. Matchavariani

Fig. 5.70 The landscape in the area of Luvisols Albic formation (Project "Cadastre and Land Registration", KfW)

negative feature of this type of soil hampering the normal development of the root system and due to its water resistance, supports soil bogging.

As the analytical data suggest, the Luvisols Albic soils have an acid reaction (pH = 4.5–6.0), with the maximum acidity fixed in the eluvial horizon, which drastically

Fig. 5.71 Profiles of Luvisols Albic. Photo by B. Kalandadze

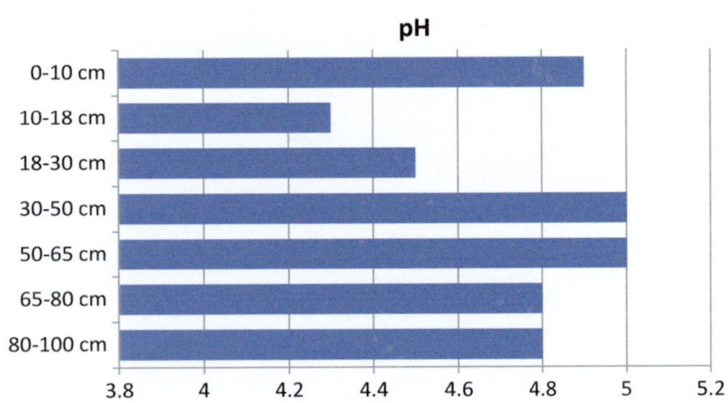

Fig. 5.72 pH distribution in profile of Luvisols Albic soil (according to data of L. Matchavariani)

decreases at greater depth (Fig. 5.72) profiles. The data of the gross chemical composition are presented in Table 5.12.

The main soil-forming processes of the Luvisols Albic soils are gleyzation. The content of soil organic matter is low or average 2.5–3% in the accumulation horizon (Fig. 5.73); the type of soil organic matter is fulvous. The texture of this type of soil is loamy or clay. Usually, the underlying horizon is particularly enriched with fine-disperse fraction. The main oxides are distributed unevenly. Eluvial horizon shows the accumulation of silica and reduced oxides. At greater depths, the amount of earth silicon increases and that of oxides decreases in the illuvial horizon. Their ratio evidences allitic weathering. The mineralogical content of silt shows dominant kaolin, chlorine, halloysite, and fine-disperse quartz. There are two maximums fixed in the distribution of individual forms of ferrum—in the upper and lower layers.

The difference between the Luvisols Albic with the Nitisols Ferralic and Acrisols Haplic soils is that the former

Table 5.12 Gross chemical composition of Luvisols Albic soil, % (according to data of L. Matchavariani)

Horizon (cm)	SiO_2	Al_2O_3	Fe_2O_3	MnO	SiO_2:Al_2O_3	SiO_2:Fe_2O_3
0–10	65.9	7.1	6.6	0.28	9.2	10.0
20–30	74.4	10.1	6.6	0.19	7.4	11.2
35–50	83.1	10.1	13.6	0.19	8.2	8.2
85–95	47.7	22.0	12.3	0.34	2.2	2.2
115–125	54.1	19.4	10.2	0.40	2.8	2.8

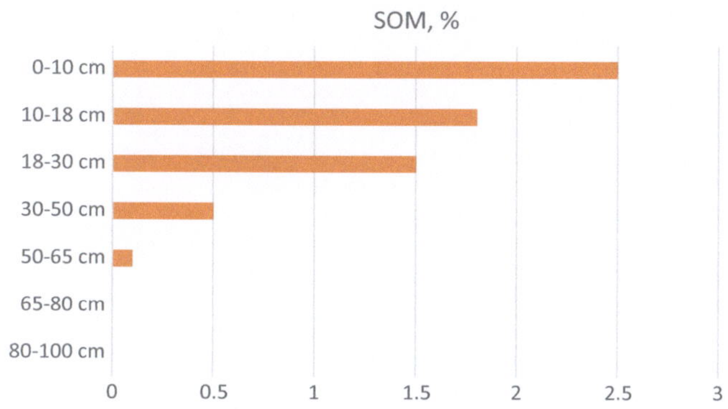

Fig. 5.73 Content of soil organic matter in Luvisols Albic soil (according to data of L. Matchavariani)

is characterized by a clearly differentiated profile, high content of Fe-concretions, and often the presence of an Ortshtein horizon, while unlike Gley-podzolic, they are characterized by slight gleyzation and less content of concretions.

According to our previous micropedological study (Matchavariani 2008), the Luvisols Albic is characterized by the following specific features of microstructure (Fig. 5.74): weak structure and porosity of the sandy-dusty mass, light color and humification, moder type of soil organic matter, isotropic plasma and bulk of ferrum concretions in the upper layers; compacted structure and fissure porosity, optical plasma orientation and high content of ferrum in the Ortshtein horizon evidenced by a joint system of microzones of large concretions and saturated with iron mass; heterogeneous structure and content, optical orientation of mosaic, sometimes current-like structure of clay and Fe-clay plasma in the lower layers. On the whole, according to many characteristics of microstructure, on the background of profile heterogeneity and strongly oriented plasma, due to the absence of intense clay migration, there is no genetic link between the profile layers to compare what is the evidence of its complex polygenetic nature.

In the relatively depressed locations of Luvisols Albic soils of humid subtropical zone, West Georgia (Fig. 5.75), are spread so-called Podzoic-gley or Yellow-podzolic-gley soils. They occupy 0.7% of the total area of the country. This soil adjoins Acrisols Haplic, Leptosols Rendzic, Luvisols Albic, and Gleysols.

Quite often, "Podzoic-gley" soils are considered as a subtype of "Subtropical Podzolic" soils. Therefore, the history of studying these soils is almost the same. "Podzolic-Gley" soils were studied by Motserelia (1954), Motserelia and Kostava (1975), Motserelia (1989), etc. As per Motserelia (1954), these soils on Colchic lowland are divided into the soils developed on the modern river terraces and on old terraces. These soils developed on modern terraces have a clear accumulative horizon with the signs of gleyzation; the degree of gleyzation increases as the depth increases and there are concretions across the whole profile. The profile of the soil formed on old terraces is clearly differentiated as genetic horizons; the accumulative horizon is of a little strength, with the gleyzation seen right from the surface, increasing as the depth increases; the concretions are observed across the whole profile and the Ortshtein layer is clearly seen.

Morphologically, the abovementioned soils are characterized by a clearly differentiated profile: A–A1A2–B1–B2–BC–CDg–G or A1A2–A2–A2B–BCg (Fig. 5.76). By its terms of development, it is quite close to the Luvisols Albic soils, but differs from them by a more intense humidification with ground and surface waters.

As the analytical data suggest, "Podzoic-Gley" soils are characterized by an acid, sometimes neutral or weak alkaline reaction what is associated with chemism of groundwaters and a moderate content of fulvous soil organic matter and deep humification. With its texture, this type of soil belongs to loams and clays. Its accumulative and eluvial horizons are depleted with a fine fraction. The main oxides are characterized by alluvial–illuvial differentiation. The content of silicate Fe usually exceeds that of a nonsilicate iron.

The mentioned soils are formed by simultaneous processes of bogging and gleyzation. As a result of bogging, gley horizon is formed and as a result of podzolization, podzolized and Ortshtein horizons are formed. A typical hydrological regime of this type of soil is noteworthy: during the period with abundant precipitations, the level of

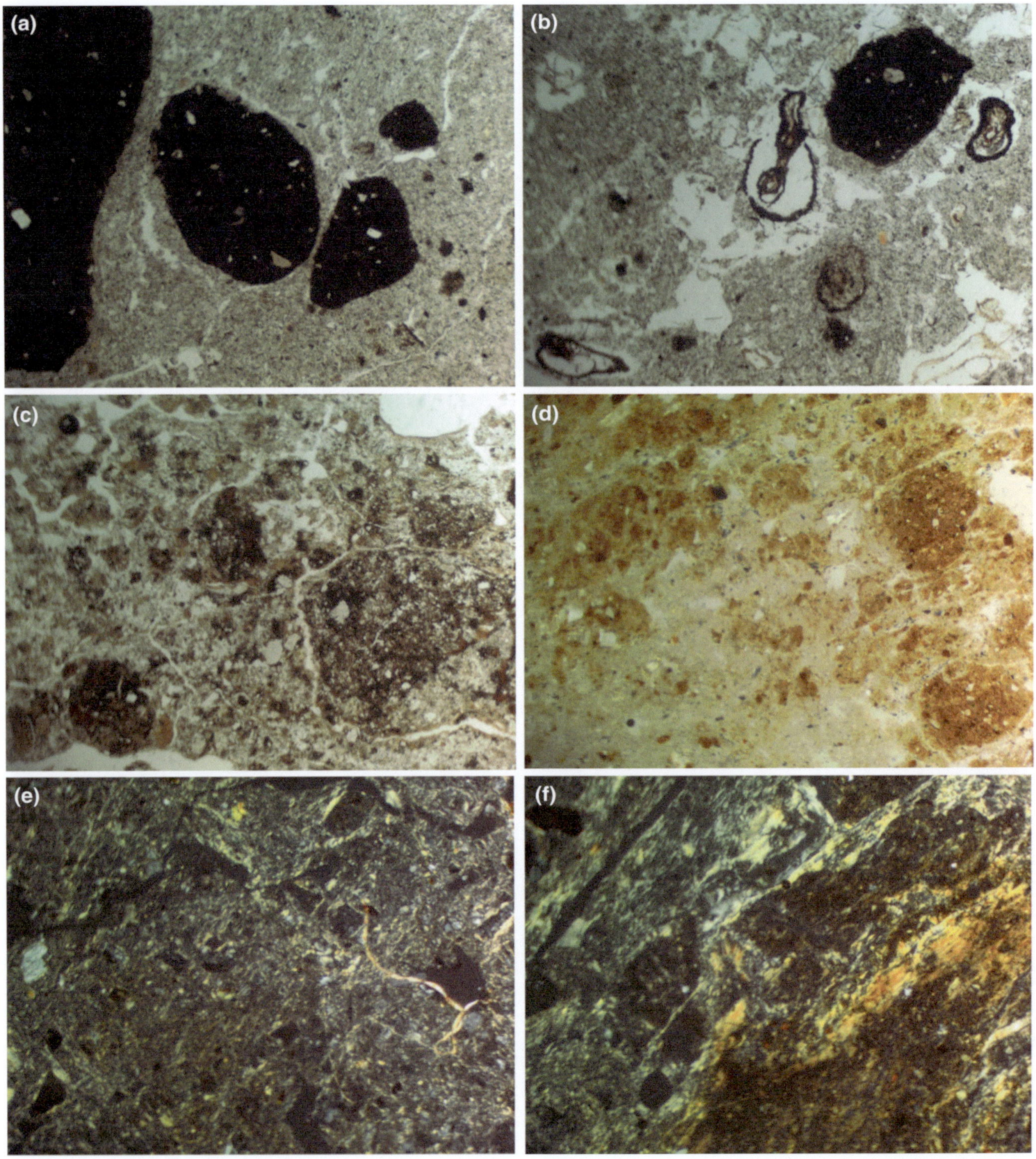

Fig. 5.74 Microstructure of Luvisols Albic soil, nic.II (**a**, **b**, **c**, **d**), nic.+ (**e**, **f**); Horizons: **a**–**b** 0–18 cm; **c** 18–33 cm; **d** 33–48 cm; **e** 48–75 cm; **f** 75–100 cm. Photos by L. Matchavariani

Fig. 5.75 The landscape in the area of "Podzoic-gley" soil formation (Project "Cadastre and Land Registration", KfW)

groundwater rises and the soil pores are filled with water, while during the droughts, the groundwater level decreases. As a result of alternating aerobic and anaerobic conditions, the oxidation–reduction processes take place and oxide is transformed into peroxide causing gleyzation and Fe-segregation.

Due to the abundant moisture, soil cultivation is associated with a number of difficulties. Therefore, its use to grow perennial crops without preliminary drainage melioration is often impossible. This type of soil is mostly used to sow corn and other annual crops.

According to our previous micropedological study (Matchavariani 2008), the "Podzoic-gley" soils are characterized by a compact microstructure of its clay mass, dusty-plasma and sandy-dusty-plasma elementary microstructure, bulk of plasma on skeleton, raw-moder or moder-raw light-colored soil organic matter and intense plasma saturation with ferrum-hydroxides, presence of the vegetation remains and carbonized particles, with the participation of fine Fe-concretions, uneven distribution of skeleton and fine-scale structure and optical orientation of plasma (Fig. 5.77).

5.14 Gleysols

Gleysols (called in national classifications as bog/boggy, swampy, marshy soils) are mainly spread in West Georgia, on Colchic Lowland (which is a kind of triangle with its shape, with its top stretching from the Black Sea coast toward east); fragments of swampy soil are also spread in the East and South Georgia (Fig. 5.78).

The area occupied by marsh and swampy soils in Georgia is 2200 km^2 making about 3% of the total territory of the country. Bog-silted (1304 km^2) and bog-peat (706 km^2) soils can be distinguished in the Gleysols.

One of the first researchers of the Bog soils on Colchic Lowland was D. Gedevanishvili. Later, these soils were studied by Zakharov (1924), Motserelia (1954), Motserelia and Kostava (1975), Kostava (1976), etc.

The climate in Colchic Lowland is warm, humid, and mild, with an average annual temperature of +14 °C and the average amount of precipitations of 1500 mm (with the minimum in summer and maximum in autumn and winter). Average annual relative humidity is as high as 71–82%.

Fig. 5.76 Profile of “Podzoic-gley” soils (Project “Cadastre and Land Registration”, KfW)

Colchic Lowland is a delta-accumulative horizon plain and is filled with alluvial material, containing the decomposition products of the constituent rocks of the Great Caucasus and Transcaucasian southern mountains. The sediments are mostly carbonated with a high content of clay in their upper layers. The dominant type of vegetation is plain forests mixed with marsh vegetation (Fig. 5.79).

The Gleysols have the following morphological structure: A–B–BC (Fig. 5.80). They have a strong profile, cloddy and block structure, heavy mechanical content, and signs of gleyzation. The bog-silted soils are characterized by weak alkaline or neutral reaction (Fig. 5.81), with a high content of soil organic matter (Fig. 5.82), heavy texture through the profile, intense dispersion, bulk exchange calcium in the absorbed cations, high content of different forms of iron (with amorphous Fe accumulated in the upper profile and crystal ferrum accumulated at greater depths), and with an uneven distribution of main oxides evidencing the alluvial nature of this type of soil. The data of the gross chemical composition are presented in Table 5.13.

Gleysols contain a wide spectrum of clay minerals: montmorillonite, kaolin, halloysite, illites, etc., with uneven distribution of different forms of. In addition, the content of silicon Fe much exceeds that of a non-silicon ferrum.

There are different views about swamping of Colchis Lowland. On the one hand, swamping is thought to be associated with the atmospheric precipitations and action of the surface waters flowing from the riverbeds; on the other hand, the process of swamping is associated with the ground and soil-and-groundwater actions.

The fund of Gleysols is the reserve, which, if dried, will give many thousands of hectares of land for agriculture in the subtropical zone.

According to our previous micropedological study (Matchavariani 2008), the Gleysols are characterized by a compact, dense, nonhomogenous, and fine-disperse mass, with partly fragmental microstructure and are saturated with ferrum, with raw and partly moder-raw type of soil organic matter, dusty-plasmic microstructure, with bulk of plasma on the skeleton, clay and clay-Fe content of isotropic plasma, with micro-sites with intense optical micro-zonal orientation, uneven saturation of the main mass with ferrum material and with the presence of organogenic remains, carbonized particles and neogenic Fe-formations and fine clay films on the pore walls. In general, marsh soils have a nonhomogeneous profile, weak skeleton, with an increasing content of primary minerals at greater depths, high ferrum content, and stratification of the profile depending on the plasma content and microstructure (Fig. 5.83).

5.15 Fluvisols

Fluvisols, which correlate with the national classification as Alluvial soil—an azonal type, are spread along the rivers in different natural zones. Consequently, alluvial soil is spread all over the territory of Georgia and covers 5% of the total area of the country (Fig. 5.84). The alluvial soils in different regions of Georgia are explored by Zakharov (1924), Sabashvili (1948, 1965), Motserelia (1954), and others.

Fluvisols are characterized by a regular flooding and sedimentation of new alluvion layers on the surface. Their properties mainly result from the nature of the basin where they develop, and consequently their regimes, structure, and properties are much diversified.

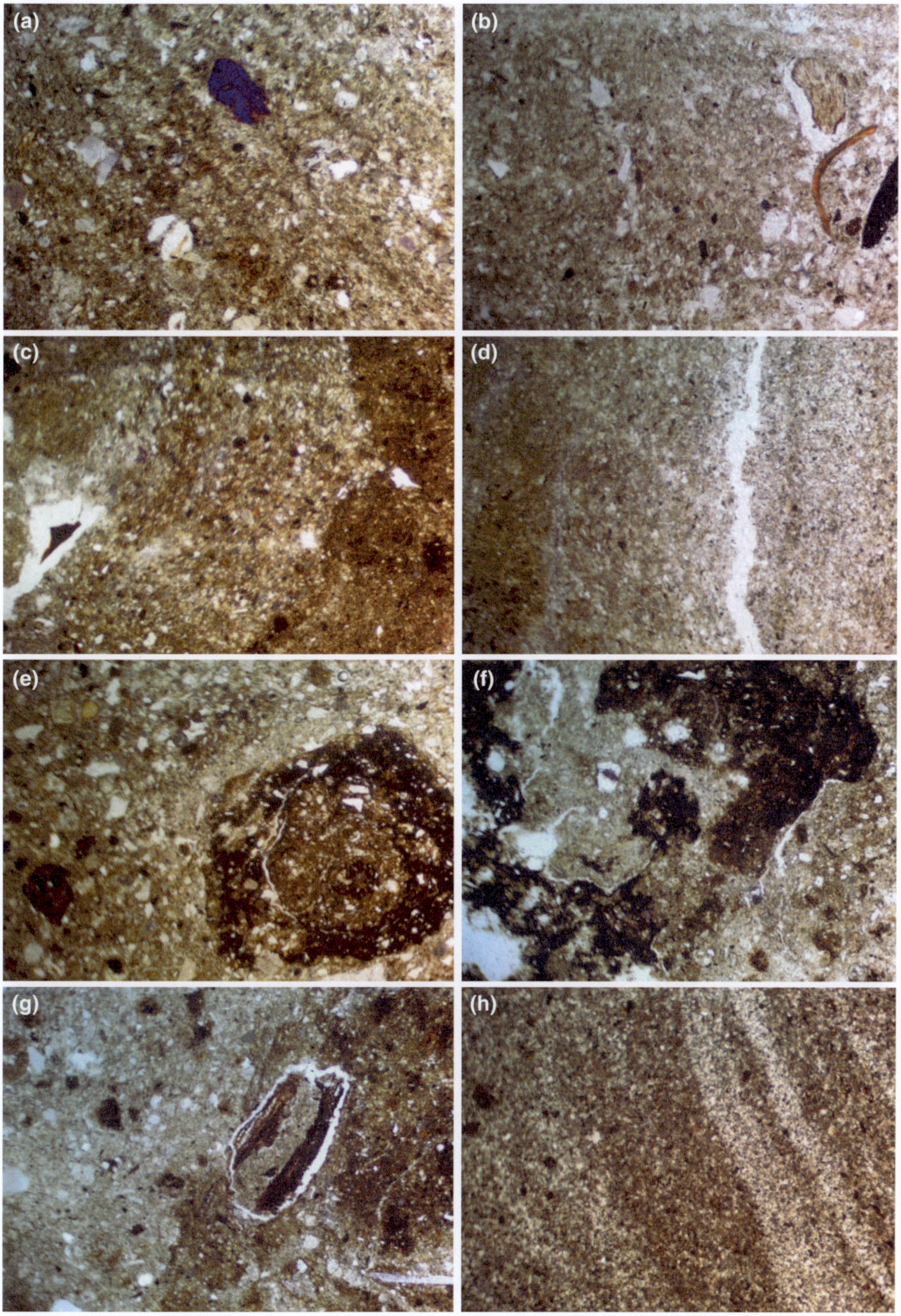

Fig. 5.77 Microstructure of "Podzoic-gley" soils, nic.II; Horizons: **a–b** 0–15 cm; **c–d** 33–60 cm; **e** 60–95 cm; **f** 95–120 cm; **g** 120–150 cm; **h** 150–180 cm. Photos by L. Matchavariani

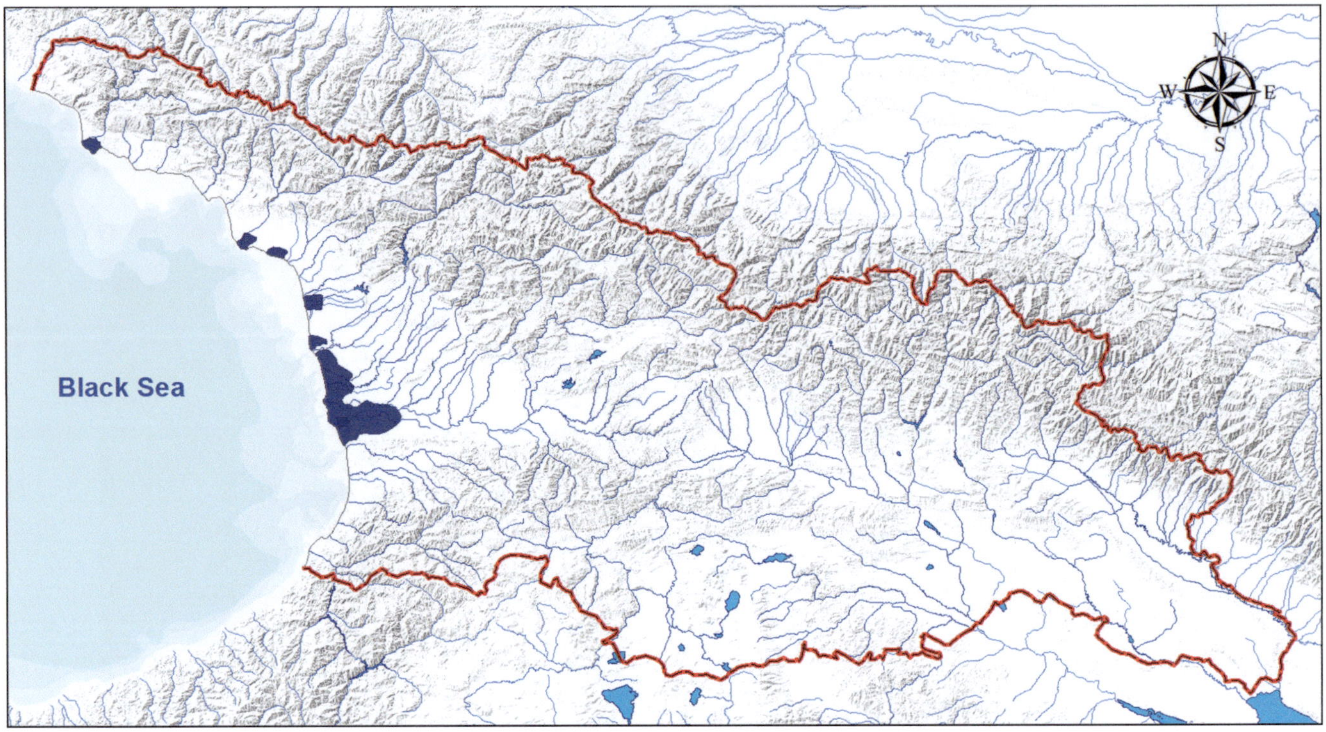

Fig. 5.78 Location of Gleysols. This map is created by D. Svanadze, based on the data of L. Matchavariani

Fig. 5.79 The landscape in the area of Gleysols formation (Project "Cadastre and Land Registration", KfW)

As the Fluvisols develop over the alluvial sediments in different natural zones, they are influenced by the climatic conditions of the zone where they were formed. The material of the alluvion over which they are formed is also much diversified. The natural biological cover is presented by floodplain vegetation (Fig. 5.85). Large areas of these territories are used to grow different agricultural crops.

The profile of the Fluvisols has the following structure: A–BC–C–CD. Morphologically, one of the most typical diagnostic properties is profile stratification (with its

Fig. 5.80 Profile of Gleysols. Photo by B. Kalandadze

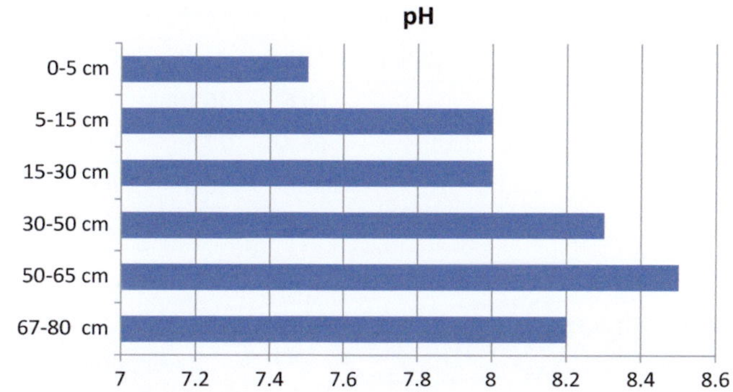

Fig. 5.81 pH distribution in profile of Gleysols (according to data of T. Ramishvili)

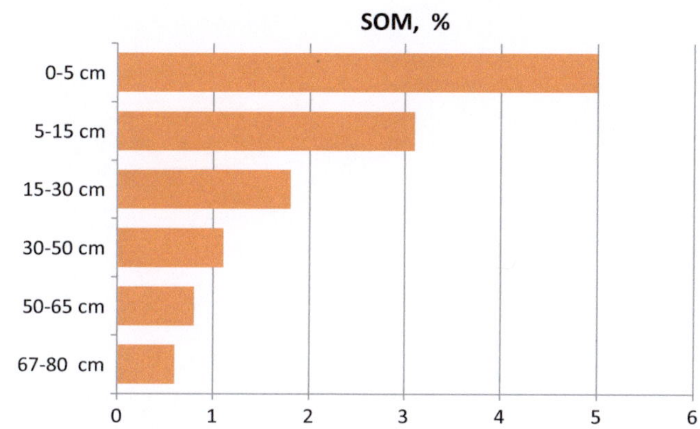

Fig. 5.82 Content of soil organic matter in Gleysols (according to data of T. Ramishvili)

mechanical structure first of all), weak structural properties and skeletal nature (Fig. 5.86). This type of soil is distinguished by a strong profile, clearly seen accumulative horizon and intense humification.

As the analytical data suggest, the reaction in Fluvisols is acid, neutral, or alkaline depending on the type of basin the concrete profile was formed in (Fig. 5.87). The content of soil organic matter is average or less; the profile is deeply humified (Fig. 5.88); the nitrogen content is high or average; and the absorption capacity is low or average. The distribution of the main oxides is more or less even both, in the soil and silt fraction. As for the gross chemical composition, the data are presented in Table 5.14.

Fluvisols differ from the zonal types of soil with a weaklier developed profile, stratified structure, and gleyzation. This soil incorporates two types: turf acid and turf saturated. Mostly alluvial turf-acid soil is formed under the herb vegetation in the high-mountainous and forest zones distinguished for the youngest age. They are usually low productive and are used as arable land or hay meadows. Turf-saturated soil is mainly spread in the steppe zone of East Georgia and is often of a carbonate content.

According to our previous micropedological study (Matchavariani 2008), Fluvisols are characterized by a dusty mass with a compact microstructure, which is large channel-like and chamber-like, branched and with figurative porosity at greater depths, vegetation remains decomposed to

Table 5.13 Gross chemical composition of Gleysols, % (according to data of T. Ramishvili)

Horizon (cm)	Loss on ignition	SiO_2	Al_2O_3	Fe_2O_3	TiO_2	CaO	MgO	SiO_2:R_2O_3	SiO_2:Al_2O_3	SiO_2:Fe_2O_3
0–16	11.0	60.6	16.1	7.3	1.09	3.1	2.0	5.65	6.9	27.8
28–45	9.6	65.8	17.6	8.6	1.11	1.8	2.0	5.0	6.4	21.8
45–70	11.3	69.6	16.0	6.6	1.09	1.8	2.0	5.8	7.3	29.1
70–100	11.0	78.8	7.9	3.3	1.40	1.3	0.9	13.1	16.4	65.5

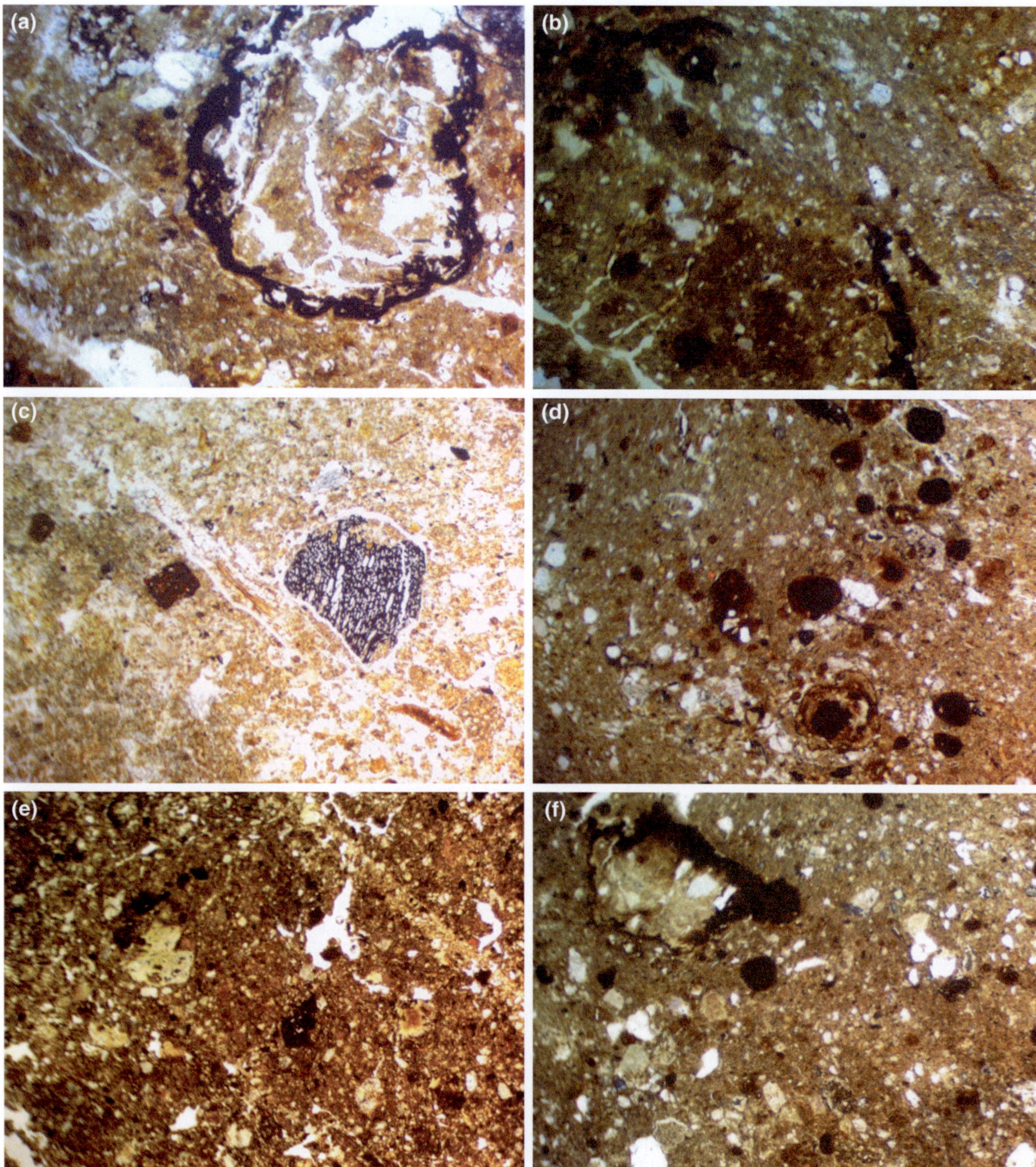

Fig. 5.83 Microstructure of Gleysols, nic.II; Horizons: **a** 0–16 cm; **b** 16–28 cm; **c** 28–45 cm; **d** 45–70 cm; **e–f** 100–140 cm. Photos by L. Matchavariani

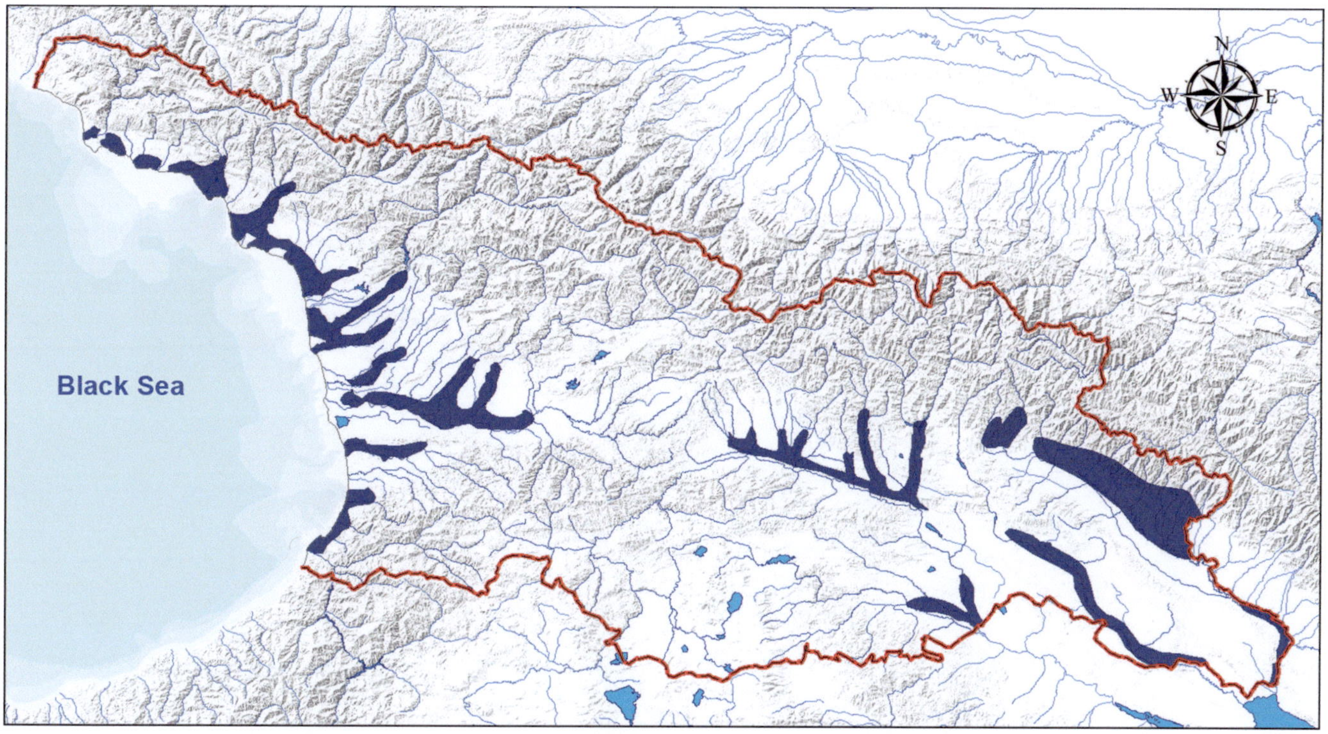

Fig. 5.84 Location of Fluvisols. This map is created by D. Svanadze, based on the data of L. Matchavariani

Fig. 5.85 The landscape in the area of Fluvisols formation (Project "Cadastre and Land Registration", KfW)

Fig. 5.86 Profiles of Fluvisols. Photo by B. Kalandadze

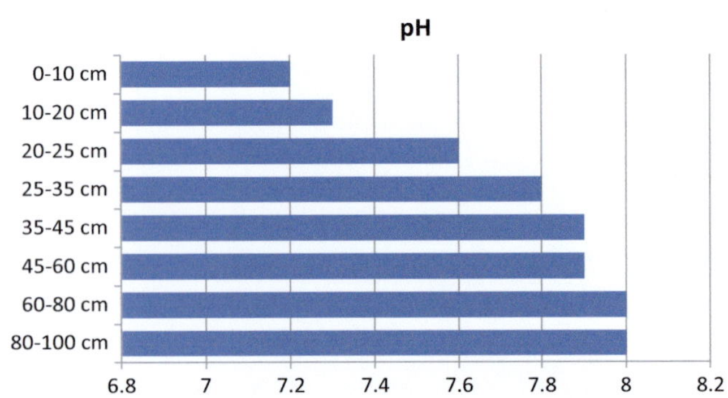

Fig. 5.87 pH distribution in profile of Fluvisols (according to data of T. Ramishvili)

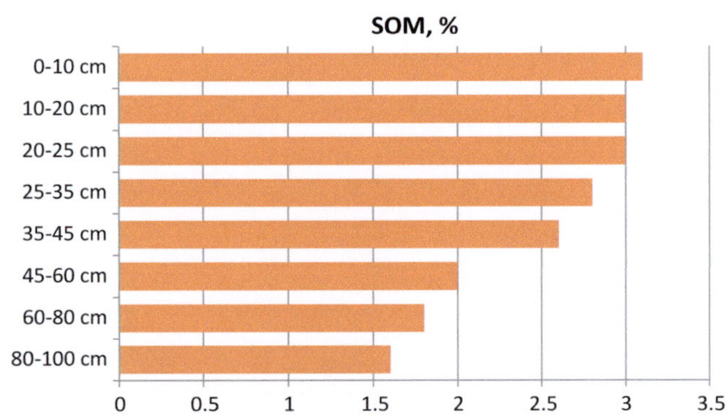

Fig. 5.88 Content of soil organic matter in Fluvisols (according to data of T. Ramishvili)

different degrees in the voids, light soil organic matter color, increased ferrum content in upper horizons, average and coarse-dusty content of skeleton, with plasma content more than skeleton content, weak optical orientation and organic-Fe-clay content. In general, Fluvisols are distinguished for a moder type of soil organic matter, with the maximum content of vegetation remains in the lower horizons, increased porosity and elementary sandy-dusty-plasmic microstructure (Fig. 5.89).

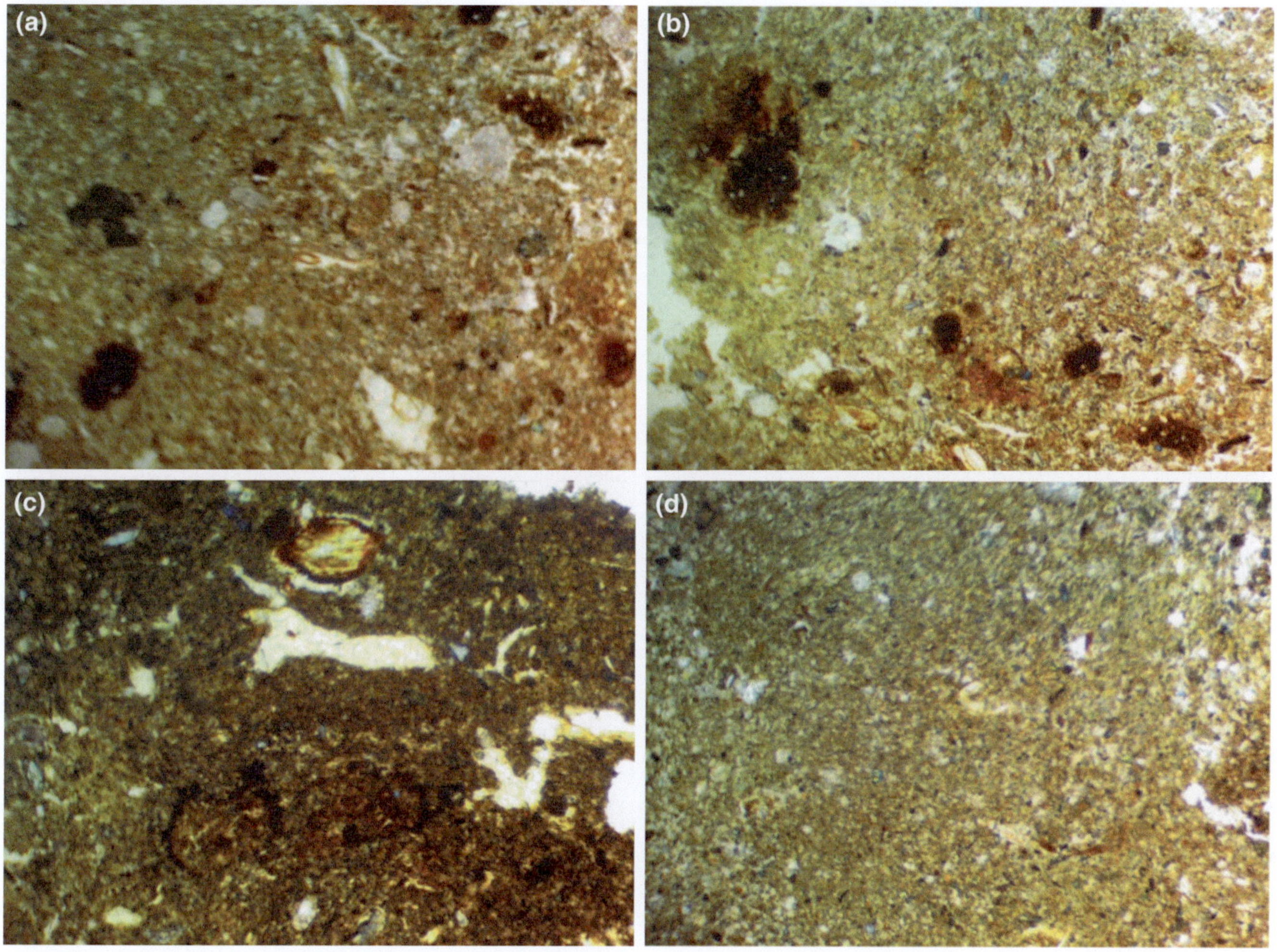

Fig. 5.89 Microstructure of Fluvisols, nic.II; Horizons: **a–b** 0–37 cm; **c** 37–72 cm; **d** 72–120 cm. Photos by L. Matchavariani

Table 5.14 Gross chemical composition of Fluvisols, % (according to data of T. Ramishvili)

Horizon (cm)	Loss on ignition	SiO_2	Al_2O_3	Fe_2O_3	TiO_2	CaO	MgO	K_2O	Na_2O	SiO_2: R_2O_3	SiO_2: Al_2O_3	SiO_2: Fe_2O_3
0–14	8.9	65.2	15.6	7.4	0.6	3.5	2.7	2.2	2.2	5.5	7.1	23.6
14–52	5.9	65.9	15.4	7.7	0.4	3.4	2.6	2.1	2.1	5.5	7.3	22.9
52–105	5.3	65.2	15.6	7.6	0.6	3.2	3.0	2.3	2.3	5.4	7.1	23.1
105–125	5.4	64.7	15.8	7.7	0.7	3.3	2.5	2.3	2.3	5.3	7.0	22.5

References

Akhvlediani GK (1973) The soil cover of Samgori Valley (Samgoris velis niadaguri safari). Sabchota Sakartvelo, Tbilisi, p 205

Anjaparidze I (1979) Cinnamonic soils of Georgia (Saqartvelos yavisferi niadagebi). Sabchota Sakartvelo, Tbilisi

Chkheidze T (1977) Raw humus calcareous soils of Georgia (Sakartvelos kordian-karbonatuli niadagebi). Sabchota Sakartvelo

Chkhikvishvili V (1970) The salt soils of Georgia and their melioration (Sakartvelos damlashebuli niadagebi da mati melioracia). Sabchota Sakartvelo

Daraselia M (1939) Water regime of red soils (Tsitelmitsa niadagebis wylis rezjimi). Metsniereba, Tbilisi

Daraselia M (1974) Dynamics of soil solutions in red soils of Georgia (Niadaguri khsnarebis dinamika saqartvelos tsitelmitsa niadagebshi). Metsniereba, Tbilisi

Daraselia MK (1949) Red and subtropical podzol soils and there use for subtropical crops (Tsitelmitsa da subtropikuli etseri niadagebi da mati gamokheneba subtropikuli kulturebis qvesh). VNIIChSK, p 445

Gerasimov IP (1966) What are the subtropical podzols of Abkhazia? (Chto takoye subtropicheskie podzoli Abkhhazii?) Pochvovedenie, vol 11, pp 21–34

Hartemink AE, Bockheim JG (2013) Soil genesis and classification. CATENA, vol 104. Elsevier, pp 251–256
Iashvili N (1987) Soils of Svaneti. Sabchota Sakartvelo, p 225
Ketskhoveli N (1935) Vegetation cover of Georgia. Literasi
Kostava G (1976) Methods for drainage and reclamation of waterlogged soils of Colchis. Sabchota Sakartvelo
Mardaleishvili R (1973) Overview of chernozem soils of the northeastern part of the Iori Upland. In: Proceedings of the Institute of Soil Science (Ioris zegnis chrdilo-agmosavleti nacilis shavmica niadagebis dakhasiateba. Niadagmcodneobis institutis shromebi), vol 14, pp 20–35
Mardaleishvili R, Pipia TS (1988) Cherozems of Iori Upland of Eastern Transcaucasia (Chernozemi Iorskogo ploskogorya Vostochnogo Zakavkazya). Agropromizdat, Moscow
Mardaleishvili R, Tvalavadze M, Lolishvili R (2005) Ways to improve the efficiency of low-fertile chernozems in Eastern Georgia (Puti povisheniya effektivnosti nizkoplodorodnikh chernozemov Vostochnoy Gruzii. Izvestiya Agrarnoy Nauki). 3(2)
Matchavariani L (2002) The Genetic significance of micro-pedological studies relatively to subtropical podzols. Geography of Georgia, Tbilisi University Publishing House, no 1, pp 18–23
Matchavariani LG (1987) On the genesis of subtropical podzolic soils (O genezise subtropicheskikh podzolistikh pochv. Trudi TGU, Sovet molodikh uchennikh, seriya yestestvennikh nauk). In: Proceedings of TSU, series of natural sciences, XIII. Tbilisi University Publishing House, pp 165–174
Matchavariani LG (2005) Morphogenetic typification of concretions in subtropical podzolic soils of Georgia. Eurasian Soil Sci 38 (11):1161–1172
Matchavariani LG (2008) Geographic paradigms of the microfabric of major soils of Georgia. Universal, Tbilisi, p 309
Motserelia AV (1954) Conversion of Colchis (Preobrazovanie Kolkhidi). Nauka, Moscow
Motserelia I (1989) Soils of the Colchis lowland—an object of melioration (Pochvi Kolkhidskoy nizmennosti—obyekt melioratsii). Tbilisi
Motserelia A, Kostava G (1975) The methods of drying melioration of colchic marsh soils (Kolkhetis cjaobiani niadagebis dashrobiti melioraciis metodebi). Sabchota Sakartvelo
Nakaidze EK (1977) Cinnamonic and medow-cinnamonic soils of Georgia (Sakartvelos khavisperi da mdelos-yavisperi niadagebi). Metsniereba, Tbilisi, p 303
Nikolaishvili DA, Matchavariani LG (2010) Humus reserves and their distribution in the landscapes of Georgia. Eurasian Soil Sci 43 (1):39–48
Palavandishvili SH (1987) Red soils of Adjara and their agroindustrial using (Adjaris tsitelmitsa niadagebi da mati agroindustriuli gamokheneba). Sabchota Adjara
Pipia TS (2008) Hydrological regime of the ordinary chernozems of Iori Upland. Ann Agr Sci (Gidrologicheskiy rezjim obiknovennikh chernozemov Iorskogo Ploskogorya. Izvestiya Agrarnoy Nauki), 10(3)
Romashkevich AI (1966) The study of red soils microstructure of Western Georgia in relation to their genesis. Micromorphological method in the study of soil genesis. (Issledovanie mikrostroenia krasnizemov Zapadnoy Gruzii. Mikromorfologicheskiy metod v issledovanii genezisa pochv). Nauka, Moscow, pp 42–57
Romashkevich AI (1974a) Soils and weathering crusts in humid subtropics of West Georgia (Pochvi i kori vivetrivania vlazjnikh subtrpikov Zapadnoy Gruzii). Nauka, Moscow, p 218
Romashkevich AI (1974b) Soils and weathering crusts in humid subtropics of West Georgia (Pochvi i kori vivetrivania vlazjnikh subtrpikov Zapadnoy Gruzii). Nauka, Moscow, p 218
Romashkevich AI (1979) Subtropical pseudo-podzolic (Eluvial-surface-gley) Soils. Genetic soil types of subtropical Transcaucasia (Subtropicheskie psevdopodzolistye (eluvial nj-poverkhnostno-gleevie) pochvi. Geneticheskie tipi pochv Yuzhnoy Gruzii). Nauka, Moscow, pp 52–114
Sabashvili MK (1939) Soil Map of Georgia (1:500 000), Tbilisi
Sabashvili MK (1948) Soils of Georgia. Akad. Nauk GSSR, Tbilisi
Sabashvili MK (1965) Soils of Georgia. Metsniereba, Tbilisi
Sabashvili MK (1970) Soil science. Metsniereba, Tbilisi
Shevardnadze M (1963) The mountain-and-forest soils of Adjara. In: TSU Proccedings, vol 60, pp 89–104
Talakhadze G (1962) The black soils of Georgia. Tbilisi
Talakhadze G (1964) The principal types of soils in Georgia. Ganatleba, Tbilisi
Talakhadze G, Mindeli K (1980) The high mountainous soils of Georgia (Sakartvelos magalmtianetis niadagebi). Sabchota Sakartvelo, Tbilisi
Tarasashvili GM (1956) Mountainous-forest and mountainois-medow soils of East Georgia (Gorno-lesnie i gorno-lugovie pochvi Vostochnoy Gruzii). AS GSSR, Tbiliai
Urushadze T (1987) The soils of mountain forests of Georgia. Metsniereba, Tbilisi, p 243. (Pochvi Gornikh Lesov Gruzii, 243 c.)
Urushadze T (1997) The principal soils of Georgia (Sakartvelo Dziritdi Nidagebi). Metzniereba, Tbilisi, p 268
Urushadze T (2013) Soils classification. Tbilisi University Publishing House, p 199
Urushadze T, Blum W (2011) Soil geography with the basics of soil science. Tbilisi University Publishing House, p 425
Urushadze T, Sanadze E, Kvrivishvili T (2010) Soil morphology (niadagis morfologia). Mtsignobari, Tbilisi
World Reference Base for Soil Resources 2014 (2015). International soil classification system for naming soils and creating legends for soils maps. Update 2015. Word Soil Resources Report, 106. FAO UN, Rome, p 192
Zakharov SA (1924) About the pivotal results and the basic-problems of studying of Georgia's soils. Polytechnic Institute, Tiflis, issue 3
Zonn SV (1974) Soil science and soils of subtropics and tropics (Pochvovedenie i pochvi subtropikov i tropikov). Moscow
Zonn SV (1987) Soil cover and problems of transformation the nature and economy of the USSR subtropics (Pochvenniy pokrov i problemi preobrazovania prirodi i khoziaystva subtropikov SSSR). Nauka, Moscow
Zonn SV, Shonia NK (1971) Pseudopodzolizationn in subtropical soils of western Georgia. Pochvovedenie, pp 103–115

Land Use

6

Gizo Gogichaishvili

Abstract

An important part of the structure of the agricultural lands of the country is the areas of arable land, and their multiyear dynamics is very important. The largest areas of arable land were registered in 1932. Since then, the reduction of the arable lands of Georgia had been an irreversible process, and in 1983, amounted to 53% of its maximum value. The sown area of winter wheat exceeded 200 thousand ha in all year. Starting from 1955, this area decreased permanently and by 1980, amounted twice as less as compared to 1927–1955. In 1913–1955, the sown corn areas amounted to 350–400 thousand ha. Since 1955, the sown corn areas drastically reduced year after year (3.5 times less the same index of 1945). Potato with its sown areas varying from 3.4 to 16.8 thousand ha in 1921–1937. Starting from 1937, the areas of potato increased year after year and reached its maximum in 1980.

Keywords

Arable land • Sown area • Wheat production • Potato production

Agriculture is one of the most important and traditional branches of economy for Georgia. Historically, in Georgia, in the country of "Georgians" (ground tillers), land cultivation was highly developed. One opinion suggests that one of the areas of vine and wheat origin was on the territory of Georgia. That is why Georgia is recognized worldwide as the "homeland of wine". In Aspindza and Akhaltsikhe regions in South Georgia, there are stone terraces of the tenth to twelfth centuries survived, which were located over the slopes of the gorges of the rivers Mtkvari and Akhalkalaki Tskali. It should also be noted that these ancient terraces, despite the great slope inclination, were irrigated, which is the evidence of a high culture of land cultivation in Georgia in the ancient times.

An important part of the structure of the agricultural lands of the country is the areas of arable land, and their multiyear dynamics is very important. The largest areas of arable land (1263.4 thousand ha) were registered in 1932 (Table 6.1). In the following years, the area of the arable lands in Georgia decreased gradually and by 1940, it was reduced to 1048.3 thousand ha. In 1945, by the end of the Second World War, the area of the arable lands increased again and made 1126.9 thousand ha. Since then, the reduction of the arable lands of Georgia had been an irreversible process, and in 1983, amounted to 665.3 thousand ha making only 53% of its maximum value. Such a permanent reduction of arable land year after year was mainly caused by the agricultural specialization, meaning introducing the southern perennial agricultural crops (citruses, tea) to Georgia. Until 1921, the perennial plantings in Georgia were limited to vine and fruit (such as apple, peach, apricot, and wild apricot) gardens occupying 8.5 thousand ha area. Later, before the onset of the Second World War, the areas of perennial crops drastically increased and reached their maximum in 1940 (293.3 thousand ha). In 1940–2003, the area of perennial plantings reduced a little and varied from 213.0 thousand ha (in 1965) to 218.2 thousand ha (in 1983). Since 1921, the areas of the homestead farms steadily increased in Georgia from 77.1 to 192 thousand ha caused by an increasing rural population in that period (Table 6.1).

The dynamics of the sown areas in Georgia is of particular interest (Table 6.2). It is known that the peak of the economic upsurge of the Russian Empire, with Georgia being a part of it by that time, was fixed in 1913. In that year, the sown areas of winter wheat amounted to 168.9 thousand ha. In 1927, winter wheat areas declined as compared to 1913. It is known from the history that the process of the general collectivization started in Georgia at the end of the 1920s and continued to the first half of the 1930s. From 1927 to 1955, the sown area of winter wheat changed from

G. Gogichaishvili (✉)
Department of Hydrometeorology, Agrometeorological Division, National Environmental Agency, Tbilisi, Georgia
e-mail: gizo.gogichaishvili@yahoo.com

L. Matchavariani (ed.), *The Soils of Georgia*, World Soils Book Series,
https://doi.org/10.1007/978-3-030-18509-1_6

Table 6.1 Structure and dynamics of cultivated lands (Iashvili 1965; Charkseliani et al. 1988)

Years	Unit	Arable lands		Perennial plantations		Household plots		Total		The deviation from the absolute maximum (Arable lands, 1932)	
		thousand ha	%	thousand ha	%	thousand ha	%	thousand ha	%	thousand ha	%
1921	thousand ha	728.7	57.7	98.5	33.6	77.1	40.2	904.3	58.4	−534.7	−42.7
	%	80.6	–	10.9	–	8.5	–	100	–	–	–
1928	thousand ha	1185.8	93.8	132.9	45.3	80.7	42.0	1399.4	90.4	−77.6	−6.1
	%	84.7	–	9.5	–	5.8	–	100	–	–	–
1932	thousand ha	1263.4	100	150.9	51.4	133.1	69.3	1547.4	100	0.0	100
	%	81.6	–	9.8	–	8.6	–	100	–	–	–
1937	thousand ha	1109.2	87.8	215.4	73.4	116	60.4	1440.9	93.1	−154.2	−12.2
	%	77.0	–	14.9	–	8.1	–	100	–	–	–
1940	thousand ha	1048.3	83.0	293.3	100	166.2	86.6	1507.8	97.4	−215.1	−17.0
	%	69.5	–	19.5	–	11.0	–	100	–	–	–
1945	thousand ha	1126.9	89.2	280.0	95.5	131.8	68.6	1538.7	99.4	−136.5	−10.8
	%	73.2	–	18.2	–	8.6	–	100	–	–	–
1950	thousand ha	1119.6	88.6	252.8	86.2	118.6	61.8	1491.0	96.3	−143.8	−11.4
	%	75.1	–	17.0	–	7.9	–	100	–	–	–
1955	thousand ha	1049.7	83.1	260.6	88.8	103.0	53.6	1413.3	91.3	−213.7	−16.9
	%	74.3	–	18.4	–	7.3	–	100	–	–	–
1960	thousand ha	885.5	70.1	259.4	88.4	106.5	55.2	1245.4	80.5	−377.9	−29.9
	%	71.1	–	20.3	–	8.6	–	100	–	–	–
1965	thousand ha	718.4	56.9	213.0	72.6	–	–	931.4	–	−545.0	−43.1
	%	77.1	–	22.9	–	–	–	100	–	–	–
1980	thousand ha	673.2	53.3	288.5	98.4	–	–	961.7	–	−590.2	−46.7
	%	70.0	–	30.0	–	–	–	100	–	–	–
1983	thousand ha	665.3	52.7	281.2	95.9	192.0	100	1138.5	73.6	−598.1	−47.3
	%	58.4	–	24.7	–	16.9	–	100	–		
1997	thousand ha	781.1	61.8	284.6	97.0	–	–	1065.7	–	−482.3	−38.2
	%	73.3	–	26.7	–	–	–	100	–	–	–
2000	thousand ha	792.9	62.7	269.3	91.8	–	–	1062.2	–	−485.2	−31.4
	%	74.6	–	25.4	–	–	–	100	–	–	–
2003	thousand ha	801.8	63.5	263.8	89.9	–	–	1065.6	–	−481.8	−31.1
	%	75.2	–	24.8	–	–	–	100	–	–	–

203.8 to 262.4 thousand ha. In these years, the sown area of winter wheat exceeded 200 thousand ha in all year. Starting from 1955, this area decreased permanently and by 1980, amounted to 114.4 thousand ha only what is on average, twice as less as compared to 1927–1955.

Corn production (the forage and industrial product in cattle breeding) has always been extremely important for the agrarian sector of the country's economy. In 1913–1955, the sown corn areas amounted to 350–400 thousand ha, except the year of 1917 when this figure was 245.9 thousand ha only. Since 1955, the sown corn areas drastically reduced year after year and amounted to 125.3 thousand ha by 1980, which was 3.5 times less the same index of 1945 (Table 6.2).

Potato is another very important food product, with its sown areas varying from 3.4 to 16.8 thousand ha in 1921–1937. Starting from 1937, the areas of potato increased year after year and reached its maximum in 1980 (33.9 thousand ha). The same is true with the areas planted with vegetables amounting to only 1.2 thousand ha in 1913. Later, the areas with vegetables increased gradually and by 1980, reached their maximum of 35.4 thousand ha (Table 6.2).

Table 6.2 Dynamics of the structure of sown area (Iashvili 1965; Charkseliani et al. 1988)

Years		1913		1917		1923		1927		1932		1937		1940		1945	
		Abs.	%	Abs.	%	Abs.	%	Abs.	%	Abs.	%	Abs.	%	Abs.	%	Abs.	%
Winter wheat	Abs.	168.9	22.6	111.6	22.4	126.9	24.1	231.0	28.9	203.8	20.4	243.2	23.9	232.8	26.0	237.6	25.1
	%	100	–	66.1	–	75.1	–	136.8	–	120.7	–	144 .0	–	137.8	–	140.7	–
Spring barley	Abs.	65.5	8.8	47.1	9.5	38.2	7.2	64.6	8.1	80.1	8.0	88.4	8.7	69.1	7.7	60.9	6.4
	%	100	–	17.9	–	58.3	–	98.6	–	122.3	–	135.0	–	105.5	–	92.9	–
Maize	Abs.	352.2	47.1	245.9	49.4	296.2	56.2	396.9	3.6	387.1	38.7	406.8	40.0	355.3	39.7	435.9	46.1
	%	100	–	69.8	–	84.1	–	112.7	–	109.9	–	115.5	–	100.8	–	123.8	–
Legumes	Abs.	5.3	0.7	3.0	0.6	0.8	0.1	2.5	0.3	4.5	0.4	12.0	1.1	13.3	1.5	16.7	1.8
	%	100	–	56.6	–	15.1	–	47.2	–	84.9	–	226.4	–	250.9	–	315.1	–
Total Grain Crops	Abs.	706.7	94.8	475.8	95.5	508.9	96.5	758.5	94.8	831.5	83.2	873.5	85.9	748.4	83.6	827.6	87.5
	%	100	–	67.3	–	72.0	–	107.2	–	117.7	–	123.6	–	105.9	–	117.1	–
Tobacco	Abs.	10.5	1.4	11.9	2.4	4.2	0.8	14.1	1.8	21.7	2.2	21.1	2.1	20.9	2.3	15.7	1.7
	%	100	–	113.3	–	40.0	–	134.3	–	20.7	–	201.0	–	199.1	–	149.5	–
Total technical crops	Abs.	20.7	2.8	14.7	3.0	7.9	1.5	22.5	2.8	99.9	10.0	52.7	5.2	51.6	5.8	42.5	4.5
	%	100	–	71.0	–	38.2	–	108.7	–	482.6	–	254.6	–	249.3	–	205.3	–
Potatoes	Abs.	7.2	0.8	3.4	0.7	4.7	0.9	8.0	1.0	16.8	1.7	23.1	2.3	28.7	2.8	22.2	2.3
	%	100	–	47.2	–	65.3	–	111.1	–	233.3	–	320.8	–	341.7	–	308.3	–
Vegetables	Abs.	1.2	0.1	1.9	0.4	5.3	1.0	4.5	0.6	18.3	1.8	44.3	4.4	14.4	1.6	17.2	1.8
	%	100	–	158.3	–	441.7	–	375.0	–	1525	–	3692	–	1260	–	1433.0	–
Potatoes and vegetables	Abs.	9.1	1.2	5.3	1.1	10.8	2.0	14.0	1.8	38.1	3.8	44.3	4.4	43.1	4.8	42.5	4.5
	%	100	–	58.2	–	118.7	–	153.8	–	418.7	–	485.8	–	473.6	–	467.0	–
Forage crops	Abs.	5.7	0.7	2.1	0.4	0.6	0.14	2.7	0.3	29.9	3.0	46.3	4.6	52.6	5.9	32.7	3.5
	%	100	–	36.8	–	10.5	–	47.4	–	524.6	–	812.3	–	922.8	–	573.7	–
Total area under crops	Abs.	747.4	–	498.1	–	527.4	–	800.3	–	999.4	–	1016.8	–	895.7	–	945.3	–
	%	100	–	66.6	–	70.6	–	107.1	–	133.7	–	136.0		119.8	–	126.5	–

Years		1950		1955		1960		1965		1970		1975		1980	
		Abs.	%	Abs.	%	Abs.	%	Abs.	%	Abs.	%	Abs.	%	Abs.	%
Winter wheat	Abs.	231.8	254.4	262.4	27.6	166.2	20.0	186.2	24.2	126.9	17.2	140.0	18.4	114.4	15.5
	%	137.2	–	155.4	–	98.4	–	110.2	–	75.1	–	82.9	–	67.7	–
Spring barley	Abs.	52.3	5.7	49.3	5.2	26.6	3.2	39.0	5.1	24.5	3.3	19.7	2.6	17.8	2.4
	%	79.8	–	75.3	–	40.6	–	59.5	–	37.4	–	30.1	–	27.2	–
Maize	Abs.	38.9	42.6	323.5	34.0	226.5	27.3	215.8	28.0	184.0	25.0	155.8	20.5	125.3	17.0
	%	110.4	–	91.8	–	64.3	–	61.3	–	52.2	–	44.2	–	35.6	–
Legumes	Abs.	161.0	1.8	14.0	1.5	11.9	1.4	15.3	2.0	7.9	1.1	9.9	1.3	13.9	1.9
	%	303.8	–	264.1	–	224.5	–	288.7	–	149.1	–	186.8	–	262.3	–
Total grain crops	Abs.	759.7	83.3	714.6	68.3	471.6	56.9	500.8	65.0	388.5	52.7	373.0	49.2	317.0	42.9
	%	107.5	–	101.1	–	66.7	–	70.8	–	55.0	–	52.8	–	44.9	–
Tobacco	Abs.	12.7	1.4	14.5	1.5	14.7	1.8	13.9	1.8	12.4	1.7	12.3	1.6	11.3	1.5
	%	121.3	–	137.6	–	140.0	–	132.7	–	118.3	–	117.5	–	107.9	–
Total technical crops	Abs.	42.7	4.7	41.0	4.3	39.9	4.8	44.0	5.7	39.5	5.4	43.7	5.8	48.2	6.5
	%	206.3	–	198.1	–	192.7	–	212.6	–	190.8	–	211.1	–	232.8	–
Potatoes	Abs.	24.4	2.7	30.9	3.3	22.2	2.7	24.1	3.1	24.6	3.3	28.3	3.7	33.9	4.6
	%	338.9	–	429.2	–	308.3	–	334.7	–	341.7	–	393.0	–	470.8	–
Vegetables	Abs.	15.2	1.7	19.4	2.0	22.5	2.7	24.1	3.1	29.4	4.0	32.7	4.3	35.4	4.8
	%	1266.7	–	1616.7	–	1875.0	–	2008.3	–	2450.0	–	2725.0	–	2950.0	–

(continued)

Table 6.2 (continued)

Years		1950		1955		1960		1965		1970		1975		1980	
		Abs.	%	Abs.	%	Abs.	%	Abs.	%	Abs.	%	Abs.	%	Abs.	%
Potatoes and vegetables	Abs.	43.0	4.7	54.5	5.7	48.3	5.8	51.2	6.6	57.1	7.7	63.2	8.3	76.2	10.3
	%	472.5	–	589.9	–	530.8	–	562.6	–	62.7	–	694.5	–	837.4	–
Forage crops	Abs.	67.0	7.3	140.4	14.8	269.6	32.5	1274.1	22.6	251.5	34.1	276.2	36.4	297.9	40.3
	%	117.5	–	2463.2	–	4730.0	–	3054.4	–	4412.3	–	4845.6	–	5226.3	–
Total area under crops	Abs.	912.4	–	950.5	–	829.4	–	770.1	–	736.6	–	758.7	–	739.3	–
	%	122.1	–	127.2	–	110.0	–	103.0	–	98.5	–	101.5	–	98.9	–

It is known that forage crops (silage corn, root crops, etc.) are widely used in cattle breeding. Their areas were insignificant until 1927 and covered 0.1–5.7 thousand ha (1913). From 1982, these areas increased drastically and varied from 29.9 to 67.0 thousand ha in 1932–1950. The areas occupied with forage crops increased drastically following the further development of the cattle forage base showing an almost 10-fold increase, and as a result, the areas occupied with forage crops reached 297.9 thousand ha by 1980 (Table 6.2) (Tourmanidze et al. 1999).

In the 1990s, the country's economy, including the agrarian sector, was badly shocked. The land plots, hayfields, and pastures were made a private property of the people. This process is still continuing.

In 2000–2003, no significant changes were seen with the distribution of the agricultural plots of field of the country (Table 6.3). However, if in 2000, the arable lands occupied 792.9 thousand ha, the sown areas covered 610.8 thousand ha making 77% of the arable lands (Table 6.4), and in 2003, the arable lands increased to 801.8 thousand ha and the sown areas constituted 70% of the arable lands. In 2000–2009, the sown areas were reduced by 50%.

The areas occupied by winter cereals (wheat, barley, and rye) in the country by 2000 reduced by 49.9% in 2009 and by 54% in 2007 (Table 6.4). The areas of spring crops (wheat, barley, rye, oats, maize, and legumes) in 2000–2009 from 150.8 thousand ha (in 2006) decreased by 41.5%. The same is true with the areas of potato, melons, and watermelons, having decreased by 52.4% from 2000 to 2009. The areas with food crops (perennial and annual grasses, food root crops) decreased from 61.5 thousand ha (in 2000) to 9.6–9.3 thousand ha (in 2007–2008) (Table 6.4).

Table 6.3 Distribution of crops by types (thousand ha)

Agricultural crops	Years			
	2000	2001	2002	2003
Total	3019.7	3022.7	3023.5	3025.8
Arable lands	792.9	795.3	789.7	801.8
Perennial plantations	269.3	267.9	264.9	263.8
Hayfields and pastures	1938.1	1939.7	1940.1	1940.4
Buildings and yards	19.4	19.8	19.8	19.8

The goal of plant growing, one of the major agricultural branches, is growing vegetation production. From 2000 to 2009, in Georgia, the production of wheat, one of the most important foodstuffs changed from 53.9 (in 2009) to 306.5 thousand tons (in 2001). Wheat harvest decreased drastically since 2005. Average wheat harvest in 2000–2005 was 199.4 thousand tons and it was 69.7 thousand tons in 2006–2009, i.e., showed a 65% reduction on average (Table 6.5). From 1913 to 1980, an average wheat harvest of winter wheat in the country was 175.4 thousand tons. The least harvest was registered in 1917 (111.6 thousand tons) and in 1980 (114.4 thousand tons) exceeding the harvest of 2009 by 52–53%.

Thus, the wheat harvest was extremely low in 1913–2009. Average annual corn harvest in 2000–2005 was 379.7 thousand tons and decreased by 25% in 2006–2009. Average annual potato harvest in 2000–2005 was 402.7 thousand tons and decreased by 50% in 2006–2009. Average annual harvest of vegetables in 2006–2009 was 56% less than in 2000–2005 (Table 6.5).

Average harvest of annual crops per hectare in Georgia is two or three times less than that in the developed countries (Table 6.6).

The main producers of cereals and pulse plants are the regions of Kakheti, Imereti, Samegrelo-Zemo Svaneti, Shida Kartli, and Kvemo Kartli (Table 6.7).

The main wheat-growing regions are Kakheti, Shida Kartli, and Kvemo Kartli. In 1998–2009, the average annual wheat harvest in Kakheti was 77.5 thousand tons; it was 35.6 thousand tons in Shida Kartli and 27.8 thousand tons in Kvemo Kartli (Table 6.8). Unprecedentedly, low wheat harvest (8.5–4.7 thousand tons) was gained in Shida Kartli in 2006–1007 and in Kvemo Kartli (5.5 thousand tons) in 2007. Even less wheat was harvested in other regions (Imereti, Mtskheta-Mtianeti, and Meskheti–Javakheti) in 2007–2009.

The average annual harvest of corn, the main spring crop, was 398.6 thousand tons in 1998–2005 and it decreased by 29% in 2006–2009 (Table 6.9). 29% of the total corn harvest is gained in Imereti region, 27% is gained in

Table 6.4 Crop areas of winter and spring crop farms of all categories (thousand ha) (Statistical Yearbook of Georgia 2005, 2006, 2007, 2008, 2009)

Agricultural crops	Years									
	2000	2001	2002	2003	2004	2005	2006	2007	2008	2009
Sown area, total	610.8	564.5	577.0	561.7	534.0	539.6	330.2	297.2	329.3	308.3
Winter crops										
Cereals (wheat, barley, rye)	119.5	131.9	156.4	133.0	113.6	99.0	76.6	55.0	58.6	60.5
Spring crops										
Cereals and beans (wheat, barley, rye, maize, beans)	257.9	242.0	237.9	234.2	237.0	252.1	150.8	151.0	176.9	156.3
Potato and melons	93.1	85.9	84.7	87.2	87.0	94.4	56.4	58.9	54.8	44.3
Perennial grass hay, annual grass of hay, food root crops)	61.5	51.8	49.9	55.0	51.7	50.7	17.9	9.6	9.3	17.0
Other crops							28.5	22.7	29.7	30.2

Table 6.5 Production of annual crops farms of all categories (thousand tons) (Statistical Yearbook of Georgia 2005, 2006, 2007, 2008, 2009)

Agricultural crops	Years									
	2000	2001	2002	2003	2004	2005	2006	2007	2008	2009
Wheat	89.4	306.5	199.7	225.4	185.8	190.1	69.7	74.9	80.3	53.9
Barley	30.2	98.9	57.5	48.3	61.3	65.4	30.6	40.3	49.3	19.9
Rye	–	–	–	–	–	–	0.0	0.1	0.1	0.0
Oats	–	–	–	–	–	–	1.3	1.6	2.9	4.2
Corn for grain	295.9	288.6	400.1	461.9	410.6	421.3	217.4	295.8	328.2	291.0
Beans	–	–	–	–	–	–	7.6	10.5	11.6	10.2
Potatoes	302.0	422.2	415.3	425.2	419.5	432.2	168.7	229.2	193.4	216.8
Vegetables	354.2	396.0	405.6	430.1	400.5	436.7	179.7	190.3	165.0	170.3
Melons	80.0	83.9	125.1	125.0	109.5	119.6	37.8	73.5	52.8	43.7
Food root crops	–	1.2	–	–	4.4	2.9	1.5	1.2	0.2	3.2
Perennial grass hay	32.4	110.3	86.1	130.8	100.3	100.7	25.8	8.8	30.2	23.0
Annual grass of hay	19.1	43.9	48.8	48.6	46.6	37.4	26.5	20.5	5.0	14.6

Samegrelo-Zemo Svaneti region, 12% is gained in Kakheti, 9% is gained in Kvemo Kartli, and 8% is gained in Guria. Other regions of the country gain 15% of the total corn production of the country.

The yield and raw material base of perennial agricultural crops play an important role in the agrarian sector and processing industry of Georgia.

In 2006–2009, the grape yield in the country varied between 150.1 thousand tons (in 2009) and 227.3 thousand tons (in 2007), while the average harvest amounted to 178.9 thousand tons. The highest grape harvest is usually gained in Kakheti (95.4 thousand tons) making over 53% of the total grape production of the country. Kakheti is followed by Imereti with 23% of the total grape production.

On average, 12.8 thousand tons of grapes are produced in Shida Kartli making 7.2% of the total average harvest of Georgia; the share of the average harvest of other regions of the country is 16.5% (Table 6.10).

Apple growing plays an important role in the agriculture and processing industry of Georgia with its average harvest of 64.1 thousand tons in 2006–2009. Half of the apple harvest (50.1%, making 32.1 thousand tons) is gained in Shida Kartli, the region specializing on the given kind of production (Table 6.11). Apple yield in Meskheti–Javakheti is 12.6%.

In the last decade of the twentieth century, the areas with nuts increased intensively. With this indicator, Georgia ranks fourth in the world following Turkey, Italy, and USA. In recent years, the average annual nut yield in Georgia has been 21.4 thousand tons, with 54% of it produced in Samegrelo-Zemo Svaneti, 21% produced in Guria, and 15.9% produced in Imereti (Table 6.12).

Citruses are grown only in the humid subtropical zone of West Georgia. In 2006–2009, an average annual citrus production was 75 thousand tons, with the largest share of the total harvest (74.4%) gained in Adjara and 17.9% gained in Guria (Table 6.13).

Table 6.6 Average productivity of annual crops in farms of all categories (tons/ha) (Statistical Yearbook of Georgia 2005, 2006, 2007, 2008, 2009)

Agricultural crops	Years									
	2000	2001	2002	2003	2004	2005	2006	2007	2008	2009
Cereals and beans—total	1.3	2.0	1.8	2.1	2.0	2.0	1.6	–	–	–
Among them										
Winter wheat	1.0	2.7	1.7	1.9	1.9	2.1	1.3	–	–	–
Spring wheat	1.1	1.9	1.9	2.0	1.5	1.7	1.1	–	–	–
Wheat (mean)	1.1	2.3	1.8	2.0	1.7	1.9	1.2	1.7	1.7	1.1
Winter barley	0.7	2.2	1.2	1.2	1.7	1.4	1.2	–	–	–
Spring barley	1.2	2.2	1.8	1.4	1.7	1.4	1.1	–	–	–
Rye	–	–	–	–	–	–	1.0	1.0	1.5	1.5
Oat	–	–	–	–	–	–	0.7	2.0	1.0	1.4
Maize	1.6	1.6	2.1	2.4	2.2	2.2	1.8	2.4	2.3	2.4
Sunflower	0.23	0.97	0.58	0.56	0.69	0.60	0.55	–	–	–
Soybean	0.66	0.52	0.90	0.50	0.60	0.71	0.34	–	–	–
Potato	8.9	11.6	11.4	11.7	11.4	10.9	7.4	10.8	8.0	11.5
Vegetable	9.3	10.2	10.6	11.0	10.3	9.9	6.6	6.1	5.9	6.8
Melons	9.8	11.2	18.6	14.2	14.1	12.5	11.1	13.2	13.6	14.8
Food root crops	–	21.0	–	–	16.2	10.3	9.1	2.0	5.5	7.4
Perennial grass hay	1.1	36.1	2.7	3.2	3.0	2.9	2.8	2.7	3.9	2.3
Annual grass of hay	1.0	3.0	2.5	3.5	2.8	3.2	3.7	3.8	3.8	2.6

Table 6.7 Production of cereals and bean crops by regions (thousand tons) (Statistical Yearbook of Georgia 2005, 2006, 2007, 2008, 2009)

Regions	Years								
	1998	1999	2000	2001	2002	2003	2004	2005	2006
Georgia, total	597.8	780.5	420.5	713.6	672.2	754.1	679.3	702.9	326.8
Among them									
Imereti	134.5	139.9	83.6	37.6	135.0	137.9	129.0	135.5	63.4
Samegrelo–Upper Svaneti	108.2	121.0	118.9	68.3	66.7	130.6	94.2	109.9	78.1
Shida Kartli	77.3	95.6	46.2	91.5	95.7	94.9	76.7	74.0	17.7
Kakheti	90.4	182.9	62.1	297.8	154.8	162.4	172.1	189.5	70.3
Kvemo Kartli	70.2	105.6	36.1	92.0	95.3	103.2	85.9	74.8	41.4
Samtskhe–Javakheti	23.0	48.5	19.9	57.5	52.7	28.7	42.3	39.9	17.1
Other regions	94.2	87.0	53.7	68.9	72.0	96.4	79.1	79.3	38.8

Table 6.8 Maize production by regions (thousand to ns) (Statistical Yearbook of Georgia 2005, 2006, 2007, 2008, 2009)

Regions	Years											
	1998	1999	2000	2001	2002	2003	2004	2005	2006	2007	2008	2009
Georgia, total	420.2	490.5	295.9	288.6	400.1	461.9	410.6	421.3	217.4	295.8	328.2	291.0
Among them												
Imereti	132.8	137.7	83.0	36.6	132.1	134.3	122.7	127.6	61.8	85.8	97.2	95.4
Samegrelo–Upper Svaneti	108.1	120.9	118.8	68.0	66.5	130.0	93.8	108.3	77.2	83.9	95.0	93.3
Guria	45.8	26.1	30.0	16.5	13.8	36.9	22.1	25.9	23.9	34.3	37.7	29.8
Kakheti	37.1	81.2	12.6	58.3	53.0	38.4	62.2	61.6	14.8	23.5	51.6	32.4
Kvemo Kartli	43.7	45.3	16.6	40.1	51.6	40.7	30.7	24.2	21.1	38.0	17.3	14.6
Other regions	52.7	79.3	34.9	69.1	83.1	81.6	79.1	73.7	18.6	30.3	29.4	25.5

Table 6.9 Wheat production by regions (thousand tons) (Statistical Yearbook of Georgia 2005, 2006, 2007, 2008, 2009)

Regions	Years											
	1998	1999	2000	2001	2002	2003	2004	2005	2006	2007	2008	2009
Georgia, total	144.6	226.1	89.4	306.5	199.7	225.4	185.8	190.1	69.7	74.9	80.3	53.9
Among them												
Shida Kartli	56.3	59.8	28.3	49.7	60.9	57.7	37.0	34.3	8.5	4.7	11.3	19.0
Kakheti	49.0	92.5	42.7	193.2	84.0	105.0	87.3	96.6	42.7	62.0	52.7	22.2
Kvemo Kartli	22.0	51.9	12.2	38.8	35.6	47.7	43.5	41.7	14.1	5.5	12.4	8.7
Other regions	17.3	21.9	6.2	24.8	19.2	15.0	18.0	17.5	4.4	2.7	3.9	4.0

Table 6.10 Grape production by regions (thousand tons) (Statistical Yearbook of Georgia 2006, 2007, 2008, 2009)

Regions	Years			
	2006	2007	2008	2009
Georgia, total	162.5	227.3	175.8	150.1
Among them				
Imereti	36.3	54.5	43.7	30.3
Shida Kartli	10.9	16.0	8.1	16.4
Kakheti	80.2	118.6	100.0	82.7
Other regions	35.1	38.2	24.0	20.7

Table 6.11 Apple production by regions (thousand tons) (Statistical Yearbook of Georgia 2006, 2007, 2008, 2009)

Regions	Years			
	2006	2007	2008	2009
Georgia, total	32.8	101.3	41.5	80.7
Among them				
Imereti	6.1	4.9	4.0	2.6
Samegrelo–Upper Svaneti	7.2	5.5	3.9	1.4
Shida Kartli	9.0	55.6	11.3	52.6
Kvemo Kartli	1.4	9.1	3.3	2.0
Samtskhe–Javakheti	0.8	10.7	6.7	14.3
Other regions	8.3	15.5	12.3	7.8

Table 6.12 Nut production by regions (thousand tons) (Statistical Yearbook of Georgia 2006, 2007, 2008, 2009)

Regions	Years			
	2006	2007	2008	2009
Georgia, total	23.6	21.2	18.7	21.8
Among them				
Imereti	3.2	3.2	3.9	3.2
Samegrelo–Upper Svaneti	13.5	12.0	9.3	11.4
Guria	5.7	4.5	4.2	3.7
Other regions	1.1	1.5	1.3	3.5

Table 6.13 Citrus fruit production by regions (thousand tons) (Statistical Yearbook of Georgia 2006, 2007, 2008, 2009)

Regions	Years			
	2006	2007	2008	2009
Georgia, total	52.2	98.9	55.2	93.6
Among them				
Ajara	31.9	80.5	32.6	78.3
Samegrelo–Upper Svaneti	9.0	5.3	5.9	2.8
Guria	11.3	13.1	16.7	12.5

It should be noted that no permanent reduction of the perennial agricultural crop harvest was fixed in 1998–2009 what is perceived as a positive trend.

In general, cattle breeding is one of the principal branches of agriculture. The cattle population in Georgia changed from 1050.0 (in 1998) to 1260.4 (in 2005) (Table 6.14). Most cattle are in the Imereti Region (245.7 thousand heads) making 21.4% of the total number of cattle in the country.

Then, they rank Samegrelo-Zemo Svaneti Region with 194.2 thousand heads (16.9% of the total number) and Kvemo Kartli with 157.7 thousand heads (13.7% of the total number).

In 1993–2005, the number of pigs in the country was 440.6 thousand (Table 6.15). From 2005, due to the spread of swine flu (H_1N_1) in the country, the number of pigs in Georgia decreased by 112.2 thousand (24.6%) in 2006, by 233.2 thousand (32.0%) in 2007, and by 23.5 thousand (21.3%) in 2008. The reduction of the number of pigs in 2005–2008 was 367.5 thousand making 83.4% of the average pig population (440.6 thousand) in the country in 1993–2005. In 2009, the pig population in the country increased by 48.8 thousand and reached 135.2 thousand. This number of pigs is 30.7% of the average annual pig population in 1993–2005, while it is 11% more than the pig population in 2008. As the average data of 1998–2009 suggest, the greatest pig population was fixed in Samegrelo-Zemo Svaneti Region (105.9 thousand making 28.7% of the total pig population of the country), Imereti Region (77.7 thousand making 21.1% of the total pig population),

Table 6.14 Cattle breeding (including buffalo) by regions (thousand heads) (Statistical Yearbook of Georgia 2005, 2006, 2007, 2008, 2009)

Regions	Years											
	1998	1999	2000	2001	2002	2003	2004	2005	2006	2007	2008	2009
Georgia, Total	1050.0	1122.1	1177.4	1180.2	1216.0	1242.5	1250.7	1260.4	1163.6	1048.5	1045.5	1014.7
Among them												
Ajara A.R.	108.6	117.8	123.2	122.7	126.1	128.4	119.5	113.2	115.9	103.5	87.7	87.5
Imereti	241.3	258.6	267.7	266.6	278.3	287.3	269.5	272.0	223.8	186.8	204.2	192.7
Samegrelo–Upper Svaneti	166.4	174.9	192.2	202.2	197.1	198.3	200.8	208.3	190.8	196.2	204.0	198.8
Inner Kartli	78.4	79.9	84.6	81.7	85.0	85.4	85.0	85.7	86.4	69.9	65.5	75.0
Kakheti	91.3	110.7	117.8	116.0	122.8	130.2	130.7	135.8	112.3	87.5	89.2	82.8
Lower Kartli	120.3	129.4	129.5	144.9	150.7	149.0	177.2	183.1	177.8	186.2	186.7	158.1
Samtskhe–Javakheti	97.9	98.5	98.6	99.5	100.9	106.1	106.0	104.2	121.0	101.6	88.6	103.0
Other regions	146.7	152.3	163.8	146.6	155.1	157.8	162.0	158.1	135.6	116.8	119.6	116.8

and Kakheti Region (59 thousand making 16% of the total pig population) (Table 6.15).

Sheep breeding is a traditional branch of cattle breeding in Georgia and is best developed in East Georgia. In 1998–2009, the sheep population was stable in the country and varied from 521.7 thousand (in 1998) to 719.8 thousand (in 2005). It should be noted that in 2009, as compared to 2008, the sheep population in the country decreased by 87.7 thousand, i.e., 12.7% which is quite a high value. Such a decrease in the sheep population was caused by the mass export of Georgian selective sheep species to Iran for its high-quality wool.

Average sheep population in Kakheti is 255.9 thousand making 40.7% of the total sheep population of Georgia. Then, they rank Kvemo Kartli with 170.9 thousand sheep (27.2% of the total sheep population of the country) and Meskheti–Javakheti with 90.6 thousand sheep (14.4% of the total sheep population) (Table 6.16).

Besides sheep breeding, poultry raising is another commercially profitable branch in Georgia. In 1998–2009, the average number of all species of poultry in Georgia was 7780.5 thousand poultry in the country (Table 6.17).

The maximum number of poultries was fixed in 2004 (9836.2 thousand) and the minimum number was fixed in 2006 (5331.7 thousand). In general, from 2004 to 2009, the number of all poultry species shows quite a falling trend. Average annual numbers of all poultry species in 1998–2009 in different regions were as follows: 1711.1 thousand in Samegrelo-Zemo Svaneti Region (22% of the total number of poultry), 1615.8 thousand in Imereti (20.8%), 1369 thousand in Kvemo Kartli (17.6%), and 1103.2 thousand in Kakheti (14.2%) (Table 6.17).

As the above-cited statistical data of the agrarian sector of Georgia evidence, it is obvious that the country has good perspectives to develop the given branch of economy evidenced by the following facts:

- The areas of nut plantations increase in West Georgia every year;

Table 6.15 Pig stock by region (thousand heads) (Statistical Yearbook of Georgia 2005, 2006, 2007, 2008, 2009)

Regions	Years											
	1998	1999	2000	2001	2002	2003	2004	2005	2006	2007	2008	2009
Georgia, Total	365.9	411.1	443.4	445.4	446.1	473.8	483.9	455.3	343.1	109.9	86.4	135.2
Among them												
Imereti	87.9	93.7	93.5	95.6	97.3	105.1	103.2	99.7	58.1	34.6	27.4	35.7
Samegrelo–Upper Svaneti	108.0	117.9	129.4	134.3	131.9	136.8	145.1	151.0	122.7	37.2	23.2	33.0
Guria	22.0	20.8	43.0	36.5	33.1	33.7	33.4	27.5	25.5	–	–	–
Racha–Lechkhumi and Lower Svaneti	16.0	18.2	17.0	20.9	20.7	21.2	24.9	20.0	31.4	–	–	–
Mtskheta–Mtianeti	20.9	22.7	24.4	24.4	25.1	24.9	31.4	23.6	14.2	–	–	–
Kakheti	57.2	79.1	81.3	73.9	78.3	91.3	84.7	74.6	46.7	7.4	10.4	22.8
Lower Kartli	20.0	23.4	21.3	24.5	23.6	24.9	25.7	31.5	19.9	8.5	4.8	13.3
Other regions	33.9	35.3	33.5	35.3	36.1	35.9	35.5	27.4	24.6	22.2	20.6	30.4

Table 6.16 Sheep population by region (thousand heads) (Statistical Yearbook of Georgia 2005, 2006, 2007, 2008, 2009)

Regions	Years											
	1998	1999	2000	2001	2002	2003	2004	2005	2006	2007	2008	2009
Georgia, total	521.7	553.2	546.9	567.5	611.2	628.8	689.2	719.8	696.8	711.0	690.0	602.3
Among them												
Mtskheta–Mtianeti	56.7	54.2	54.5	57.2	58.0	60.1	57.4	60.0	57.1	67.0	79.8	50.0
Kakheti	193.4	221.1	206.8	230.0	256.2	262.3	277.6	273.9	266.1	313.9	300.2	269.4
Lower Kartli	127.0	129.5	122.6	145.5	145.9	146.3	212.4	242.8	230.0	210.1	206.8	131.8
Samtskhe–Javakheti	102.5	102.3	114.9	87.2	94.4	96.9	89.0	87.8	90.0	72.8	61.7	87.4
Other regions	42.1	46.1	48.1	47.6	56.7	63.2	52.8	55.3	53.6	47.2	41.5	63.7

Table 6.17 All types of poultry, according to regions (thousand wings) (Statistical Yearbook of Georgia 2005, 2006, 2007, 2008, 2009)

Regions	Years											
	1998	1999	2000	2001	2002	2003	2004	2005	2006	2007	2008	2009
Georgia, Total	8239.7	8473.3	7825.5	8495.9	8905.2	9200.6	9836.2	7482.2	5331.7	6149.7	6682.2	6674.8
Among them												
Imereti	1903.5	1931.5	1766.4	1871.2	1802.1	1801.0	1967.7	1470.5	1211.6	1159.4	1318.3	1186.3
Samegrelo–Upper Svaneti	1606.6	1785.7	1812.0	1919.5	2008.4	2164.8	2152.0	2032.2	1013.9	1471.0	1359.2	1207.8
Inner Kartli	619.0	506.9	416.4	543.3	562.5	537.0	551.6	394.9	265.1	266.3	314.7	446.8
Kakheti	1190.0	1278.4	1114.9	1181.1	1158.4	1272.7	1296.8	969.5	878.7	804.8	**1004.4**	1088.5
Lower Kartli	847.2	893.9	768.7	915.7	1541.1	1726.2	2203.1	1461.8	1211.7	1572.5	1641.4	1644.9
Other regions	2073.4	2076.9	1947.1	2065.1	1832.7	1698.9	1665.0	1153.3	819.7	875.7	**1044.2**	1100.5

- East Georgia, Lagodekhi Region, in particular, has vast reserves to develop nut growing;
- The country had over 60 thousand ha of tea bushes, which are mostly abandoned at present, with some exceptions. Rehabilitation of tea plantations will undoubtedly bring a significant benefit both to the population and to the country budget;
- The climate change and warming processes will lead to the shift of the natural zones. Consequently, the areas of growing various agricultural crops (wheat, corn, fruit, and vine) over the mountainous relief will increase. If considering that 70% of the territory of Georgia is mountainous, the scale of additional areas used in agriculture as a result of such an increase at the expense of warming in terms of the climate change is easily imaginable;
- The climate warming will promote the introduction of the commercially profitable agricultural crops to the country. For instance, it is obvious that growing olives both in the East and in West Georgia is extremely perspective;
- Table grape variety was traditionally imported to Georgian agrarian market from Central Asia and neighboring countries. The demand for this product is high in August, in autumn, winter, and spring. Only 5–10% of the demand for this product is met by Georgian agriculture. As the tourism develops in the country, the demand for this product will increase further. The climate warming will help produce the table grape variety in Kakheti, Kvemo Kartli, and Shida Kartli, which will promote the repletion of the agrarian market with local produce increasing the budgetary incomes of the local population (farmers) and country;
- On the background of the warmth provision of the agricultural crops both on the lowland and in the mountains, the development of irrigation systems and introduction of the modern irrigation technologies will become necessary, which is possible if considering the available water resources of the country. It is known that highly productive annual and perennial crops on the irrigation lands yield twice or thrice richer harvest a year than those growing on nonirrigated lands.

Thus, based on the statistical data on the agrarian sector of the Georgian economy, the country has extremely favorable prospects to develop agriculture. In West Georgia, the areas of nut grow year after year. There are reserves available to expand the same areas in East Georgia, in the Lagodekhi Region in particular. Tea plantations occupied over 60 thousand ha of the country territory, but at present, they are in fact abandoned. The rehabilitation of the tea plantations will bring significant benefit both to the population and country budget. By considering the climate change and global warming processes, they predict an altitudinal shift of the natural zones. As a result, the areas to grow wheat, corn, fruit, and vine will expand in the mountainous regions, and new agricultural crops will be possible to introduce. As one can see, the recent attempt to grow olives in the country was successful. Consequently, in terms of reasonable and rational land management in Georgia, the field of agriculture will have favorable prospects.

References

Charkseliani A, Petriashvili R, Kipiani M (1988) Land inventory of the Georgian SSR. Tbilisi, Sabchota Sakartvelo, p 735

Iashvili NS (1965) The land fund of the Georgian SSR. In: Natural resources of the Georgian SSR, vol IV, Agricultural resources. Moscow, Nauka, pp 219–273 (Zemelnyi Fond Gruzinskoi SSR. V knige "Prirodnye resursi Gruzinskoy SSR", tom IV, Selskokhozyaistvennie resursi, Moskva, Nauka)

Statistical Yearbook of Georgia (2005) Department of statistics under ministry of economic development of Georgia. Tbilisi, p 329

Statistical Yearbook of Georgia (2006) Department of statistics under ministry of economic development of Georgia. Tbilisi, p 329

Statistical Yearbook of Georgia (2007) Department of statistics under ministry of economic development of Georgia. Tbilisi, p 328

Statistical Yearbook of Georgia (2008) Department of statistics under ministry of economic development of Georgia. Tbilisi, p 326

Statistical Yearbook of Georgia (2009) Department of statistics under ministry of economic development of Georgia. Tbilisi, p 322

Tourmanidze T, Shelia V, Gogichaishvili G (1999) Agroecological monitoring of soil fertility. Perspectives of modern information and communication systems in agriculture, food and production and environmental, vol B. Bonn, Germany, pp 829–833

7 Soil Erosion

Gizo Gogichaishvili

Abstract

As the data suggest, 224 thousand ha of arable lands and 569 thousand ha of pastures and hay meadows are eroded. 95 thousand ha out of 162 thousand ha of eroded melioration lands is slightly eroded, 51 thousand ha is averagely eroded and 16 thousand ha is strongly eroded. The eroded areas of melioration lands in East Georgia are 3 times more than those in West Georgia, and slightly eroded areas are 5 times more and the averagely eroded areas are twice as more. 76 thousand ha of melioration land in Georgia is damaged due to the deflation processes, with 30 thousand ha being slightly deflated, 22 thousand ha being averagely deflated, and 24 thousand ha being strongly deflated. The studies evidence that 1004 thousand ha of 1134 thousand ha of melioration fund of Georgia is hazardous in respect of the development of erosive processes, with 289 thousand ha being slightly erosive hazardous, 50 thousand ha being averagely erosive hazardous, while 665 thousand ha is extremely erosive hazardous. Among them, 12 thousand ha of irrigation lands is hazardous in respect to erosion.

Keywords

Soil erosion • Erosion by water • Erosion by wind • Erosive hazardous

7.1 Introduction

Among the exogenous factors contributing to the formation of the Earth relief, particularly important are erosive processes, with water and wind erosive actions being outstanding. Under the action of these two types of erosion, erosive and deflation relief forms are formed. The action of the two relief-forming processes is episodic and does not result in any obvious changes of the natural (intact) environment. The situation is absolutely different with the territories used by people. The natural vegetation in such areas is transformed or degraded artificially, by the human's action. In such a situation, depending on the intensity or degree of development, the erosive-deflation processes may lead to catastrophic outcomes. Large-scale migrations of peoples, grand wars of conquest, and even the destruction of a number of states in the world history were often provoked by such processes.

Despite the significant progress in the erosion research and practical use of anti-erosive measures, there is still much to do attempt and exploit the environment in an optimal and predictable manner in the future. No widely accepted or physically grounded models of soil erosion, which could be used to solve the practical problems of slope erosion and erosion, in general, e.g., approved selection of anti-erosive measures for optimal development of individual slopes and territories is developed. At present, numerous different empirical equations are used for these purposes. However, their areas of use are local and narrow and often lack relevant reference materials.

At present, particularly stressed is the ecology of agricultural plots of field where due to erosion, great amounts of weed and pest-killer chemicals (pesticides, herbicides, insecticides, etc.) capable of provoking even ecological catastrophes on certain territories are washed down to the water reservoirs.

The situation with the exploration of deflation processes is similar. Deflation is considered as a process of destruction of the agricultural soil cover under the action of strong winds. Often, the erosive and deflation processes of the soil cover take place in the same area. In addition, both processes have the same physical basis. Therefore, in such areas of concern, the practical issues of soil protection must be considered by taking both, erosion by water and deflation, processes into account.

G. Gogichaishvili (✉)
Agrometeorological Division, Department of Hydrometeorology, National Environmental Agency, Tbilisi, Georgia
e-mail: gizo.gogichaishvili@yahoo.com

L. Matchavariani (ed.), *The Soils of Georgia*, World Soils Book Series,
https://doi.org/10.1007/978-3-030-18509-1_7

The subject of erosion studies is the regularities and reasons for occurrence of soil erosion, erosion-prone and eroded soils, anti-erosion measures for soil, and the melioration of eroded soils.

The development of erosion studies as that of a theory is the concern of different branch specialists. A wide concept of erosion was developed by geographers and geologists, who considered erosion, mostly as exogenic processes of the earth surface development (Penck 1894; Davis 1898, 1902; Lazarevich 1973).

Pedologists started to explore erosion more specifically. E. Wollny, a German pedologist, was the first to set experiments to study the influence of atmospheric precipitations on soil wash down (Wollny 1895). The first person to set special experiments to study the depth and sheet erosion was Kozmenko (1909, 1928).

American scientists Bennett and Chaplin (1928) were the first to pay attention to the erosion as to the catastrophic phenomenon for the mankind.

The foundation to erosion studies was laid by the scientists from different branches: hydrologists, geologists, geomorphologists, pedologists, glaciologists, geobotanists, agronomists, foresters, and others. This phenomenon was studied by hydrologists in view of development of rivers and lakes (Lopatin 1952; Makkaveev 1953), while glaciologists studied the aspects of erosion associated with the formation of different forms of the earth surface and soil destruction by the action of snow and ice; geobotanists studied erosion in view of the association between the locations and the vegetation; agronomists gave a particular importance to the protection of agricultural lands and so on.

A fundamental role in the development of the science of soil erosion was played by the definition of soil erosion. Term *soil erosion* was first used in England by McGee in 1911. Later, the first monographs about the subject, including *Soil erosion and its control* by Ayres (1936), were published, where it discusses the role of the many different factors and agronomic treatments that affect soil loss. In 1937, famous collection of *Soil Erosion* (*Eroziya pochv*) was published in Russia. Editor and one of the authors of the collection was a famous scientist, Prof. A. Pankov. In Georgia, in 1932, a work by Prof. T. Kvaratskhelia *Soil wash-down in tea plantations* was published. A US erosion specialist H. Bennett is considered the founder of erosion studies. In 1939, he published his famous monograph *Soil conservation* and in 1955, he published his work *Fundaments of soil conservation*. In addition, a number of works were published in the English language, including some very remarkable monographs by Archer (1956), Stallings (1957), Kohnke and Bertrand (1959) and Hudson (1971).

On the initiative of US Professor H. Bennett, a Soil Protection Department was established as a unit of the Ministry of Agriculture. Aiming at studying the intense deflation and erosion processes in the 1930s and protecting agricultural plots of field and lands against them, first, USLE and later RUSLE1 and RUSLE2 were designed. In the same years, by the order of the Ministry of Agriculture of the USA, a new physical model of soil erosion WEPP—Erosion by water Prediction Project (Flanagan et al. 1997) was developed. In the 1960s, erosion by wind prediction project WEQ—Erosion by Wind Equation (Woodruff and Sidoway 1965) was developed, and in 1998, its revised version RWEQ (Fryrear et al. 1998) was finalized. In 2007, a new model based on the new, physical regularities of erosion by wind WEPS—Erosion by Wind Prediction System (Project Leader L. Wagner) was made available. At the same time, soil erosion studies were accomplished in Europe, with a number of models developed (MMF, G2, EUROSEM—European Soil Erosion Model), whose practical use has a number of limitations.

It is known that Georgia is an ancient agricultural country. The fact of Georgia being considered as one of the homelands of viticulture and wheat and its residents being called Georgians, i.e., ground tillers, evidences the presence of highly developed ancient agriculture on the territory of present Georgia.

It is natural that developed agriculture means protecting soil and various agricultural plots of field against the harmful impact of erosion. One of the evidences of the highly developed agrarian culture of Georgia is ancient artificial terraces survived to date in Georgia. Terrace agricultural system was used in Meskheti, Mtiuleti, Kartli, Racha-Lechkhumi, Adjara, and Abkhazeti. In this respect, artificial terraces in Meskheti built with stone are very interesting. They are spread on the slopes of the gorges of the river Mtkvari and Akhalkalaki-Tskali, starting from Vardzia (village Saro, Khizabavra, Toki, Tmogvi, Khertvisi, Toloshi, Kisatibi, etc.), as well as over smaller areas in the city of Akhaltsikhe, in village Atskuri, and at other locations. These terraces were built in the middle centuries and are still operable.

In the environs of Tbilisi, the forest-and-melioration works of landscaping, urban economy, protection of streets, and premises against erosion started in 1891. Similar works were undertaken in the environs of resort Abastumani.

Since the 1920s, they started to massively grow tea and other subtropical agricultural crops in the humid subtropical zone of West Georgia. Specialized farms were established. In line with the abovementioned, the Scientific-Research Institute of Tea and Subtropical Crops studied the soil erosion processes in view of protecting soil against erosion, improving the fertility of eroded soils, and optimal cultivation of greatly inclined slopes. In this respect, particularly large-scale works were accomplished by Chakvi branch of the said Institute and at the base station of village Keda.

Since 1946, the Mountain Forestry Institute of Georgia has accomplished fundamental research to study soil erosion and develop forest-melioration measures to protect soils against erosion in the forest-and-mountain zones of Georgia.

Due to the intense water and erosion by wind of the agricultural plots, important scientific research studies were accomplished at Georgian Institute of Farming in the 1960s. For the erosion-prone regions, such soil-protective practices as specific soil cultivation, annual crops sowing, and growing and crop rotation for different natural zones of the country were developed.

Large-scale works to cultivate greatly inclined and intensely eroded slopes by terracing were accomplished at Georgian Scientific-Research Institute of Horticulture, Viticulture, and Oenology.

As early as since the beginning of the 1930s, the research for the purpose of engineering exploration of the soil-erosive processes as that of a physical phenomenon had been accomplished at Georgian Scientific and Research Institute of Hydrotechnics and Melioration. The institute accomplished such works as physical modeling of the erosive processes developed over the irrigation areas, development of prognostic models, and hydrotechnical measures to protect soils. In 1960–1970, under the leadership of Prof. Ts. Mirtskhulava, a hydromechanical prediction model of erosion by water (rain) and erosion by wind prediction model were developed at the institute.

At Georgian Institute of Soil Science, Agrochemistry, and Melioration, a Soil Erosion Department was established in 1946 immediately after foundation of the Institute. The goal of the department was to study the reasons and regularities of soil erosion development based on the study of the properties of soil anti-erosion and agricultural production properties of eroded soils. Another goal of the department was to develop the measures to improve soil productivity through practical realization of the works accomplished by the institute specialists.

7.2 Erosion by Water

Term *erosion* is of the Latin origin and means *eating away*. Concept *erosion* is used in different branches of science, such as geology, geomorphology, hydrology, soil science, medicine, techniques, etc. In soil science, the term has many different meanings, and one can observe various terms associated with erosion: water, wind, irrigation, chemical, military, and other kinds of erosion. The opinions about the explanation of term *erosion* also differ. The researchers studying erosion as hydrodynamic and aerodynamic processes mean a set of processes of detachment, transportation, and deposition of the particles of soil and sometimes, bedrocks or underlying parent materials under the action of temporal surface water currents or wind.

Erosion by water is caused by the action of rain, snowmelt, or irrigation water (Fig. 7.1), but it does not mean

Fig. 7.1 Eroded relief, Lilo, surrounding of Tbilisi, East Georgia. Photo by B. Kalandadze

erosion of seashores, river banks, or shores of lakes or water reservoirs, because in this case, the process is caused by permanent and not temporal water currents.

The conditions for erosion by water to develop occur when surface slope currents or surface flow is formed. They distinguish three kinds of surface flow: rainwater, snowmelt water, and irrigation water flow corresponding to three kinds of erosion: (1) erosion caused by rain (downpour) water, (2) erosion caused by snowmelt water, and (3) irrigation erosion. Naturally, these kinds of erosion differ with the sources of flow formation, mechanism of erosion process, and scales of the material damage caused by them.

The erosion processes caused by snowmelt water are absolutely different in the mountainous and plain areas. On the territories of Northern Europe and Russia, erosion caused by snowmelt water is less intense and lasts longer than the erosion by water caused by rainwater. Snow melting over the slopes in Georgia differs much depending on the slope exposition and altitude and mostly varies from 1 week to one or one and a half month. The snow cover in West Georgia may melt even in 1–3 days. The duration of the erosion processes caused by rainwater is much less than those caused by snowmelt and usually varies from several minutes to several hours. Usually, the erosion caused by rainwater results in more soil loss than the one caused by the snowmelt.

Depending on the size and falling velocity of raindrops, the destruction intensity of the surface soil layer varies. After a raindrop falls on the soil surface, the drop disintegrates together with very small amount of soil affected by the raindrop. The product of soil disintegration splashes. A part of the splash falls not on the soil surface, but in temporal water furrows and is transported from the site of its origination by the water current. Consequently, rain contributes to the "enrichment" of the temporal water currents flowing down the slope with a hard phase (soil particles). Besides, the rainwater, after falling on a thin layer of concentrated water current flowing down the slope, increases the current turbulence a lot and drastically improves its washing and transportation capacity.

Erosion caused by irrigation waters (called irrigation erosion) can be divided into the following subtypes depending on the methods of irrigation: erosion caused by gravity, row and cell irrigation, as well as overhead irrigation.

It is known that depending on the morphological features, they distinguish between the surface erosion and the linear erosion. Both kinds of erosion are accompanied by either soil wash down or wash away, or often both, depending on the location of the study section on the slope.

On its turn, surface erosion is divided into plane and stream erosion, with only a conventional difference between them. It is considered that erosion is caused by the motion of a continuous thin water layer over the slope surface. Such type of currents occurs extremely rarely, and the soil is mostly washed down by stream currents. The boundary between the surface and linear erosions is conventional. It is considered that if the trace of erosion is removed on the plot by ordinary soil cultivation, the erosion is surface, but if the trace of erosion survives, the erosion is linear.

Climatic conditions, relief, rules to grow agricultural crops and agricultural techniques, soils, and other conditions determine the intensity of erosive processes on the agricultural plots, particularly on arable lands. As per the data of the Land Management Design Institute, 205.7 thousand ha of land making 26.2% of the total arable land are eroded in Georgia to different degrees (791.2 thousand ha). 110.5 thousand ha of them are slightly eroded (53.7%), 74.4 thousand ha show medium erosion (36.2%), and 20.8 thousand ha of them are intensely eroded (10.1%). The area of slightly and intensely eroded arable land in Georgia amounts to 95.2 thousand ha (46.3% of total arable lands). This constitutes quite a large area if considering that Georgia is a land-poor country. It should also be considered that soil erosion reduces the soil productivity a lot evidenced by a 30–35% falls in the harvest yield. Besides, erosion leads to the damage to the whole infrastructure (roads, bridges, premises, water pipes, etc.), flooding and silting the slope bottoms and adjacent areas, and the agricultural plots of field and sown areas are destructed.

7.3 Erosion by Wind

Term *deflation* can be used in lieu of term *erosion by wind*. Deflation, as a process, in the geographical literature, means only sweeping the solid particles from the earth (soil) surface by the wind and taking them up in the air. The process of transportation of the soil particles suspended in the air is also a part of deflation. The final outcome of this process is the sedimentation of the soil particles suspended in the air on the earth surface as the wind dies down. The sediments originated due to the deflation processes are called eolic sediments in the geographic literature. Consequently, term *erosion by wind* describes the said process more thoroughly, as it includes both deflation and sedimentation processes. Therefore, when we talk about the soil loss due to the wind, term *deflation* can be used, while when talking about the sediments made up of soil particles, term *sedimentation* is to be used.

A necessary condition for erosion by wind to occur is wind with its speed sufficient to move soil particles. Depending on such external features, as intensity, duration, and inflicted damage, they distinguish between the daily erosion by wind and the dust hurricanes. This difference is conventional in this case too. A distinguishing feature of daily erosion by wind is relatively low wind speed, which is slightly more than the critical speed of the given kind of soil.

Daily erosion by wind is often limited by one or more adjacent plots with all phases of the process starting from soil sweeping through particle sedimentation developed on them.

In case of high wind velocities, much exceeding the critical soil speed, the soil particles are transported up in the air to much greater distances reaching several hundreds of meters, while the traveling distance of the soil particles suspended in the air reaches hundreds to thousands of kilometers. In meteorology, the transportation of large amounts of dust by strong wind is called dust hurricane. Dust hurricane is a dangerous natural phenomenon with its scales often being so large that it takes a form of a natural calamity. Erosion by wind is quite widely spread in Georgia. The major areas of erosion by wind in Georgia cover Shida Kartli Plain, Iori Plateau, and Shiraki Plain in particular. It should be noted that virtually, the experimental studies of erosion by wind or deflation have never been accomplished in Georgia. However, cases of negative impact of erosion by wind on agriculture have been fixed for many times. As the witnesses state, deflation phenomena occur annually on Shiraki Plain. The typical period of deflation in Georgia is observed in February or March. In case of high wind velocities, the crops of winter wheat are often physically damaged or swept away from the field making it necessary to resow the fields what entails high costs.

102.5 thousand ha of arable lands are damaged by the erosion by wind in Georgia making 21.1% of the arable lands of East Georgia. This means that every fifth hectare of the arable land is under the negative impact of erosion by wind. It is clear that the problem of erosion by wind is important for Georgia, but taking anti-deflation measures is quite limited due to various factors (little soil depth, great slope inclination, small plot areas, multi-contour shapes, etc.), and therefore soils are less protected.

The lands or areas with the prospect of origination and further development of erosion following the certain combination of natural conditions when used for farming or economic purposes without taking anti-erosion measures are considered hazardous in respect of erosion.

The main factors causing erosion can be divided into relatively stable (climate, relief, soil genetic type, and grain-size distribution) and dynamic (vegetation cover, economic use of lands, water permeability, soil structure, etc.) factors. The existing methods to evaluate the hazard of erosion and forecast the ground washout can be divided into the following four groups:

- qualitative (mostly landscape geographical),
- semiquantitative (points),
- quantitative-analytical, and
- quantitative-empirical dependencies.

At present, best elaborated to forecast erosion are four quantitative models. First three of them (Methodical recommendations… 1978; Wischmeier and Smith 1965, 1978; Instructions for… 1979) are developed for practical use, while the fourth model (Shvebs 1974), despite being theoretically elaborated quite thoroughly, lacks concrete recommendations and is not used for practical purposes as a result. As there is no long series of observation over the soil wash down, it is in fact impossible to examine the accuracy of the abovementioned models. The question of giving preference to one of the three quantitative models must be decided by looking at the model based on the mechanism of erosive process and the degree to which it considers the major factors causing erosion. Besides, the labor-intensiveness and simplicity of a model must be considered.

7.4 Brief Description of the Soil Erosion Mechanism

As the modern views suggest, the erosive process is considered as two independent processes: (1) detachment of the soil particles from the total soil mass and (2) their transportation with the water current. In order to prevent the detachment of a soil particle from the total soil mass, the force of adherence and particles' own weight must be overcome. During the transportation of a particle, which is in a free, non-adhered state, the water current overcomes only the force of friction, which is less than the sum of the weight of the particle and force of adherence. With a sufficient speed, the water current is capable of detaching the particles from the soil mass. If considering that the detachment of the particles from the total soil mass and its further transportation occurs only at the expense of the living force of surface currents of a slope, then we must introduce a concept of a nonerosive or non-developed erosion zone of the slope (Horton 1948; Mirtskhulava 1970; Makkaveev 1971). However, in real terms, the soil wash down even in the surface currents with little discharge takes place immediately at the watershed crest, as the living force of minor-depth surface currents of a slope is not the only and often main agent in the process of detachment of the soil particles from the whole soil mass and its transportation. Raindrops play a great role in the development of this process. It is proved that the amount of particles detached from the whole soil mass and splashed in the air due to the action of raindrops is quite great (Laws 1940; Hobs and Kezweeny 1967; Mutchler and Hansen 1970; Mutchler and Larson 1971; Siscoe and Levin 1971; Tound and Painter 1974; Green and Houk 1980; Poesen and Savat 1981; Park et al. 1983; Zaslavsky et al. 1981) and is often equal or more the soil mass washed down from the small area (area of overhead irrigation machine or flow ground) due to erosion.

The role of raindrops is not limited to their participation in the first phase of erosion development, associated with the detachment of the particles from the total soil mass, but they improve the transportation ability of the surface current of the slope as they arouse impact waves in them (surface currents) and give additional turbulence to the current (Makkaveev 1971, 1973). It is the raindrops playing the main role in suspending the sediment in the surface slope current. If excluding the impact of raindrops on the slope currents of small depths (up to 1–2 cm), the transportation capability of the currents will diminish by 10–20 times. Besides, the raindrops break ground clods, solidify the soil surface, form a certain kind of plug on it, and smooth the micro-relief strongly reducing the soil infiltration ability. In some cases, the water permeability of the compacted crust of the soil surface may be 200 times less than that of the soil (Mc`Intyre 1958). By considering the properties described above, the rainwater currents have strong eroding properties. If ignoring the energy of raindrops in the erosive processes, the soil wash down will diminish by 10–20 times (Hudson 1974).

The slope currents also have a number of hydraulic peculiarities increasing the eroding ability of the currents. As the depth of the surface slope currents is little, the movement velocity of the unit waves is usually less the current movement velocity, i.e., their regime is mostly turbulent and intense. Consequently, unevenness in the surface currents created by every particle causes the division of the current depending on its value and changes its direction. The impact on the resistance of a turbulent current is concentrated in nature, unlike still particles. Besides, the water level in a turbulent current increases a lot in front of the resistance and sharply falls behind it. This is how the hydrostatic pressure is formed, which coincides with the direct (frontal) pressure and which is developed under the impact of the water current on the soil clod lug. The relative impact of hydrostatic pressure on the detachment of the particles from the soil surface and their transportation increases as the ratio between the transported particle size and the current depth approximates (Makkaveev 1973).

7.5 Selecting the Methods to Forecast Soil Erosion and Evaluate the Hazard of Erosion

None of the abovementioned models considers all peculiarities of erosion by water (rain) mechanism. The hydromechanical model by Prof. Ts. Mirtskhulava (Methodical recommendations… 1978) is based on the mechanism of detachment of soil particles under the impact of current living force, but does not consider the role of the energy of raindrops in the development of erosion. H. Wischmeier and D. Smith equation (Wischmeier and Smith 1965, 1978) shows the impact of rain kinetic energy on erosion, while the impact of other hydraulic parameters of the slope currents is determined through the empirical dependencies. The model of Russian State Hydrological Institute (Instructions for… 1979) is based on the empirical relation between the volumes of liquid and solid flows from the small water catch basins. The first two models consider the conditions determining the erosion intensity more thoroughly. Soil erosion resistance, phases of plant development, and yield of crops, as well as crops growing agricultural techniques and soil cultivation technologies, slope inclination and shape, and impact of the anti-erosion measures on erosion intensity are defined quantitatively in H. Wischmeier and D. Smith equation (Wischmeier and Smith 1965, 1978). The model by Ts. Mirtskhulava (Methodical recommendations… 1978) considers the same factors as H. Wischmeier and D. Smith equation, but the anti-erosion coefficients of agricultural crops are a reduced option of the recalculated coefficients of H. Wischmeier and D. Smith's instruction.

Calculation of the soil loss caused by erosion with the last two models—the models of H. Wischmeier and D. Smith and State Hydrological Institute—is much simpler than it is suggested by Ts. Mirtskhulava's hydromechanical equation. By considering the degree of consistence with the real processes of soil wash down and detailed treatment, Ts. Mirtskhulava's hydromechanical equation and H. Wischmeier and D. Smith's empirical-statistical models must be preferred.

H. Wischmeier and D. Smith's model is mainly based on a climatic parameter. Soil erosion resistance parameter on its turn is closely connected to the climatic parameter. In addition, the impact of agricultural crops and agricultural techniques to grow them and yield on the intensity of erosion and its development are calculated depending on the phases of erosion development. Due to this, H. Wischmeier and D. Smith's model, so-called USLE—Universal Soil Loss Equation, was used to evaluate and forecast the hazard of erosion (Wischmeier and Smith 1965, 1978). Soil loss was calculated to compare the two models. The deposit erosive index for each rain was calculated by using H. Wischmeier and D. Smith's methods, which is directly proportional to the soil erosion value. For this purpose, the environs of Akhaltsikhe Basin were selected. The length of the slope was taken at 150 m and 11% was taken as its gradient. The coefficient of determination between the precipitation erosion index (potential) and washed-down soil mass calculated with a hydromechanical model amounted to 0.911.

A similar relation was established based on the data of 13 weather stations located in the European part of the former Soviet Union: in Baltic countries, Byelorussia, Central black-soil regions, Ukraine, North Caucasus, and Lower Volga Federal District. Based on the correlation between the soil loss calculated with a hydromechanical model and the precipitation erosion index (potential) calculated with H. Wischmeier and D. Smith's model, the determination coefficient amounted to 0.922–0.946 (Zaslavsky et al. 1981).

H. Wischmeier and D. Smith's model, also known as the Universal Soil Loss Equation, is used to calculate the long-term and multiyear average soil loss caused by erosion. Therefore, it is not recommended to use it to calculate the soil loss in a concrete year or after concrete rain. The equation to calculate the soil loss due to erosion following the concrete rain will give the value of average soil loss in case of similar rains. An indicator used to calculate any average value may have significant variance. This regularity is proved by the results of the station experiments held in village Zendidi, Adjara. The factual soil loss caused by erosion was compared to the soil loss calculated by USLE. The variance of the factual values of 9-year-long observations over the soil wash down in different years is quite big and is from 4,23 to 36,86 t/ha. Deviation between the average factual and calculated values for 9 years is up to −2%. The deviation between the value of soil loss calculated by using an average annual erosion index (potential) (amounting to 15.61 t/ha) and the average factual soil loss for 9 years is +12%. It should be noted that when making a quantitative forecast of such a complex, multifactorial process, as soil erosion is, a 12% deviation (error) from the factual values is absolutely acceptable.

The universal equation of soil loss due to erosion is as follows (Wischmeier and Smith 1965, 1978; Salnikov 1965; Larionov 1973; Hudson 1974; Zaslavskii 1979): A = R K L S C P, where

- A is the soil loss, t/ha/year;
- R is the factor of precipitations represented by the units of erosion indices, t/m/ha;
- K is the soil eroding factor, with its value equaling to 22,13 m (72.6 ft) and amount of the soil washed down by erosion from the plot with the inclination of 4.5° (9%) divided by the precipitation erosion index (potential), i.e., the amount of soil washed down from the said plot per unit precipitation erosion index (potential). At the same time, the given plot is black fallow land (a physical surface) all over the year, t/ha;
- L is the slope length factor, with its value equaling to the ratio between the amount of soil washed down from the slope of a given length and the amount of the soil washed down from the slope with the length of 22.3 m in terms of the same slope inclination, a dimensionless value;
- S is the slope inclination factor, with its value equaling to the ratio between the amount of soil washed down from the slope of a given inclination and the amount of soil washed down from the slope with the inclination of 4.5° (9%), a dimensionless value;
- C is the factor of vegetation, crop rotation, agricultural techniques, and soil cultivation system. Its value equals to the ratio between the amount of soil washed down from the system of crop rotation and soil cultivation and the amount of soil where the soil eroding factor is determined, a dimensionless value;
- P is the value considering the impact of anti-erosion measures on the soil wash down. Its value equals to the ratio between the amount of soil washed down from the plot where anti-erosion measures are taken and the amount of soil washed down from the slope where the soil is being cultivated and sowing (planting) takes place in the direction of the slope inclination, a dimensionless value.

A simple mathematical expression of USLE allows calculating the optimal values of some parameters by simple manipulation, which is as follows:

$$L_{opt.} = A_{Acceptable}/R \times K \times S \times C \times P$$
$$S_{opt.} = A_{Acceptable}/R \times K \times L \times C \times P$$
$$C_{opt.} = A_{Acceptable}/R \times K \times S \times L \times P$$
$$P_{opt.} = A_{Acceptable}/R \times K \times S \times L \times C$$

where

A is the admissible soil loss due to erosion, or soil tolerance, T, t/ha/year;

R, K, L, S, C, and P are the parameters of USLE discussed above;

$L_{opt.}$, $S_{opt.}$, $C_{opt.}$, and $P_{opt.}$ are optimal values of USLE in case of average annual soil losses caused by erosion.

7.6 Marginal Admissible Norms of Soil Erosion

Human impact on natural phytocenoses, their substitution with cultural phytocenoses, or excess and intense exploitation of natural cenoses (e.g., pastures and hay meadows) are the reasons for the increased areas of eroded soils, gullied slopes, reduced agricultural areas, areas of the agricultural crops diminished by one-third (30%) on average, etc. The cultivation of millions of hectares of land has resulted in extreme activation of erosive processes, and has put the question of developing anti-erosion measures and using them in the branch of agriculture on the agenda. As the erosive phenomena are virtually impossible to prevent, the anti-erosion measures must reduce the soil loss to a certain limit. How can this limit be calculated? Surely, it must be identified depending on the regularities of the development of natural processes. Erosion occurs in parallel to the soil-formation processes: the genetic horizons, as they form, get washed away because of erosion. The soil-formation process on the background of complex biochemical, physical–chemical, and other processes means the transformation and accumulation of the products of weathering of organic and organic mineral compounds and soil-forming parent materials and resultant formation of soil genetic horizons. It should be noted that the rate of soil formation is greater as a result of transformation of organic mass from the surface than in case of parent materials weathering. It is the soil accumulation horizon subject to the impact of erosive processes in the first instance. Therefore, if the rate of formation of accumulation horizon does not exceed the rate of the soil wash down owing to erosion, the erosion is considered normal. Otherwise, the erosion is accelerated, i.e., the soil formation, or the formed soil amount is less the soil loss due to erosion.

A limited or admissible amount of soil loss due to erosion must consider the social and economic conditions of the society. There are various kinds of anti-erosion measures available: organizational-economic, agrotechnical, forest-and-melioration, and hydrotechnical measures. The practical realization of any anti-erosive agrotechnical measure is associated with great capital investments and long payoff period. When designing and practically realizing anti-erosion measures, the priorities of social–economic development and material standing of the society must be considered. Besides, quite often, a large-scale and thorough use of anti-erosion measures makes the development of certain branches of agriculture on a local area or generally, land cultivation virtually nonprofitable. Therefore, in agriculture, it is necessary to use the set of anti-erosion measures to a limited extent and in a differentiated manner. Consequently, it is based on the introduction of inexpensive and easily realizable anti-erosion measures to life, with organizational-economic and agrotechnical measures as the most important ones.

The definition of the limit soil loss due to erosion accepted in 1950 states that soil tolerance (maximum admissible loss) is the amount of soil loss caused by erosion without any significant diminution of the nutrition elements necessary for the plant and by preserving the existing level of harvest of agricultural crops.

We use anti-erosion measures for eroded soils, but the erosion processes still continue with certain intensity. The wash down of the most fertile soil layer is followed by the loss of not only the nutrition chemical elements of the plants, but other fundamental soil properties having a direct impact on its fertility deteriorate a lot. In addition to losing the nutrition elements owing to erosion, by harvesting over 70% of the harvest (organic mass) from the plot, we lose the amount of nutrition elements needed to form almost the same amount of harvest. Due to the abovementioned, the amount of nutrition elements necessary to maintain the soil fertility and agricultural harvest on non-washed down or eroded grounds must be balanced annually. As a result of development of erosion processes on the eroded soils, the intensity of use of fertilizers increases year after year, as in this instance, there occur the following kinds of losses:

- Soil loss caused by erosion processes meaning the greatly reduced natural soil fertility and reaching great values. In order to prevent the deterioration of soil fertility of this kind, it is necessary to use small, but increasing doses of mineral and organic fertilizers every year.
- Soil fertility loss owing to the removal of the nutrition chemical elements needed for the plants. This kind of loss can be compensated by using organic and mineral fertilizers.
- Evaporation and wash down of the applied mineral fertilizers into the lower soil layers or most importantly, their wash down due to erosion.
- The loss of mineral portion of soil due to erosion leads to the deterioration of its aqueous, physical and aeration properties, and strong resultant inhibition of microbiological and biochemical processes. This, on its turn, has a negative impact on soil fertility. It is difficult to restore or improve the physical properties of the eroded soils.

Following the abovementioned, it can be concluded that quite great amounts of mineral fertilizers and chemical plant-protection means are used on eroded soils, which, in terms of active erosion processes, result in the stressed ecological state over vast areas with possible adequate outcomes. The production harvested from the eroded lands must be economically profitable.

In 1956, the Committee of Joint Conference of scientists of the Department of Agricultural Studies and Department of Soil Protection, aiming at identifying the value of soil tolerance (limit loss), set the following requirements:

- The soil depth must be maintained.
- The quantities of the nutrition chemical elements necessary for the plants must be maintained.
- The control over the floods and sedimentation must be maintained.
- The gully-formation process must be prevented.
- The harvest reduction caused by the loss of 1-inch-thick (2,54 cm) surface soil layer must be prevented.
- The water loss from the fields must be prevented.
- The loss of sowings (resowing and the like) must be prevented.

Table 7.1 Admissible soil loss caused by erosion (T-Tolerance for the soils of different depths)

Soil depth		Permissible soil loss (T-Tolerance)			
		Renewable soil[a]		Nonrenewable soil[b]	
inch	cm	t/acre	t/ha	t/acre	t/ha
0–10	0–25	1.0	2.2	1.0	2.2
10–20	25–50	2.0	4.5	2.0	2.2
20–40	51–100	3.0	6.7	3.0	4.5
40–60	101–150	4.0	9.0	4.0	6.7
>60	>150	5.0	11.2	5.0	11.2

[a]Soils with favorable substrate, which can be renewed by means of treatment, application of mineral or organic fertilizers, and other measures
[b]Soils with unfavorable substrate, such as stones, solid, and very solid parent materials. The renewal of such soils is associated with great material expenses and is not economically practical

Consequently, the Committee resolved that the soil loss caused by erosion must not exceed 5 t/acre (1 acre = 0,405 ha) a year, making 11 t/ha a year. At present, the definition of soil tolerance is mostly the same and means "the annual value of soil erosion allowing reaching a high level of agricultural production economically and without limitations" (Wischmeier and Smith 1978).

The indicators of soil loss caused by erosion and soil tolerance are more elaborated as compared to the previous option (Table 7.1). With them, the soils are divided into renewable and nonrenewable soils. The renewable soils are developed on soft soil forming or strongly transformed, weathered parent materials (e.g., moraine sediments, clays, intensely weathered sandstones, aeolian soils, loess-like parent materials, etc.). Nonrenewable or practically nonrenewable are the soils developed on solid parent materials. Soil tolerance, both of the renewable and the nonrenewable ones, varies depending on the depth of soil, from 2.2 to 11.2 t/ha a year.

7.7 Evaluation of the Soil Erosion Hazard to Corroborate the Content of Anti-Erosion Measures

Within the scope of the Plan of Melioration and Water Economy Development before 2000, the soil erosion–deflation corroboration was developed for the Melioration Fund of Georgia. The water (rain) erosion within the scope of this corroboration was done by using the Universal Soil Loss Equation (USLE), while the hazard of deflation intensity was forecasted by using the methods developed by the Soil Erosion and Bed Processes Problems Laboratory of Lomonosov Moscow State University (Methodological guidelines… 1982). The forecast of irrigation erosion was done by the methods of the same Problems Laboratory by the specialists of project institute "Saktskalproekti". This job was done with scale 1,500,000 and a map of soil erosion and deflation hazards of the melioration Fund of Georgia was compiled.

At the first stage, we did the geomorphological zoning of Georgia; at the second stage, a map of precipitation erosion potential was compiled; at the third stage, the factor of erosion of individual soil types and subtypes was calculated on the soil map based on many of our own and literary materials; and at the fourth stage, the factor of peculiarities of agricultural plants and techniques, as well as soil cultivation factor, was calculated in different administrative regions of Georgia. By matching the abovementioned factual maps, the separate elementary erosive areas were identified. With the first approximation, the acceptable soil loss was fixed at the value from 2.5 t/ha a year. In respect of erosion, the soil loss not exceeding the admissible loss under the existing soil use model is considered nonhazardous. In respect of erosion, the soils with the loss of 2.5–5.0 t/ha a year are considered slightly hazardous; the soils with the loss of 5–10 t/ha a year are considered averagely hazardous; and the soils with the loss exceeding 10 t/ha a year are considered extremely hazardous. The intensity of the development of erosive processes is determined by considering the areas occupied by eroded soils at different degrees. As the data of project institute "Saktskalproekti" suggest, 224 thousand ha of arable lands and 569 thousand ha of pastures and hay meadows are eroded. 95 thousand ha out of 162 thousand ha of eroded melioration lands is slightly eroded, 51 thousand ha is averagely eroded, and 16 thousand ha is strongly eroded. The eroded areas of melioration lands in East Georgia are 3 times more than those in West Georgia, and slightly eroded areas are 5 times more and the averagely eroded areas are twice as more. Particularly large areas are eroded on the territory of Samachablo, exceeding the eroded areas in Abkhazeti by almost 2.5 times and by 5 times the eroded areas in Adjara. 76 thousand ha of melioration land in Georgia is damaged due to the deflation processes, with 30 thousand ha being slightly deflated, 22 thousand ha being averagely deflated, and 24 thousand ha being strongly deflated.

The studies evidence that 1004 thousand ha of 1134 thousand ha of melioration fund of Georgia is hazardous in respect to the development of erosive processes, with 289 thousand ha being slightly erosive hazardous, 50 thousand ha being averagely erosive hazardous, while 665 thousand ha is extremely erosive hazardous. Of them, 12 thousand ha of irrigation lands are hazardous in respect to erosion.

The elementary areas marked on the map (Fig. 7.2) are divided into four categories: with less 5% of the I-category lands; 5–20% of the II-category lands; 20–40% of the III-category lands; and over 80% of the IV-category lands needing anti-erosion measures. By considering the mentioned categories, 10 thousand ha of 103 thousand ha of melioration lands in Abkhazeti is slightly erosive, 20 thousand ha of land is averagely erosive, and 73 thousand ha is strongly erosive. 41 thousand ha of 50 thousand ha of melioration fund of Adjara (82%) is prone to erosion, with 5 thousand ha (13%) being slightly erosive hazardous, 10 thousand ha (24%) being averagely erosive hazardous, and 26 thousand ha (63%) being extremely erosive hazardous. 40 thousand ha of 64 thousand ha of the melioration fund of Samachablo (62.5%) is prone to erosion, with 13 thousand ha (32%) being slightly erosive hazardous, 8 thousand ha (20%) being averagely erosive hazardous, and 19 thousand ha (48%) being extremely erosive hazardous.

304 thousand ha of the melioration fund of Georgia (19%) is slightly erosive hazardous, 80 thousand ha (5%) is averagely erosive hazardous, and 1194 thousand ha (76%) is strongly erosive hazardous. 12 thousand ha of irrigation lands are also prone to erosion (Gogichaishvili 2007).

510 thousand ha of the melioration lands of Georgia is hazardous in respect of deflation development, with 220 thousand ha (43%) being slightly erosive hazardous, 224 thousand ha (44%) being averagely erosive hazardous, and 66 thousand ha (13%) being strongly deflation hazardous. As deflation hazard is determined by using deflation indexes, the evaluations given above are qualitative and are approximate.

The erosion hazard of the melioration land fund in Georgia is based on the quantitative model of erosion prognosis. 128 thousand ha of 137 thousand ha of potentially hazardous melioration fund of East Georgia (i.e., 93%) is slightly erosive hazardous and 9 thousand ha (7%) is averagely erosive hazardous. In West Georgia, 47 thousand ha of the melioration fund is erosive hazardous to different degrees, including 39 thousand ha (83%) slightly erosive hazardous, and 8 thousand ha (17%) averagely erosive hazardous. It should be noted that no strongly erosive-hazardous areas of the melioration fund of Georgia are identified by using these methods (Gogichaishvili 2016).

When evaluating the erosion hazard of the melioration land fund of Georgia, the potential land loss was mainly associated with high values of precipitation erosion indices and relief factors. In addition, factor C of erosion hazard, vegetation, and its growing (agricultural techniques), which were calculated based on the phases of the agricultural plants development, precipitation erosion index, and soil cultivation system, changed from 0.40 to 0.65 across the municipalities. The calculations of prognostic soil loss due to erosion did not consider the impact of anti-erosion measures on the erosive processes. When compiling the map of the given scale (1,500,000), it is impossible to identify the effects of all anti-erosion measures on some or other plots. The soil cultivation system and crops growing techniques imply the need for and the possibility of taking anti-erosion measures. The analysis of the considered map evidences that the plant itself, the agricultural techniques to grow it, and the available soil cultivation system in Georgia fail to duly protect the soil against erosion. The question as to how much it is possible to reduce the soil loss caused by erosion to the desirable minimum by selecting and using the relevant anti-erosion measures and to maintain the soil fertility at the acceptable level is less studied. It is known that particularly complex conditions in view of erosion are formed when growing annual row crops over the slopes. They are the major food, forage, and technical crops of the country, occupying large areas of the arable land (maize, potato, sunflower, tobacco, watermelons, and vegetables). The issue was to minimize the soil loss caused by erosion by selecting the relevant anti-erosion measures without reducing the areas of the plots of field.

79% of the arable land on the hills bordering Colchic Lowland are located on the slopes with Relief Erosion Index (REI) of over 4.0 units. If taking the value of the annual admissible erosion index of 3–3.5 tons, it will become clear that the areas occupied by maize (Table 7.2) need additional anti-erosion measures, such as furrowing or mulching, as even such anti-erosion measure as making buffer zones, fails to reduce the erosive processes to the acceptable level. The situation is similar to growing the tobacco over the slopes. Planting tobacco across the slope as compared to inclined planting (at an angle) reduces the soil loss caused by erosion, but in the final run, even making the rows or buffer zones does not yield the desirable result. It should be noted that making the buffer zones of grass, leguminous grass particularly seems a much-needed measure in the mountainous areas, as it helps protect the soil against erosion on the one

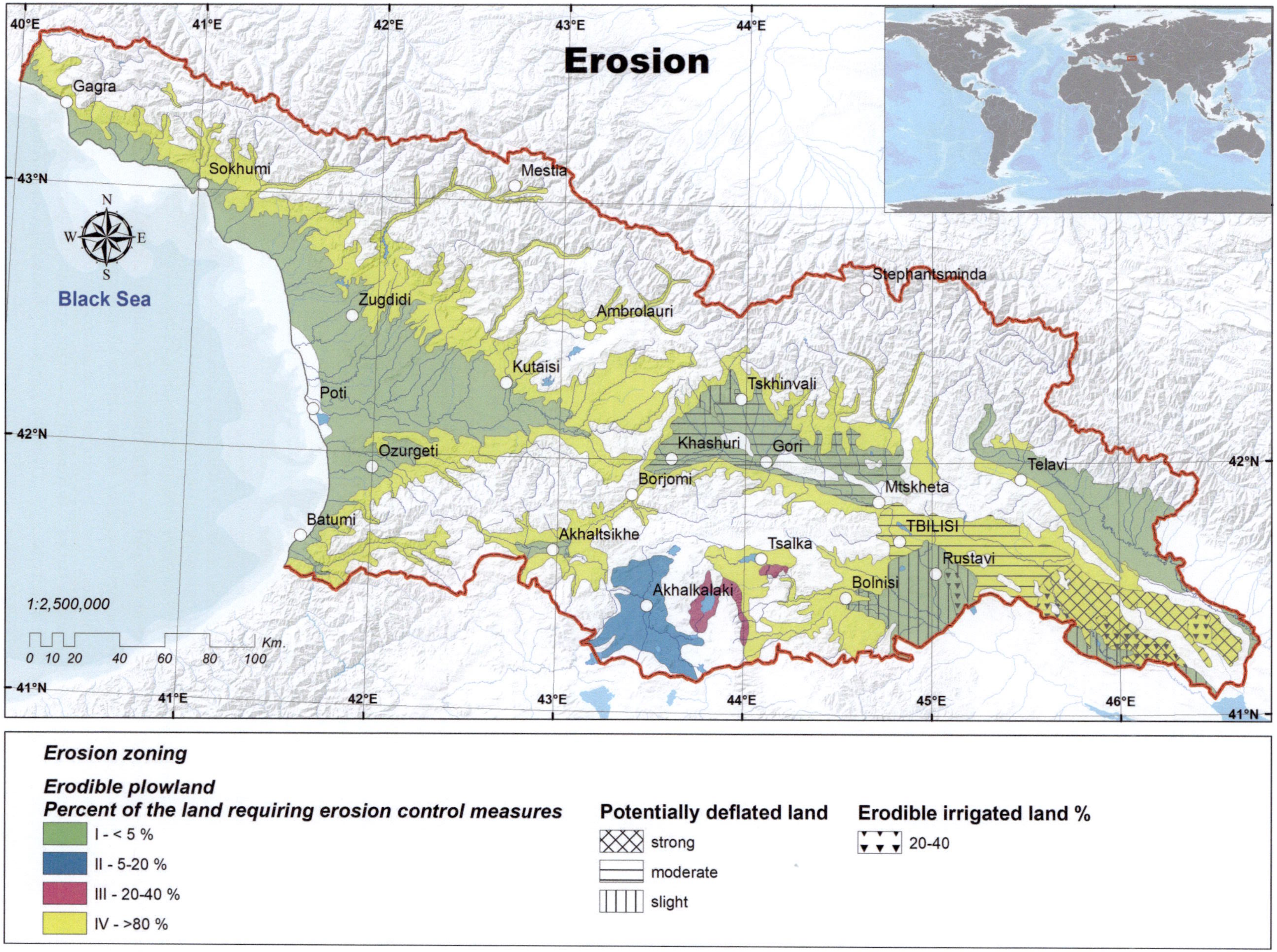

Fig. 7.2 Erosion zoning. This map is created by D. Svanadze, based on the data of G. Gogichaishvili

Table 7.2 Potential loss of soil by erosion on arable lands with a set of anti-erosion (erosion control) measures on terraced hills bordering the Colchic Lowland (Vicinities of Sukhumi)

Set of erosion control		R × K = 53.54 × 1.4 = 54.94									
		P = 0.75–1.0	P = 1.75–2.0	P = 2.5–3.0	= 4.0–4.5	P = 5.0–5.5	P = 5.5–6.0	P = 6.5–7.0	P = 7.5–8.0	P = 8.5–9.0	P = 9.5–10.0
		41.21	96.15	137.35	219.76	274.70	302.17	357.17	412.05	466.99	521.93
		54.94	109.88	164.82	247.23	302.17	39.64	384.58	439.52	494.46	549.40
Corn	Contour processing of slopes	12.98	30.29	43.27	69.22	86.53	95.18	112.51	129.80	147.10	164.41
		17.31	34.61	51.92	77.88	95.18	103.84	121.14	138.45	155.76	173.06
	Furrow	5.19	12.12	17.31	27.69	34.61	38.07	45.00	51.92	58.84	65.76
		6.92	13.84	20.77	31.15	38.07	41.53	48.46	55.38	62.30	69.22
	Buffer strips	1.04	2.42	3.46	5.54	6.92	7.61	9.00	10.38	11.77	13.15
		1.38	2.77	4.15	6.23	7.61	8.31	9.69	11.08	12.46	13.84
Tobacco	Contour processing of slopes	8.37	19.52	27.88	44.61	55.76	61.34	72.51	83.65	94.80	105.95
		11.15	22.31	33.46	50.19	61.34	66.92	78.07	89.22	100.38	111.53
	Furrow	3.35	7.81	11.15	17.84	22.30	24.54	29.00	33.46	37.92	42.38
		4.46	8.92	13.38	20.08	24.54	26.77	31.23	35.69	40.15	44.61
	Buffer strips	0.67	1.56	2.23	3.57	4.46	4.91	5.80	6.69	7.58	8.48
		0.89	1.78	2.68	4.02	4.91	5.35	6.25	7.14	8.03	8.92
	Angle planting	10.10	23.56	33.65	53.84	67.30	74.03	87.51	100.95	114.41	127.87
		13.08	26.92	40.38	60.57	74.03	80.76	94.22	107.68	121.14	134.60
	Buffer strips	0.81	1.88	2.69	4.31	5.38	5.92	7.00	8.08	9.15	10.23
		1.05	2.15	3.23	4.85	5.92	6.46	7.54	8.62	9.69	10.77

hand and restore the soil fertility on the other hand. However, as an alternative measure, it is possible to cultivate the area with row or broadcast crops growing in rows. The soil protection coefficients (properties) of broadcast crops are twice as high as those of row crops. The situation with the terraced hills bordering Colchic Lowland in Imereti, in the environs of Tskaltubo is the same.

Despite quite a high value of the precipitation erosion index, 65% of the arable lands in the Okriba Basin (Tkibuli, R = 50.10) are located where the relief erosion index is less than 3.0. In such a situation, making the rows in maize plots and buffer zones is the means to control the intensity of erosion development at the admissible level. Over the slopes with the REI of 3.0–4.0 or more, the prognostic loss of soil as a result of erosive processes in case of taking the above-listed measures will increase to 8 t/ha. In such a case, it is admissible to use any additional measure of the set of anti-erosion measures or substitute the new measures for the old ones. Soil furrowing reduces the risk of development of erosion ten times as much as fallowing. Water seeps down the ditches made on the soil surface impeding or significantly reducing the surface flow. As a result, the soil loss caused by erosion diminishes. This measure is beneficial to take in a mechanized manner, but its use is limited as the slope inclination increases. In such a case, another hampering factor is the soil depth. This measure yields maximum effect for soils of a sufficient depth, soils developed over strongly weathered parent materials, aeolian soil or loess-like parent materials. Soil mulching with organic substances is another favorable soil-protecting means reducing the risk of development of erosion by 6 or 7 times. However, with mulching, soil cultivation becomes difficult or even impossible. A major portion of arable lands in Racha Basin is found over the slopes with high REI values, but the low precipitation erosion index greatly reduces the risk of erosion (Table 7.3). Despite this, contour cultivation or row making in maize plots fail to reduce the risk of development of erosive processes to the admissible level, making it necessary to make sown grass buffer zones at the expense of reduced sown areas in order to protect them against erosion or making broadcast crop rows of different widths over the slopes as necessary. A quite rare vine species grows as a traditional crop in Racha-Lechkhumi Region. Contour cultivation of slopes fails to reduce the risk of erosion to the admissible level. Rather, sowing grass, mulching, or making terraces is needed in the spaces between the vine rows.

The situation is similar in Zemo Imereti, but unlike Racha, the buffer zone in all maize plots reduces the soil loss caused by erosion to the admissible level (Table 7.4). The same effect is reached by sowing grass in the spaces between the rows.

Over the piedmonts bordering Shida Kartli, where there are 53% of arable lands found, with REI of over 4.0, in addition to the contour cultivation of maize slopes and row

Table 7.3 Potential loss of soil by erosion on arable lands and vineyards with a set of anti-erosion (erosion control) measures in Racha (Vicinities of Oni)

Set of erosion control		R × K = 13.64 × 1.4 = 19.10							
		P = 1.25–1.5	P = 1.75–2.0	P = 1.25–3.0	P = 3.5–4.0	P = 4.0–5.0	P = 5.5–6.0	P = 6.5–7.0	P = 8.0–8.5
		23.88	33.43	47.75	66.85	76.40	105.05	124.15	152.80
		28.65	38.20	57.30	76.40	95.50	114.60	133.70	162.35
Corn	Contour processing of slopes	7.52	10.53	15.04	21.06	24.07	33.09	39.11	48.13
		9.02	12.03	18.05	24.07	30.08	36.10	42.12	51.14
	Furrow	3.01	4.21	6.02	8.42	9.69	13.24	15.64	19.25
		3.61	4.81	7.22	9.63	12.03	14.44	16.85	20.46
	Buffer strips	0.60	0.84	1.20	1.68	1.93	2.65	3.13	3.85
		0.72	0.96	1.44	1.93	2.41	2.89	3.37	4.09
Vineyard	Contour processing of slopes	5.68	7.96	11.36	15.91	18.18	25.00	29.55	36.37
		6.82	9.09	13.64	18.18	22.73	27.27	31.81	38.64
	Sowing annual grasses between rows	0.57	0.80	1.14	1.59	1.82	2.50	2.96	3.64
		0.68	0.91	1.36	1.82	2.27	2.73	3.18	3.86

Table 7.4 Potential loss of soil by erosion on arable lands and vineyards with a set of anti-erosion (erosion control) measures in upper Imereti (Vicinities of Sachkhere)

Set of erosion control		R × K = 13.07 × 1.4 = 18.30							
		P = 1.75–2.0	P = 2.5–3.0	P = 3.5–4.0	P = 5.0–5.5	P = 6.0–6.5	P = 7.0–7.5	P = 8.0–8.5	P = 9.0–9.5
		32.03	45.75	64.05	91.50	109.80	127.10	146.40	134.70
		36.60	54.90	73.20	100.65	118.95	137.25	155.55	173.85
Corn	Contour processing of slopes	10.09	14.41	20.18	28.82	34.59	40.35	46.12	51.88
		11.53	17.29	23.06	31.70	37.47	43.23	49.00	54.76
	Furrow	4.04	5.76	8.07	11.53	13.83	16.14	18.45	20.75
		4.61	6.92	9.22	12.68	14.99	17.29	19.60	21.91
	Buffer strips	0.81	1.15	1.61	2.31	2.77	3.23	3.69	4.15
		0.92	1.38	1.84	2.54	3.00	3.46	3.92	4.38
Vineyard	Contour processing of slopes	7.62	10.89	15.24	21.78	26.13	30.49	34.84	39.20
		8.71	13.07	17.42	23.95	28.30	32.67	37.02	41.38
	Sowing annual grasses between rows	0.76	1.09	1.52	2.18	2.61	3.05	3.48	3.92
		0.87	1.31	1.74	2.40	2.83	3.27	3.70	4.14

Table 7.5 Potential loss of soil by erosion on arable lands with a set of anti-erosion (erosion control) measures on foothills bordering Inner Kartli (Vicinities of Akhalgori)

Set of erosion control		R × K = 11.95 × 1.4 = 16.73									
		P = 2.0–2.5	P = 2.5–3.0	P = 3.0–3.5	P = 3.5–4.0	P = 4.5–5.0	P = 5.5–6.0	P = 6.0–6.5	= 7.0–7.5	P = 8.0–8.5	P ≥ 10(12)
		33.46	41.83	50.19	58.56	75.29	92.02	100.38	117.11	133.84	200.70
		41.83	50.19	58.56	66.92	83.65	100.38	108.75	125.48	142.21	
Wheat	Contour processing of slopes	5.86	7.32	8.78	10.25	13.18	16.10	17.57	20.49	23.42	56.20
		7.32	8.78	10.25	11.71	14.64	17.57	19.03	21.96	24.89	
	Furrow	2.34	2.93	3.51	4.10	5.27	6.44	7.03	8.20	9.37	22.48
		2.93	3.51	4.10	4.68	5.86	7.03	7.61	8.78	9.96	
	Buffer strips	0.47	0.59	0.70	0.82	1.05	1.29	1.41	1.64	1.87	4.50
		0.59	0.70	0.82	0.94	1.17	1.41	1.52	1.76	1.99	
Corn	Contour processing of slopes	10.54	13.18	15.81	16.40	23.71	28.99	31.63	36.89	42.16	63.22
		13.18	15.81	16.40	21.08	26.35	31.62	34.26	39.53	44.80	
	Furrow	4.22	5.27	6.32	6.56	9.49	11.59	12.65	14.76	16.86	25.29
		5.27	6.32	6.56	8.43	10.54	12.65	13.70	15.81	17.92	
	Buffer strips	0.84	1.05	1.26	1.31	1.90	2.32	2.53	2.95	3.37	5.06
		1.05	1.26	1.31	1.69	2.11	2.53	2.74	3.16	3.58	

Table 7.6 Potential loss of soil by erosion on arable lands with a set of anti-erosion (erosion control) measures on foothills bordering Lower Kartli (Vicinities of Tetritskaro)

Set of erosion control		R × K = 12.36 × 1.4 = 17.23									
		P = 1.75–2.0	P = 2.0–2.5	P = 2.5–3.0	P = 3.0–3.5	P = 4.0–4.5	P = 5.5–6.0	P = 6.5–7.0	P = 8.0–8.5	P = 8.5–9.0	P ≥ 10(12)
		30.15	34.46	43.08	51.69	68.92	94.77	111.99	137.84	146.45	206.76
		34.46	43.08	51.69	60.31	77.54	103.38	120.61	146.45	155.07	
Wheat	Contour processing of slopes	5.36	5.03	7.54	9.05	12.06	16.59	19.60	24.12	25.63	36.18
		6.03	7.54	9.05	10.55	13.57	18.09	21.11	25.63	27.14	
	Furrow	2.14	2.41	3.02	3.62	4.82	6.63	7.84	9.65	10.25	14.47
		2.41	3.02	3.62	4.22	5.43	7.24	8.44	10.25	10.85	
	Buffer strips	0.43	0.48	0.60	0.72	0.96	1.33	1.57	1.93	2.05	2.90
		0.48	0.60	0.72	0.84	1.09	1.45	1.69	2.05	2.17	
Corn	Contour processing of slopes	9.50	10.86	13.57	16.28	21.71	29.85	35.28	43.42	46.13	65.13
		10.86	13.57	16.28	19.00	24.43	32.56	37.99	46.13	48.85	
	Furrow	3.80	4.34	5.43	6.51	8.68	11.94	14.11	17.37	18.45	26.05
		4.34	5.43	6.51	7.60	9.77	13.03	15.20	18.45	19.54	
	Buffer strips	0.76	0.87	1.09	1.30	1.74	2.39	2.82	3.47	3.69	5.21
		0.87	1.09	1.30	1.52	1.95	2.61	3.04	3.69	3.91	

making, it is necessary to take some kind of additional anti-erosion measure sufficient to reduce the soil loss caused by erosion to the admissible level (Table 7.5). As mentioned above, wheat, as a broadcast crop, protects the soil against erosion much better than row crops (maize, tobacco, sunflower, and potato). Therefore, wheat, in case of contour cultivation and row making, over the slopes with REI of 3.5–4.0, ensures soil protection with the admissible level of erosion (3.5 t/ha/year). Over the slopes with higher values of REI, the buffer zones or furrowing is a necessary measure. The situation is similar with the piedmonts bordering Kvemo Kartli. However, the hazard of erosion as compared to the arable lands over the piedmonts bordering Shida Kartli is less (Table 7.6). Here too, like in Shida Kartli, in addition to the contour cultivation and row making of slopes, additional anti-erosion measures are necessary.

Tianeti Basin is characterized by quite high values of REI. Here, 40% of arable plots are located over the slopes with 1.25 REI. The maximum value of REI with the arable land is 5.5. Despite the low value of REI, the contour cultivation or row making the slopes with wheat and maize crops fails to reduce the soil loss caused by erosion to the admissible level. In the arable plots, furrowing and making buffer zones or some other anti-erosion measure is necessary. In Ertso Basin, despite the value of REI lower than in Tianeti Basin, the value of REI is still twice as much, and the probability of erosion development increases consequently. Potential soil loss, depending on the degrees of REI, from the fallow land (as calculated for the physical surface) varies from 16 to 255 t/ha/year. 70% of the arable lands are located over the slopes with REI of over 3.0. The soil loss caused by erosion in wheat sowings, which are located over the slopes with REI of up to 3.0, corresponds to the admissible erosion level. The maize fields will meet the same level of erosion in case of contour cultivation and row making provided they are located over the slopes with the value of REI up to 1.25. In all other cases, additional anti-erosion measures are necessary.

75% of arable lands over the piedmonts of Gombori Ridge are located over the slopes with REI of 3.0. The potential soil loss caused by erosion varies from 9 to 350 t/ha/year. The highly erosive state is the result of the location over the slopes with relatively high erosion index of the arable lands and relatively high precipitation erosion index. Contour cultivation of arable plots and row making of wheat plots yield an anti-erosive effect only for the slopes with REI of up to 1.5 and for the slopes with up to 0.5 REI for maize fields. In all other cases, additional measures are necessary to reduce erosion to the admissible level. Neither contour growing nor cultivation of vine can protect the slopes against erosion unless there is a turf formed in the spaces between the rows.

75% of the arable lands over the denudation plain of Iori Plateau are located over the slopes with REI of 2.5–6.0, while other 25% are located over the slopes with REI of 1.0–2.0. Due to relatively high precipitation erosion index, despite relatively low values of REI, the potential soil loss for pure fallow land varies from 39 to 234 t/ha. Like at any other location, the broadcast crops (wheat, oats, and barley) and row crops (maize), in terms of contour cultivation and row making of the slope, fail to protect the soil against erosion, and additional anti-erosion measures are needed. Contour vine growing, cultivation, and mulching in the spaces between the rows fail to protect the slopes with REI of 2.5–3.0. Over the slopes with REI more than 3.0, it is necessary to sow the grass in the spaces between the rows and to retain a part of the green mown mass as mulch.

Akhaltsikhe Basin is not distinguished for abundant precipitations. The precipitation erosion potential is quite

Table 7.7 Potential loss of soil by erosion on arable lands with a set of anti-erosion (erosion control) measures in Akhaltsikhe hollow (Vicinities of Akhaltsikhe)

Set of erosion control		R × K = 12.36 × 1.4 = 17.23									
		P = 0.5–0.75	P = 0.75–1.0	P = 1.0–1.25	P = 1.5–1.75	P = 1.75–2.0	P = 2.5–3.0	P = 3.5–4.0	P = 4.5–5.0	P = 5.5–6.0	P = 6.5–7.0
		5.74	8.60	11.47	17.21	20.07	28.68	40.15	51.62	63.09	74.56
		8.60	11.47	14.34	20.07	22.94	34.41	45.88	57.35	68.82	80.29
Wheat	Contour processing of slopes	1.00	1.50	2.01	3.01	3.51	5.02	7.03	9.03	11.04	13.05
		1.50	2.01	2.51	3.51	4.01	6.02	8.03	10.03	12.04	14.05
	Furrow	–	–	–	–	–	2.01	2.81	3.61	4.42	5.22
		–	–	–	–	–	2.41	3.21	4.01	4.82	5.62
Corn	Contour processing of slopes	1.81	2.71	3.61	5.42	6.32	9.03	12.65	16.26	19.87	23.49
		2.71	3.61	4.52	6.32	7.23	10.84	14.45	18.07	21.68	25.29
	Furrow	–	–	–	2.17	2.53	3.61	5.06	6.50	7.95	9.40
		–	–	–	2.53	2.89	4.34	5.76	7.23	8.67	10.12
	Buffer strips	–	–	–	–	–	–	1.01	1.30	1.59	1.88
		–	–	–	–	–	–	1.15	1.44	1.73	2.02
Potato	Contour processing of slopes	2.16	3.25	4.34	6.50	7.59	10.84	15.18	19.51	23.85	28.18
		3.25	4.34	5.42	7.59	8.67	13.01	17.34	21.68	26.01	30.35
	Furrow	0.87	1.30	1.73	2.60	3.03	4.34	6.07	7.80	9.54	11.27
		1.30	1.73	2.17	3.03	3.47	5.20	6.94	8.67	10.41	12.14
	Buffer strips	–	–	–	–	–	0.87	1.21	1.56	1.91	2.26
		–	–	–	–	–	1.04	1.39	1.73	2.08	2.43

low here and makes 7.17. This leads to the relatively low erosion hazard. Maximum potential soil loss caused by erosion is 80 t/ha (Table 7.7). 50% of the arable lands are located over the slopes with REI of up to 3.0, while other 50% are located over the slopes with REI of 3.0–7.0. In terms of contour cultivation and row making of the areas occupied by row crops, the soil loss caused by erosion when REI is 2.0 is less the admissible level of erosion. In other cases, various anti-erosion measures are necessary. For wheat sowings, in terms of contour cultivation and row making, the washed down soil mass will be less than the admissible level of erosion would cause provided they are located over the slopes with REI of 4.5. With the slopes of higher values of REI, additional anti-erosion measures are necessary.

75% of the arable lands in the piedmonts of South Georgia are located over the slopes with REI of 3.0–6.5. Despite quite high values of REI, this area has the lowest precipitation erosion index in Georgia amounting to 2.60 resulting in a very low erosion hazard. The maximum value of the potential soil loss for a pure fallow land is approximately 17 tons per hectare a year. Therefore, growing potato in this area in terms of contour cultivation and row making of the slope and growing wheat only with contour cultivation does not pose any hazard of erosion. Following the above-mentioned, one can conclude that the soil loss caused by erosive phenomena can be brought to minimum when growing any of the agricultural crops in Georgia. This goal can be reached mainly by optimal planning of the slope cultivation works and taking anti-erosion measures based on the quantitative forecast of erosion. Taking agrotechnical measures is much easier and is not associated with huge capital investments.

7.8 Evaluation of the Hazard of Erosion in Mountains

Relatively thorough works were undertaken to evaluate the erosion hazard of the lands in Zemo Svaneti and Adjara. In Zemo Svaneti, these works were accomplished with scale 1,25,000, and a 1,50,000-scaled map was compiled finally. A 1,25,000-scaled map of Zemo Svaneti (Mestia) showed 101 arable plots and 27 orchards. The arable plots include the cultivated homestead lands as well, making 2921.5 ha. The plots have thin contours which are typical to the mountainous regions. The number of arable plots with the length of less than 50 m is 36 with the total area of 902 ha making 31% of the total arable lands. The total number of the plots with the length of 51–100 m is 32 with the area of 939.5 ha (32%). The total number of the plots with the length of up to 100 m is 67%, while their area is 63%. The total number of the plots with the length of 101–150 m is 16 with the total area of 480 ha (16%). The total number of the plots with the length of 151–200 m is 8 with the total area of 260 ha (9%) and the total number of the plots with the length of over 200 m is 9 with the total area of 428.5 ha (12%).

Out of the total number of plots in Zemo Svaneti, only one plot with the area of 64 ha is located over a slope with the inclination of 0–2°; 9 plots with the total area of 302 ha

(10% of the total arable area) are on the slope with the inclination of 2–5°. As the slope inclination increases, the area of the arable land on it increases as well. The slopes with the inclination of 5–10° occupy the area of 934.5 ha making 32% of the total arable lands (18 plots); the arable lands with the inclination of 10–15° occupy the area of 786.0 ha (27%) constituting 30 plots. 7 plots with the area of 278.5 ha are located on very steep slopes (over 25°). These plots occupy 9% of the arable plots in Zemo Svaneti. Arable plots with the area of 547.5 ha are located on the slopes with the inclination of 15–25°, making 19% of the total arable areas and covering 7 plots. Over the slopes with the inclination of over 15° in Zemo Svaneti, there are 42 arable plots located with the area of 826 ha making 27% of the total arable land.

The arable plots in Zemo Svaneti are mostly extremely erosive hazardous (85% of them), 14% of them are slightly prone to erosion and 1% is averagely prone to erosion. Only 35 ha area in Zemo Svaneti is averagely prone to erosion. 6% of the orchards are not hazardous in respect to the development of erosion processes, 21% are slightly prone to erosion, 5% are prone to erosion, and 68% are extremely prone to erosion. Out of the arable plots in Mestia making 946 ha, 89% are eroded to different degrees what is the practical realization of the high erosion hazard of the lands in Zemo Svaneti (Gogichaishvili 2012).

References

Archer SG (1956) Soil conservation. University of Oklahoma Press, Norman

Ayres QC (1936) Soil erosion and its control. McGraw-Hill, New York, USA, p 365

Bennett HH, Chaplin WR (1928) Soil erosion—a national menace, U.S. Department of Agriculture, Circular 33

Davis WM (1898) Physical geography, Boston

Davis WM (1902) Elementary physical geography, Boston

Flanagan DC, Laflen Elliot WJ, Meyer CR, Nearing MA (1997) WEPP —predicting erosion by water using a process-based model. J Soil Water Conserv 52(2):96–102

Fryrear DW, Ali Saleh JD, Bilbro HM, Schomberg H, Stout JE, Zobeck TM (1998) RWEQ: improved erosion by wind technology. J Soil Water Conserv 55:183–189

Gogichaishvili G (2007) Quantitative assessment of soil erodibility on the ameliorated lands in Georgia. Eurasian Soil Sci 40(11): 1227–1232

Gogichaishvili G (2012) Erodibility of Arable soils in Georgia during the period of storm runoff. Eurasian Soil Sci 45(2):197–201

Gogichaishvili G (2016) Soil erosion in river basins of Georgia. Eurasian Soil Sci 49(6):696–704

Green T, Houk D (1980) The resuspension of underwater sediment by rain. Sedimentology 27:607–610

Hobs PV, Kezweeny AJ (1967) Splashing of a water drop. Science 155:1112–1114

Horton R (1948) Erosional development of rivers and catchments. Publishing House, M: Geografgiz, p 158 (Khorton R. Erozionnoe razvitie rek i vodosbornikh basseinov. 1948, M: Geografgiz)

Hudson NW (1971) Soil conservation. Batsford, London

Hudson N (1974) Soil protection and erosion control. Publishing House "Misl", Moscow (Okhrana pochvi i borba s eroziei), p 400

Instructions for the determination of computed hydrological characteristics in the design of anti-erosion measures in the European Territory of the USSR (1979) Leningrad, Publishing House Gidrometizdat (Instruktsia po opredeleniu raschotnikh gidrologicheskikh kharakteristik pri proektirovanii protivoerozionnikh meropriatii na Evropeiskoi Territorii SSSR), p 60

Kohnke H, Bertrand A (1959) Soil conservation. McGraw-Hill, New York-Toronto-London

Kozmenko A (1909) Failure, landslide and erosion formations of the North-Eastern part of the Novosilsky district of the Tula province. Geography Book 3–4

Kozmenko AC (1928) Struggle against ravines and moistening of fields in the Tula province (Borba s ovragami i uvlazhnenie poley v Tulskoy gubernii)

Larionov GA (1973) Foreign experience in assessing the potential danger of erosion. In collection, Assessment and mapping of erosion-hazardous and deflationary dangerous lands (Zarubejnii opit otsenki potentsialnoi opasnosti erozii. V sb., Otsenka I kartirovanie erozionnoopasnikh i defliatsion-noopasnikh zemel. M., Izd. MGU). Publishing House of Moscow State University, Moscow, pp 38–46

Laws JO (1940) Recent studies in raindrops and erosion. Agric Eng 21:431–433

Lazarevich R (1973) Erosion in the SFD of Yugoslavia (in Serbian), Zbornik radova 12, 105–123, Institut za šumarstvo I industriju, Beogtad

Lopatin GV (1952) River sediments of USSR, Geograpgizdat (Nanosi rek SSSR, Geografizdat)

Makkaveev NI (1953) River bed and erosion in its basin. AS of USSR (Ruslo reki i erozia v eyo basseiyne, AN SSSR)

Makkaveev NI (1971) Stock and channel processes. Texts of lectures for geomorphologists. Publishing House MSU, Moscow, p 115 (Stok b ruslovie protsessi. Teksti lektsii dlz geomorfologov. M., Izd-vo MGU)

Makkaveev NI (1973) Hydraulic typification of erosion process. In the book: soil erosion and channel processes, Issue 3. Publishing House Moscow State University, Moscow, pp 65–77 (Gidravlicheskaia tipizatsia erozionnogo protsesa. 1973, V kn., Erozia pochv i ruslovie protsesi. Vip. 3, M., Izd. MGU)

Mc`Intyre DS (1958) Permeability measurements of soil crusts formed by raindrop impact. Soil Sci 85(4):185–189

Methodical recommendations on the forecast of water (rain) erosion (1978) Publishing House VASKHNIL, GruzNIIGIM, (Metodicheskie rekomendatsii po prognozu vodnoi (dozdevoi) erozii), Moscow, p 60

Methodological guidelines for the definition of land reclamation fund. Section: Erosion-deflationary justification (1982) Ministry of Land Reclamation and Water Economy of the USSR, "Soyuz-vodproekt", Specialized Department for Research and Research Institute "Soyuzg-iprovodkhoz". Scheme of the development of land reclamation and water management of the USSR for the period up to 2000 year. (Metodicheskoe ukazania po opredeleniu meliorativnogo fonda. Razdel: Erozionno-deflatsionnoe obosnovanie. 1982, Ministerstvo melioratsii i vodnogo khoziaistva SSSR, "Soiuz-vodproekt", Specializirovannoe otdelenie po iziskaniam i issledovaniam instituta "Soiuz-giprovodkhoz". Skhema razvitia melioratsii i vodnogo khoziaistva SSSR na period do 2000 goda)

Mirtskhulava TSE (1970) Engineering methods for calculating and predicting erosion by water. Publishing House Kolos, Moscow, p 240 (Injenernie metodi raschota i prognoza vodnoi erozii. 1970, M., Izd. Kolos)

Mutchler CK, Hansen LM (1970) Splashing of a water drop at terminal velocity. Science 167:1311–1312

Mutchler CK, Larson CL (1971) Splash amounts from water drop impact on a smooth surface. Water Res 7:195–200

Park SW, Mutchler JK, Bubenzer GD (1983) Rainfall characteristics and their relation to splash erosion. Trans of the ASAE (General Edition. St. Joseph, Mich) 26(3):795–804

Penck A (1894) Morphologie der Erdoberfläche, Bd 1–2

Poesen J, Savat J (1981) Detachment and transportation of loos sediments by raindrop splash. Detachability and transportability measurements. Catena 8:19–41

Salnikov VK (1965) Equation for predicting soil losses due by erosion (Uravnenie dlia predskazania poter pochv v rezultate erozii. Viborochnii perevod s angliiskogo iazika). Selective translation from English. Pub. By Vintish, M, p 20

Shvebs GI (1974) Formation of erosion by water, sediment runoff and their assessment (on the example of Ukraine and Moldova). Leningrad, Publishing House Gidrometizdat, p 184 (Shvebs G.I. Formirovanie vodnoi erozii, stoka nanosov i ikh otsenka (na primere Ukraini i Moldavii). Leningrad, Gidrometizdat)

Siscoe GL, Levin Z (1971) Water-drop-surface—wave interactions. J Geophys Res 76:5112–5116

Stallings JH (1957) Soil conservation. Englewood Cliffs, New Jersey

Tound TP, Painter DJ (1974) The interaction between a raindrop and a shallow body of water. In: First Australian conference on hydraulics and fluid mechanics, Christchurch, New Zeland, pp 96–102

Wischmeier WH, Smith DD (1965) Predicting rainfall-erosion losses from cropland east of the Rocky Mountain. In: Agriculture handbook, vol 282, p 47 (US Dep. Agric., 1965)

Wischmeier WH, Smith DD (1978) Predicting rainfall erosion losses: a guide to conservation planning. In: Agriculture handbook, vol 537 (US Dep. Agric., 1978)

Wollny E (1895) Untersuchungen über das Verhalten der atmosphärischen Niederschläge zur Pflanze und Boden. Forschungen aus dem Gebiet der Agriculturphysik 18

Woodruff NP, Sidoway FH (1965) A erosion by wind equation. Soil Soc Amer Proc 29:602–608

Zaslavsky MN (1979) Soil erosion (Erozia pochv). Publishing House "Misl", Moscow, p 241

Zaslavsky MN, Larionov GA, Dokudovskaya OG, Tarabrin NP (1981) Map of the erosion index of rainfall in the European territory of the USSR and the Caucasus. In the collection. Erosion of soils and channel processes. 1981, Publishing House of Moscow State University, Moscow. 8, pp 17–30 (Karta erozionnogo indeksa dojdevikh osadkov Evropeiskoi territorii SSSR i Kavkaza. V. sb. Erozia pochv i ruslovie protsesi. M., MGU, vip. 8)

Soil Pollution

8

Besik Kalandadze and Lia Matchavariani

Abstract

Soils contaminated from water or air and by artificially applied toxic substances from pesticides, as well as mineral fertilizers, accumulate toxic elements, including heavy metals, having an extremely adverse impact on the living organisms. Among all kinds of economic activities, mining is the main source of pollution that deteriorates the agrophysical properties of soils and geoecological conditions in whole. The main ecological catastrophic zones in Georgia are discussed, related to the mining industry—Bolnisi metallogenic province (Kvemo Kartli Region) and Chiatura-Zestaponi manganese province (Imereti Region). Besides, pollution of soils and water with pesticides (chlorine-organic compounds) near the former chemical warehouses is considered.

Keywords

Soil pollutants • Contaminated soils • Heavy metals • Mining industry • Pesticides

8.1 Introduction

At the turn of the twentieth century, the contemporary civilization reached a high level of development. Such a development, alongside with the general benefit, brought many problems to people. The economic activity of humans causes environmental pollution with industrial waste, wastewaters, different radioactive substances, weed and pest killers used in agriculture, etc.

As it is known, the content of micro- and macro-chemical elements in the soil depends on the soil-formation processes, chemical composition of the soil-forming parent materials, and landscape conditions of soil formation: climate, waters, relief, vegetation cover, and fauna, i.e., the factors determining the processes of solution, accumulation, and migration of substances (DIN EN ISO 14688–1:2003-01 2003).

Soil is a specific component of nature. Therefore, its pollution has the gravest outcomes. Often, the soil polluted with chemical substances becomes a source of pollution of other natural components (water, air, and plant), as the soil pollution process is permanent. Therefore, the ecological state of the biosphere is closely connected to its sanitary regime. There are already natural regions identified on the Earth with highly increased concentrations of different chemical substances. Endemic diseases are frequent in such regions (Ford 2007).

Exact quantitative assessment of the impact of heavy metals on soil productivity is a complex task. Some soils are highly toxic, and their large amounts in a human body may lead to severe health problems; moreover, if considering that the half-life of heavy metals usually varies from several tens to thousands of years. Great amounts of heavy metals in a living organism/soil lead to a severe deterioration of health/degradation of soil quality. Consequently, the soil productivity falls drastically. Thus, it is very significant and urgent issue to study an accumulation and migration of heavy metals in soils. Consideration of study results will promote the environmental protection and good health of the future generations (Brümmer 2010).

Soil is an extremely complex and specific component of nature. Polluted water and air easily regain their original state in case the toxic substances are removed from them. As for soils, the issue is much more complicated. In case of pollution, the centuries-long soil balance is disturbed and takes too long to get restored. Some important soil component is disturbed, leading to the sharp deterioration of its normal functioning (Janssen et al. 1997; Friesl and Horak 2006; Friesl et al. 2006).

B. Kalandadze (✉) · L. Matchavariani
Faculty of Exact & Natural Sciences, Department of Soil Geography, Ivane Javakhishvili Tbilisi State University, Tbilisi, Georgia
e-mail: besik.kalandadze@tsu.ge

L. Matchavariani
e-mail: lia.matchavariani@tsu.ge

L. Matchavariani (ed.), *The Soils of Georgia*, World Soils Book Series,
https://doi.org/10.1007/978-3-030-18509-1_8

The upper organic soil layer, where the plant root system develops, is subject to the first technogenic impact. In this layer, micro- and macro-elements and other toxic substances also intensely accumulate. The elements accumulated in the soil from water or air are added by artificially applied toxic substances from pesticides, as well as mineral and organic fertilizers having an extremely adverse impact on the living organisms.

The main source of pollution can be industry, farming, transport, community facilities, mining, and other kinds of economic activities.

Particularly alarming is the irrigation of the soil with the waters discharged from plants and enterprises and its impact on the chemical composition of soil. The use of wastewaters with high concentrations of heavy metals in irrigation leads to severe outcomes. However, as some authors (Saet et al. 1982) think, irrigation with polluted waters is hazardous only with high concentrations of zinc, copper, and nickel, as these elements are highly phytotoxic.

The soil buffering ability is also nonhomogeneous. Buffering ability is soil property to resist to the polluting elements. This is first of all closely connected to the cation exchange process taking place in soils (DIN ISO 10693 1997-050 1997; DIN ISO 11260 1997-05 1997; DIN ISO 10390 2005-12 2005; DIN 19730: 2009-07 2009).

Soil is an open subsystem of a geochemical landscape of an even more complex natural system. It is a diversified natural body with weathering and soil-formation processes taking place simultaneously and with one another's support in time and space (such processes are humification, nitrification, denitrification, carbonization, decarbonization, podzolization, lessivage, etc.). The soils have the ability to retain many organic and polluting substances and play a role of a filter (Kandeler 2010).

With the highest stress index of environmental pollution, the pollution with heavy metals ranks the first. The problem is augmented by the fact of the half-removal periods of heavy metals being long (e.g., 1100 years for Cd; 310–1500 years for Cu, 310–5900 years for Pb, and 510 years for Zn). Other sources of soil pollution with metals are anthropogenic impacted: ore mining, power generation, industry, transport, municipal service, land cultivation, cattle breeding, etc. (Ainsworth et al. 1994; Alloway et al. 1990).

The solid waste remaining after the processing of non-ferrous metals contains many different kinds of chemical substances. The problem is aggravated by frequent cases of removing and burying the non-utilized industrial waste within the limits of the same urbanized areas, while the volume of waste may amount to thousands of tons.

In large industrial cities, the average annual amount of domestic waste and sewage sediments per man is 0.3 t and 0.4 t, respectively. The source of the environmental pollution with lead, zinc, cobalt, and benzopyrene is motor vehicles. With the vehicle emissions, 260 000 tons of lead are deposited on the Earth's surface annually, thus thrice exceeding the amount of lead from the metallurgical plants getting into the soil (Guo et al. 2006).

One of the major sources of soil pollution is agricultural activities, with the means of chemization, the mineral and organic fertilizers, pesticides, plant-growing stimulants, soil structure-forming polymers, biocides, and other compounds having an extremely negative impact on the soil flora and fauna (BBodSchV 1999).

The intensification of agriculture is followed by an increased use of phosphoric fertilizers. At present, 15–20 mln tons of phosphoric fertilizers are produced worldwide. Phosphatic raw material (apatites and phosphorites) is enriched with mixed elements. The concentration of rare metals in apatites is one order higher than in phosphorites and 3–8 times higher than in the lithosphere. 1 ton of simple superphosphate fertilizer contains 49 gr. of lead, 540 gr. of manganese, 137 gr. of zinc, 23 gr. of nickel, 18 gr. of copper, and 45 gr. of cadmium. The concentrations of fluorine, lead, copper, uranium, and strontium in phosphorites are 30 times more than Clarke. Arsenic is intensely accumulated in phosphatic raw materials. The substances used for plant protection contain such elements as mercury, arsenic, lead, fluorine, boron, copper, etc.

The soil buffering ability is also nonhomogeneous. Buffering ability is soil property to resist to the polluting elements. This is first of all closely connected to the cation exchange process taking place in soils. It should be noted that the resistance of soils of nonacid reaction rich in organic substances is much higher than that of light granular sandy soils. However, the latter, owing to good water permeability, is capable of getting rid of the polluting elements what often ends with the pollution of ground waters.

8.2 Pollution Caused by the Mining Industry[1]

As the ore-dressing industry gets more intense, the question of environmental pollution with heavy metals gets more topical. Almost similar geochemical processes take place in virtually all regions with copper and sulfide deposits—the oxidization of sulfide minerals with the participation of ground waters, oxidization of sulfide minerals with the participation of ground waters, and origination of sulfuric acid and metal sulfates followed by a number of secondary processes (Kalandadze et al. 2009).

[1]Updated Version from the Article (Matchavariani and Kalandadze 2012), Originally Published in the "Forum Geografic" (Copyright © 2012 Forum Geografic). All Rights Reserved.

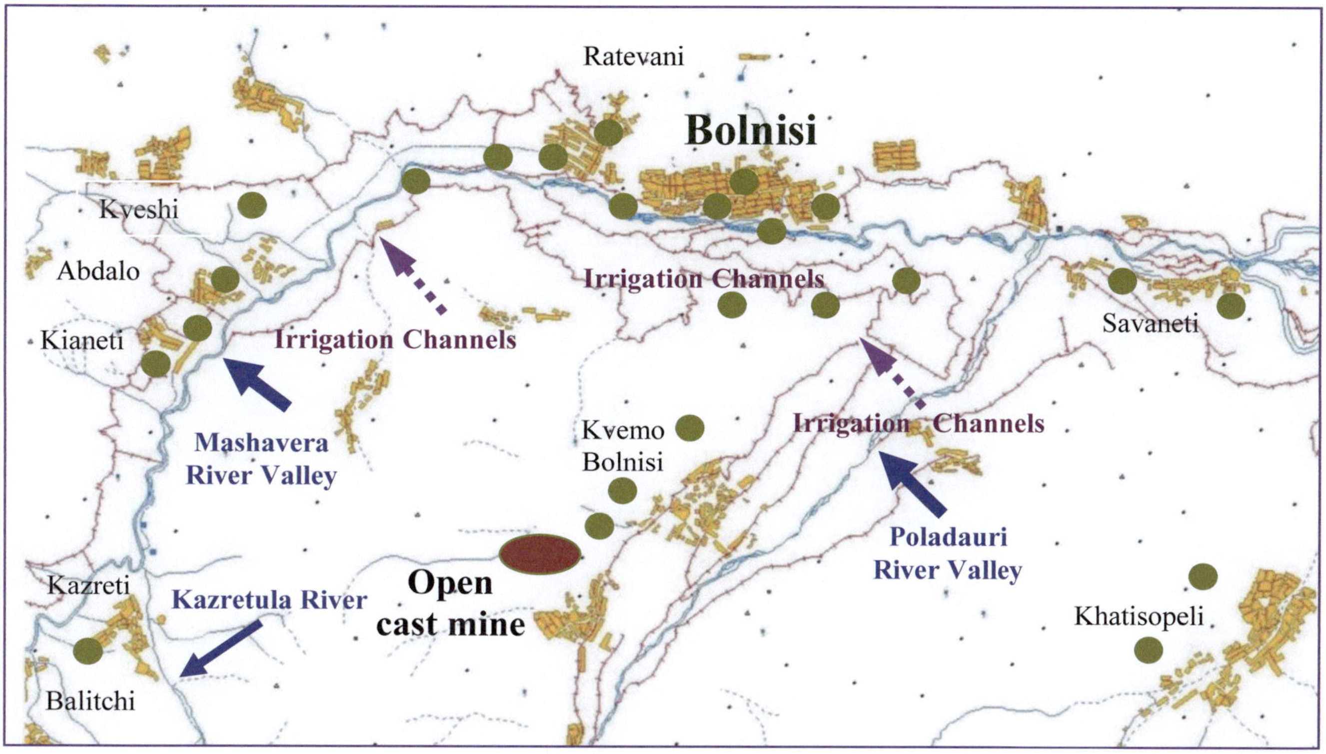

Fig. 8.1 Location of sampling soils (*"Forum Geografic", 2012, vol. XI, Issue 2, 127–137*)

The area adjacent to the town of Bolnisi (in Kvemo Kartli region) is one of the zones in Georgia with severe soil pollution caused by the mining industry.

Bolnisi municipality is one of the agricultural regions of Georgia with its plots of field located in the gorges of the rivers Mashavera and Poladauri, up to the confluence with the river Khrami (Fig. 8.1). Mostly, meadow and brown soils transient to a forest type with average and high humus content, loamy mechanical composition, and neutral to alkaline reaction are spread here.

Bolnisi, which is a northwest part of the metallogenic province of the Lesser Caucasus, is an important mining region. One of the most important ore deposits in the region is a barite-complex ore, with an ore-dressing and processing complex of enterprises operating with it. The ecological situation in this region is much severe, as the enterprises subordinate to the ore-dressing and processing complex of enterprises (copper-barite mine, ore landfills, tailing pit, copper concentrate plant, producing gold and silver alloy containing other admixtures too, from gold-containing quartzite ore stored separately by means of heap leaching, etc.) have a strong technogenic impact on the environmental components (soils, waters, and plants) which ultimately has a negative impact on the human health. Besides, the agricultural plots are irrigated with polluted irrigation waters. It is known that the pollution of natural objects due to the impact of a strong technogenic plant is seen 30–35 km from the epicenter, in the direction of movement of winds and waters (Kovda and Rozanov 1988).

The region has an average- and high-mountainous relief. The absolute height is 500–3100 m. With its genesis, the mineral ore deposit is a hydrothermal deposit.

At present, the deposit is open and the ore is extracted by open mining. The sulphuric acid waters from the mine flow directly into the river, where they are added by the wastewaters from the ore-dressing and processing enterprise, which flow first, into the river Kazretula (Fig. 8.2), and then into the river Mashavera (Fig. 8.3) and penetrate the soil through the irrigation system (Sayed 2006; Matchavariani and Kalandadze 2012; Matchavariani et al. 2015).

Besides, the storage of waste parent material, the industrial waste of the complex of enterprises is stored in so-called tailing pits, while the used waters flow into the river Kazretula, after being treated at treatment plants. A potential ecological threat is the waters sipped from the flotation waste and waters flowing from the ore-dressing and processing enterprise, which are strong sulphuric acid solutions with high contents of heavy metals. The wastewaters flow first, into the river Kazretula, and then into the river Mashavera and penetrate the soil through the irrigation system. Thus, the hydrographic network and soil cover get polluted. The rivers become the major source of pollution in the region. Besides water, the wind can also help spread the pollution; in particular, the dust originated during the explosion works

Fig. 8.2 Water pollution on the river Kazretula. Photo by B. Kalandadze

in the process of ore mining is taken to great distances. Finally, the harmful substances accumulate in soil and solid river drift (Müller 2000).

The accumulation of great amounts of heavy metals in soil and hydrosphere has quite a harmful impact on the biological world of the region what must be considered in several aspects. Besides, the growth of heavy metals in the hydrosphere and the soil may have a severe impact on the soil microflora, change its content, and negatively influence the processes of soil self-restoration. The soils that formed from the interaction of all landscape elements are the most informative component of the ecosystem. Therefore, all natural and technogenic processes are reflected in the soil (Lombi et al. 2002; Marschner et al. 2010).

The studied soils are developed over the weathering crust of the volcanogenic and sedimentary parent materials of the region, and their chemical composition depends both on the parent material content and soil-forming processes and anthropogenic factors associated with the economic activities of the human.

Tables 8.1, 8.2, 8.3, and 8.4 show the degree of pollution of the regions with heavy metals (Hanauer et al. 2011a, b; Felix-Henningsen et al. 2007).

As our data suggest, the total amounts of zinc, copper, cadmium, and sulfate ion in the Kazretula River flowing under the tanks are several times more than the maximum permissible levels established for surface waters. The Kazretula River has high concentrations of ore elements. Because of a very low water pH, these metals are mostly in a soluble form and are capable of traveling long distances. After the point where Kazretula River joins the Mashavera River, its water is diluted by 10 times. The pH of the irrigation water varies from 3 (at the mouth of the Kazretula River) to 5 (in the Mashavera River, depending on the distance of the Kazretula inflow). Besides, the pH value and turbidity of the water rise, and as a consequence the metals start floating in the water and continue to move in this form. In spite of this fact, the level of sulfate ion in the water is still quite high (DIN 11466 1995; DIN 19684-6: 1977-02 1997).

Fig. 8.3 Water pollution on the river Mashavera and tributary of Kazretula. Photo by B. Kalandadze

Table 8.1 Total amounts of heavy metals in the arable plots of Bolnisi Region, mg/kg (*"Journal of Environmental Biology", 2015, vol. 36, 85–90*)

The sampling sites (n—quantity of samples)	Cu	Zn	Mn	Pb
Ratevani-1, n = 12	60–3625	75–2250	625–1000	19–36
Abdalo, n = 5	40–6875	100–625	1125–1375	20–35
Kazreti, n = 10	35–200	85–100	750–1000	15–25
Khatisopeli, n = 4	60–100	110–212	1000–1250	14–19
Balichi, n = 8	55–85	90–120	750–1125	25–25
Savaneti, n = 4	40–115	105–135	875–1000	14–17
Ratevani-2, n = 12	55–1250	70–750	625–1000	17–31
Kveshi, n = 7	42–125	100–120	1000–1250	15–20

Table 8.2 Total amounts of heavy metals in the vineyards of Bolnisi Region, mg/kg (*"Journal of Environmental Biology", 2015, vol. 36, 85–90*)

The sampling sites (n—quantity of samples)	Cu	Zn	Mn	Pb
Ratevani, n = 20	255–3125	165–2125	750–1625	22–41
Pakhralo, n = 3	130–625	150–255	1000–1375	22–31
Kianeti, n = 2	290–305	100–110	875–1125	32–35
Bolnisi, n = 5	100–170	115–175	750–1375	17–22

Table 8.3 Average content of heavy metals in the soils of Bolnisi Region, mg/kg

Horizon (cm)	Quantity of samples	Cu	Zn	Mn	Pb
0–20	20	155	116	967	21
20–40	20	71	104	960	21
40–60	20	55	95	955	23
60–80	17	47	90	1050	21
80–100	9	48	86	930	20

Table 8.4 Content of heavy metals in soils and ore-containing rocks in the 0–30 cm depth (Chiatura Municipality), mg/kg

Prob. no	Mn	Cd	Cu	Pb	As	Ni
P1 Itkhvisi Lab. No (59001)	139,998	0.039	31.555	15.452	3.723	31.555
P2 Sareki Lab. No (58993)	101,548	0.591	67.204	22.775	10.998	281.572
P3 Mordzgveti Lab. No (58994)	1729	0.240	128.719	18.284	8.762	33.187
P6 Tiri Lab. No (58997)	127,338	0.561	83.392	33.425	11.362	346.631
P8 Martotubani Lab. No (58999)	73,522	0.448	189.960	20.109	9.061	184.004
P10 Rgani Lab. No (59008)	279,915	0.525	1069.408	87.621	22.874	479.616

The copper levels in the soils examined by us were extremely high (Matchavariani et al. 2015). The least levels were identified in about 18% cases of total samples, while 200 mg/kg or more were found in 18.3% of all samples (Tables 8.1 and 8.2)[2].

The soils, spread in Bolnisi area, are diversified both with their genesis and models of soil formation. The nature and intensity of agricultural activities lead to significant differences in the amounts of different elements. Large amounts (>200 mg/kg) of heavy metals were identified in the soils of some villages in the region. Such areas are located on the Mashavera River Basin (Hanauer et al. 2011a, b). It is noteworthy that areas with particularly high levels of metals are often found near the railway and river basin. It is natural that the levels of copper in soils of vineyard are quite higher than in the plowed land. The amounts of copper are similarly high in orchard soils. The majority of the examined land plots are found on the right bank of the Mashavera River. Vineyards and orchards occupy the largest part of the area, and about over half of the territory is grown with wheat crops. Over half the area is badly polluted by copper and zinc; copper and zinc concentration in most of the area is 200–700 mg/kg, while approximately 8–9% of the area can be termed as polluted at catastrophic levels (Amberger 1996). This area receives intense irrigation with the Mashavera River water. Virtually, we can say that we face an apparent case of anthropogenic impact, implying the impact of the irrigation system by using the water polluted following the activities of the ore-dressing and processing plant. On the other sites of the region where copper and zinc concentrations (500 mg/kg) were identified, the pollution has a fragmented nature.

The highest concentration of lead identified in the fields of Kianeti Village, distributed as follows: in vineyards was an average content of lead (33.8 mg/kg) and 24.3 mg/kg in the arable. If considering that the copper and zinc levels are minimal there, one may assume that the only source of lead is vehicles. The maximum levels of copper (875 mg/kg) are fixed in the same area and high concentration of zinc is detected (500–1600 mg/kg) as well.

So, it may be concluded that this area is subject to obvious anthropogenic impact further intensified by other factors (McBride 1989; Mench et al. 2000).

As for manganese (Mn), it is particularly noteworthy. As it is generally known, manganese is a very significant element for geochemical processes which are concerned in the water, plant, and soil. The maximum concentration of manganese is 1125–1375 mg/kg (Pollution Concentration Coefficients are 5.6–6.8, respectively), whereas minimal content is 875 mg/kg. Manganese, as one of the elements of parent materials, is actively accumulated in them. At the same time, the lowest level of Mn is in sandy and prevalent in clay soils. Therefore, maximum levels of manganese have been revealed in the basalt rocks and minimal in the alluvial soils (Table 8.3) (Stahr 2010; Usman 2004).

As per I. Vazhenin's classification (Definition Methods…, 1987), more 61% of all examined soils (70 ha) is polluted with copper, either slightly, or moderately; 30 ha (17.3%) has polluted more than average; and 24 ha (21.2%) of soils is polluted intensely or enormously.

[2]Tables 8.1 and 8.2 taken from the article (Matchavariani et al. 2015), originally published in the "Journal of Environmental Biology"

About 70 ha (61.3%) of examining soils is polluted with zinc slightly or moderately; 18.5% (20 ha) has polluted more than average; 21% (24 ha) is polluted either highly or extremely; 93 ha (81.5%) is medium or higher than average polluted; and 21 ha (18.4%) of examining soils has a high level of manganese pollution (Zeien 1995). A total indicator of concentration coefficient shows that only 19 ha of upper soil layers (0–20 cm) shows low levels of pollution; 91 ha of lands is intensely polluted (Sastre et al. 2004).

The order of chemical elements in slightly polluted soil is as follows: Mn > Zn > Cu—pollution concentration coefficient (Hc). In the upper soil layers (0–20 cm), which show average pollution with the least value of total amounts of pollution concentration, depending on the values of concentration coefficients, the chemical elements have the following order: Mn > Zn > Cu; the diminishing order of Zn(n − 1) completely changes when using maximum values (Cu > Zn > Mn).

As the concentration coefficients suggest, in the highly polluted soils with the minimum and maximum values of Zn(n − 1), the chemical elements are of different orders. With the minimum values, their order is Cu > Mn > Zn, while in case of maximum total values of pollution concentration coefficients, the order changes as Cu > Zn > Mn (Kalandadze and Matchavariani 2011, 2012).

As noted in our previous publication (Kalandadze and Matchavariani 2012), "During the last few years, the soil's characteristics have sharply deteriorated. In some places, the ground is covered with whitish/greenish waterproof coats. The porosity of the soil, as well as its productivity, is diminished … In this case, we're dealing with gypsumming. As we found out, limestone is added to the wastewater of the Madneuli enterprise in order to neutralize the acid, after that this wastewater is flowing through the sewer. In that case, gypsum is formed, which is flowing in the river water and then this water is used for irrigation of agricultural land plots. In the course of time gypsum has collected on the ground surface and it coats the soil, which worsens aeration and filtration capabilities of the soil, which subsequently causes a sharp decrease in its productivity" (Fig. 8.4).

Fig. 8.4 Contaminated soil surface, Bolnisi. Photo by B. Kalandadze

In the examined area, the main properties of averagely polluted soils, as the total values of pollution concentration coefficient suggest, are as follows: the soil is heavy clay loam, specific weight—solid-phase density in the cross section is very differentiated (higher—in the plowed horizon and deeper—lower). As the analyzed data suggest, in the lower layers of plowed fields, the soil is not so consolidated, being a much positive factor as it results in positive values of all other parameters. In particular, the general soil porosity is good there and conditions for plants are satisfactory. There is favorable aeration for plants, favorable hydrologic characteristics, and thick and capillary pores are capable of keeping the bulk of moisture as well (Kretzschmar 2010). According to our previous publication (Kalandadze and Matchavariani 2012), "This is evidenced by large quantities of productive moisture and high levels of maximum moisture volumes in the fields. The amounts of productive moisture and levels of maximum moisture volumes in the soils are 30.50–51.00% and 19.3–32.7%, which is considered to be the best values for irrigated soils." Table 8.3 shows an average content of heavy metals in soils' vertical profile.

Because of low bulk density and high general porosity, the soils have good filtration capabilities: the water can travel 1.5 m in the period of 24 h (Wilke 2010).

The soils, polluted with Zn(n − 1) and classified as the ones with average pollution, have quite satisfactory hydrophysical properties. The soil has an average value of maximum moisture volumes. The filtration ability is not so unfavorable (Knox et al. 2000).

The hydrophysical properties of intensely polluted soils are unfavorable for the plants' normal development (Fig. 8.5). The granulometric analysis suggests that soil is a light clay loam evidencing the presence of heavy metals there. The soil specific weight through the whole profile is not optimal. The soil volume weight right from the upper layer of the plowed area is quite high (Kabata-Pendyas and Pendyas 1989; Chlopecka and Adriano 1997). This fact is evidenced that the soil is much solidified, thus creating unfavorable hydrophysical conditions for vegetation. The values of maximum moisture volumes are not acceptable and the amount of productive moisture in the soil is too low (Commission regulation 2006; Contin et al. 2007; Hartley and Lepp 2008).

Vegetation, as a whole, is a determining factor of geochemical processes occurring in the soil. Plants' main ability is selective absorption, so they can receive various chemical elements from the soil, disproportionately with their composition. In fact, they are capable of choosing the chemical elements necessary for their growth. Figure 8.6 (Kalandadze and Matchavariani 2012) shows the content of heavy metals in different plants (corn, pumpkin, and grapes). Productive moisture is quite significant, which helps the plant absorb dissolved substances. The deficit of productive moisture leads to the deterioration of plants, reduced yield, and often death of the plants.

Fig. 8.5 Soil pit with calcium grains, Bolnisi. Photo by B. Kalandadze

Technogenesis is obvious in the areas, where there are several pollution factors acting simultaneously. The soils have unfavorable agrophysical properties, which cause a decrease in the filtration ability of the soil. Filtration of soil moisture is only about 0.5 m during the 24 h, that is, unfavorable for plants (Matchavariani et al. 2015).

Our research allowed to conclude that "pollutant heavy metals—copper, zinc, and manganese have an especially active negative impact on soil characteristics, its composition and soil-formation processes, which results in deterioration of hydro-physical potential of the soil. The balanced correlation between solid, liquid and air phases in the soil is violated. Characteristics and quantities of components existing in the soil are changing dramatically; the soil is degrading, vital functions of agricultural crops are disrupted and bio-productivity is falling. Summation of the agrophysical parameters of slightly, averagely and highly polluted soils provides a clear evidence of that" (Kalandadze and Matchavariani 2012).

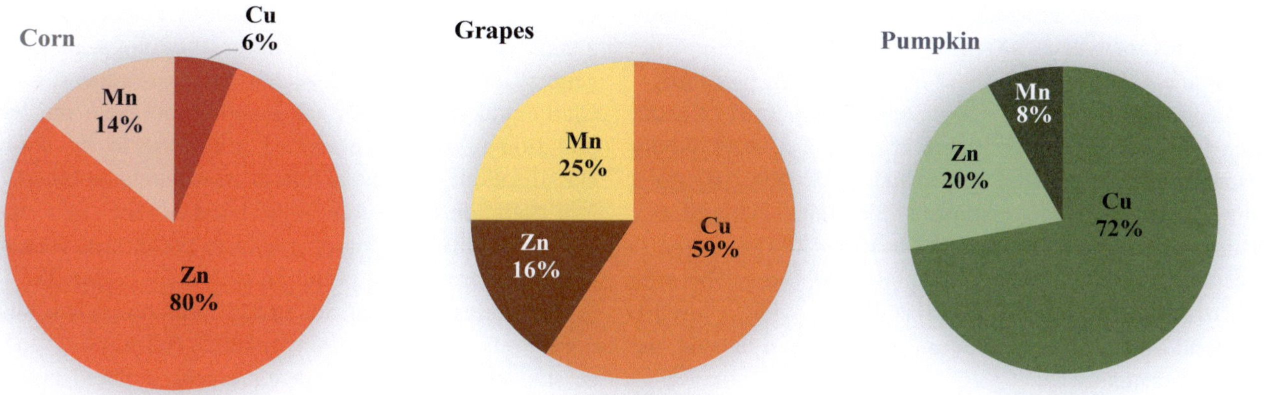

Fig. 8.6 Levels of heavy metals in plant (*"Forum Geografic", 2012, vol. XI, Issue 2, 127–137*)

Irrigation of agricultural lands with polluted water, as the crucial anthropogenic factor of soil degradation, changes pH. Heavy metals are absorbed by clay minerals just after they get into the soil. Metals are accumulated in the surface, because the soil carbonate system serves as a barrier for them.

The agricultural activities contribute to increasing the concentration of heavy metals in the soil. For instance, an intense using of copper in vineyards and orchards following the use of different copper-based chemical pesticides there. Their migration into the lower soil horizons is conditioned by relief, precipitation volumes, soil pH, and others (VDLUFA 1976, 1991, 1997).

Soil pollution with heavy metals is directly connected with soil agrophysical characteristics. In the highly polluted soils, the processes of cementing have taken place that increases the soil bulk density, deterioration of soil porosity, and greatly reduces water permeability (Kalandadze and Matchavariani 2012; Kalandadze and Trapaidze 2015).

The use and efficiency of chemical, physical–mechanical, and mechanical methods of soil restoration depend on bioclimatic, geoecological, and soil-edaphic factors such as soil texture and structure, content of soil organic matter, absorption volumes, pH, oxidation-restoration potential, etc.

Another important zone of ecological catastrophe in Georgia is Chiatura-Zestaponi manganese industrial region located in western Georgia (Imereti Region).

The relief of the area is mostly mountainous and intensely dissected. It has a complex geology dated from the Cambrian through the Quaternary Age.

The districts of the territory located over 2000 m asl are presented by Leptosols Umbric soils. The greatest areas of the mountain-and-forest zones are occupied by Cambisols Dystric. Acrisols Haplic soils are also widely spread. The areas with Rendzinas soils within the limits of the Chiatura structural plateau coincide with the areas with calcareous parent materials. Such soils occupy vast areas on the terraces of the rivers of Zemo Imereti.

The manganese deposit of Chiatura has been mined since 1860. By 1990, the amount of raw material mined from it amounted to 203 mln tons, while the amount of realizing production goods was 108 mln tons. By 1990–2005, the amount of raw material mined from it amounted to 7,734,009 tons and the amount of realized production goods was 2,713,614 tons.

In recent years, the supply of the branch with wooden materials has deteriorated, having led to the lack of fixing materials needed for the underground mining of ore. Because of this reason, the raw material is being extracted by open mining.

As the ore is dressed at the plant, the river Kvirila permanently gets polluted with manganese ore admixtures remained after the ore wash—the remained toxic substances, as a result of washing during the ore dressing, flows into the river Kvirila and pollutes it. The content of manganese ore in the river Kvirila is 10–12%, with magnesium peroxide and aluminum as main polluting ores (Kalandadze and Felix-Henningsen 2014).

The river Kvirila and its tributary Dzirula are the main arteries in the hydrographic network of the region. The waters of the river Kvirila are mainly used to irrigate the agricultural plots of field of Chiatura and Zestaponi Regions.

The major source of pollution of the city of Chiatura is slag, the waste from the metallurgical plant with its landfill is found immediately on the territory of the city, while the volume of slag amounts to 15 mln m^3. This material is dissipated all over the territory of the city by the action of wind and wastewaters. The dust penetrates the residential houses and water pipeline causing major health and living problems for people. Besides, the content of different heavy metals in the slag much exceeds the standards (Table 8.4).

The amount of atmospheric precipitations in the study area varies between 900 and 1800 mm promoting the development of severe washout and washdown processes.

Washing of one ton of raw material needs 3 m^3 of water taken from two dams over the river (technical water supply systems). The manganese ore is mainly extracted by open mining, with drill and explosion operations, without any restoration or recultivation of the territory. No measures to protect flora or fauna during the mining are taken. The used industrial waters flow into the river Kvirila without any filtration and then to the agricultural plots of field through the irrigation channels.

In addition, there is a ferroalloy plant operating in Zestaphoni, which mainly treats the manganese ore extracted in Chiatura. The production cycle of the plant uses the water of the river Kvirila, which, enriched by various heavy metals and other admixtures, returns to the river without any filtration (Table 8.4).

Various criteria and norms have been adopted for evaluating soil pollution levels, Clarke Concentration Coefficient is one of the notable examples of such norms/criteria, the coefficient is calculated with the use of the following formula: Kk = Cf/K, where Kk is Clarke Concentration Coefficient, Cf —is the actual concentration of chemical elements in the soil, and K—is the Clarke of a chemical element. That criterion shows how high or low is the concentration of a concrete chemical element in comparison with that element's Clarke.

Geoecologic condition of soils is also evaluated with Pollution Concentration Coefficient, which is calculated with the use of the following formula: Hc = Cf/level, where Hc is the pollution concentration coefficient, Cf is the actual concentration of a chemical element, f is local or general established levels of a concrete chemical element, which indicates how the concentration is increased in comparison with the general levels.

Pollution Probability Concentration Coefficient is also used for evaluating the levels of soil pollution, the formula used in that case is C = Cf/maximum permissible level, where C is the coefficient of pollution probability concentration and Cf is the actual concentration of a chemical element.

These indicators show how high is the actual concentration of a chemical element in comparison with the maximum permissible levels of concentration of that element; the higher that coefficient is in comparison with one the higher is the probability of soil pollution and negative impact of chemical elements on live organisms.

We took into consideration, correlation between the various elements of the soil and their joint impact on the soil in order to establish/evaluate the maximum levels of concentration of microelements in the soil.

Soils are grouped according to the total actual concentrations of toxic chemical elements pollutants, and this grouping is based on the methodology elaborated by Vinogradov (1957); according to that methodology, first group is general level + 1 Clarke; second group is general level + 2 Clarke, etc. When soils are grouped this way, the level of pollution of soils is classified according to the following classification: slightly polluted, moderately polluted, medium pollution (or averagely polluted), higher than medium pollution, strong pollution (or high pollution), very strong pollution (or extreme pollution/extremely polluted).

Territories adjacent to strong pollution sites such as nonferrous metal processing plants, ore-dressing and processing enterprises, etc. are extremely polluted and sometimes their level of pollution exceeds 10 Clarkes (Vinogradov 1957).

Pollution levels can be used as approximate indicators of adverse impact of chemical elements on environment. For example, at I and II (group) level pollution, the soil's biota is strongly deteriorated, biochemical processes are suppressed. At III and IV (group) level of pollution, agrochemical characteristics of the soils are worsened, vital functions of plants are disrupted, and their chemical composition is violated. At V and VI (group) level of pollution, plants become sick and they die; plant and animal products are not fit for use due to sanitary-hygienic considerations. Chemical composition of the upper layer of the soil changes and all agrochemical characteristics quickly deteriorate (Kalandadze and Matchavariani 2012).

High concentrations of heavy metals in the irrigation waters are a much problematic issue. Mostly vegetables and watermelons, melons, and gourds are grown in the region, while as it is known, these crops have a short vegetation period and great amounts of toxic substances accumulate in them. This is particularly true with the root crops. These processes are promoted by soil Hp following the climatic conditions, which is acidic in the given instance and is 5.5 (Sauerbeck 1982).

Despite the sanctions of the Environmental Inspection, the situation has not changed for better. The reason for this is minor sanctions, which cost nothing to the enterprise or the complex of enterprises reluctant to spend additional finances to put the plant to order. In the final run, it is the people who are damaged.

8.3 Soil Pollution with Pesticides

A typical feature of modern agriculture is the wide-scale introduction of the industrial techniques used to grow agricultural crops, meaning an intense use of such technical means, as pesticides. Out of 2 million people with pesticide intoxication registered annually, 50,000 are deceased.

Improper use of mineral and organic fertilizers and chemical pest and weed killers in the agricultural production leads to the environment pollution, resulting in the pollution of soil and atmosphere, vegetable and animal products, and drinking and irrigation waters with different toxic substances. The use of fertilizers in large amounts deprives the bacteria of the capability to transform into an organic substance and form the vital product necessary for plants (Schatz et al. 2015).

There are approximately 187 active substances and 400 pesticide preparations of complex compounds registered in Georgia. It is known that pesticides damage bacteria and other organisms necessary for soil. This process deteriorates the agrophysical properties of soil. The unfavorable processes in the soil have a negative impact on soil porosity and water permeability change the range of productive moisture and other processes resulting in extremely unfavorable agrobiological and agroecological conditions. Overall, this results in the deteriorated soil productivity (Lezhava and Matchavariani 1983).

It is established that over 90% of the pesticides occur in the human body through the meal. Persistent organic pollutants (POPs) are one of the severest, yet an invisible problem in Georgia. In the twentieth century, these harmful substances were used in different industries. Pesticides containing these substances were used in agriculture to yield a rich harvest. Consequently, toxic substances are deposited in water and soil posing a great threat both to the environment and human health. It is recognized that the POPs cause oncological, reproductive, endocrinal, and immunological diseases (Nurzhanova et al. 2013).

OCPs (organochlorine pesticides) actively used in agriculture several decades ago have four common features: high toxicity, resistance to degradation, evaporation and propagation to great distances (through air, water or migrating species), and accumulation in fatty tissues.

POPs were prohibited in Georgia in 1975 and their legal use was stopped in the 1980s (Table 8.5). In those years, average 30 kg of pesticides was applied per hectare. The floodplains and bogged areas in the basin of the Alazani River were treated by airplanes to fight malaria. Most pesticides remain unused in the open air.

One of the most dangerous features of the pesticides is mutagenic activity having a negative impact on the health of people and future generations. Pesticides maintain their biological activity and, therefore, they pose a permanent threat to the environment and human. Pesticides have a wide range of toxic activity. Their mutagenic activity does not show itself on a human body instantly, but gives grave outcomes later (Kumpiene et al. 2006).

The highest concentration of the environmental pollutants in the atmosphere is usually fixed near the sources of pollution, but sometimes they are transferred to thousands of kilometers by wind or water. As a result, the human impact on the biosphere is global all over the world.

Phytoremediation is one potential method for reducing risk from the pesticides (Khatisashvili et al. 2015). Genetic heterogeneity of populations of wild and weedy species growing on pesticide-contaminated soil provides a source of plant species tolerant to these conditions. These plant species may be useful for phytoremediation applications.

Organochlorine pesticides taken up by the plants are distributed unevenly in different plant tissues. The main organ of organochlorine pesticide accumulation is the root system. The accumulation rate of organochlorine pesticides was found to be a specific characteristic of plant species and dependent on the degree of soil contamination. This information can be used for technology development of phytoremediation of pesticide-contaminated soils (Nurzhanova et al. 2013).

The chlorine-containing organic pesticides, which were actively used in the Soviet economy have four common features: high toxicity, resistance to degradation, evaporation and propagation to great distances (through air, water, or migrating species), and accumulation in fatty tissues. Particularly strong pollution in soils and waters of South Caucasus was identified at the hot points used as storage or distribution areas of pesticides in the past.

Following tables show pollution of soils and water with chlorine-organic compounds near the former chemical warehouses (Tables 8.6, 8.7, 8.8, and 8.9) in Bolnisi, Gori, Karaleti, and Kareli towns.

Table 8.5 Content of dust (DDT—dichloro-diphenyl-trichloroethane) and its compounds in the soils, μg/kg

Place	Pit#	2,4DDE	4,4DDE	2,4DDD	4,4DDD	2,4DDT	4,4DDT	∑DDE	∑DDD	∑DDT	∑DDX
Schilda	P2	1.92	3.49	–	–	–	–	5.41	–	–	5.41
	P5	–	61.76	12.46	–	12.09	–	61.76	12.46	12.09	86.32
	P6	–	2.11	–	–	–	–	2.11			2.11
	P7	26.41	**72.96**	44.68	–	3.45		99.38	44.68	3.45	147.51
Khachini	P16	9.44	30.40	2.92	3.41	–	–	39.84	6.33	–	46.18
Sakobo	P17	22.15	58.51	2.17	4.72	–	–	80.66	6.89	–	87.56
Tamarisi	P33	29.50	54.18	3.04	12.84	2.90	2.90	83.68	15.88	5.80	105.36

Table 8.6 Soil and water pollution with chlorine-organic compounds in Bolnisi

Pesticide	Water (μg/l)	Soil (μg/kg)
2456 Tetrachloro M Xylene	0.05	0.038
Alpha Lindane	0.49	0.368
Beta Lindane	1.67	1.23
Delta Lindane	1.46	1.095
Heptachlor Epoxide	0.13	0.098
Endosulfan	2.11	1.583
DDE	0.01	0.008
Dieldrin	0.80	N.D
Endosulfan 2	0.58	0.435
DDD	0.03	0.023
Endrin Aldehyde	2.49	1.868
Endosulfan Sulfate	2.81	2.108
DDT	19.54	14.655

Table 8.7 Soil and water pollution with chlorine-organic compounds in Gori

Pesticide	Water (μg/l)	Soil (μg/kg)
2456 Tetrachloro M Xylene	0.05	0.038
Alpha Lindane	0.56	0.42
Beta Lindane	1.48	1.11
Delta Lindane	1.45	1.088
Aldrin	0.07	0.053
Heptachlor Epoxide	0.15	0.113
Endosulfan	1.3	0.975
DDE	0.04	0.03
Dieldrin	1.51	1.133
Endosulfan 2	1.24	0.93
DDD	1.74	1.05
Endrin Aldehyde	0.41	0.308
Endosulfan Sulfate	2.24	1.68
DDT	302.91	227.183

Table 8.8 Soil and water pollution with chlorine-organic compounds in Gori–Karaleti

Pesticide	Water (μg/l)	Soil (μg/kg)
2456 Tetrachloro M Xylene	0.03	0.023
Alpha Lindane	0.66	0.495
Beta Lindane	1.17	0.878
Delta Lindane	1.96	1.47
Aldrin	0.07	0.053
Heptachlor Epoxide	2.71	2.033
Endosulfan	1.84	1.38
DDE	3.70	2.775
Dieldrin	2.11	1.583
Endosulfan 2	1.63	1.223
DDD	0.27	0.203
Endrin Aldehyde	1.27	0.953
Endosulfan Sulfate	1.19	0.893
DDT	777.15	582.863

Table 8.9 Soil and water pollution with chlorine-organic compounds in Kareli

Pesticide	Water (μg/l)	Soil (μg/kg)
2456 Tetrachloro M Xylene	0.02	0.015
Alpha Lindane	0.21	0.158
Beta Lindane	0.58	0.435
Heptachlor	0.22	0.165
Delta Lindane	0.57	0.428
Aldrin	3.05	2.288
Heptachlor Epoxide	7.56	5.670
Endosulfan	5.64	4.230
DDE	0.04	0.030
Dieldrin	2.92	2.190
Endosulfan 2	1.01	0.758
DDD	84.71	63.533
Endrin Aldehyde	1.66	1.245
Endosulfan Sulfate	1.24	0.930
DDT	69.13	51.848

References

Ainsworth CC, Pilon JL, Grassman PL, Sluys WGVD (1994) Cobalt, cadmium and lead sorption to hydrous iron oxide: residence time effect. Soil Sci Soc Am J 58:1615–1623

Alloway BJ, Jackson AP, Morgan H (1990) The accumulation of Cd by vegetables grown on soils contaminated from variety of sources. Sci Total Environ 91:223–236

Amberger A (1996) Pflanzenernährung, 4th edn. UTB, Stuttgart Bes C, Mench MJ (2008) Remediation of copper contaminated topsoils from a wooden treatment facility using in-situ stabilization. Environ Pollut 156:1–11

BBodSchV (1999) Bundes Bodenschutz und Altlastenverordnung 12 July 1999, Bundesgesetzblatt, Bundesanzeigerverlagsgesellschaft mbH, Köln

Brümmer GW (2010) Böden als Pflanzenstandort. In: Schachtschabel et al (eds) Lehrbuch der Bodenkunde, 16th edn. Spektrum, Heidelberg/Berlin, pp 379–448

Chlopecka A, Adriano DC (1997) Influence of zeolite, apatite and Fe-oxide on Cd and Pb uptake by crops. Sci Total Environ 207:195–206

Commission regulation (EC) No 466/2006. Off J Eur Union Dec 2006 Luxemburg

Contin M, Mondini C, Leita L, De Nobili M (2007) Enhanced soil toxic metal fixation in iron(hydr)oxides by redox cycles. Geoderma 140:164–175

DIN 11466 (1995) Soil quality—extraction of trace elements soluble in aqua regia. Beuth Verlag GmbH, Berlin, Deutsches Institut für Normung

DIN 19684–6: 1977–02 (1977) Methods of soil investigation for agricultural engineering—chemical laboratory tests—part 6: determination of iron soluble in oxalate solution. Beuth Verlag GmbH, Berlin, Deutsches Institut für Normung

DIN 19730: 2009–07 (2009) Soil quality—extraction of trace elements from soil using ammonium nitrate solution. Beuth Verlag GmbH, Berlin, Deutsches Institut für Normung

DIN ISO 10390 2005–12 (2005) Soil quality—determination of pH. Beuth Verlag GmbH, Berlin, Deutsches Institut für Normung

DIN ISO 10693 1997–05 (1997) Soil quality—determination of carbonate content—volumetric method. Beuth Verlag GmbH, Berlin, Deutsches Institut für Normung

DIN ISO 11260 1997–05 (1997) Soil quality—determination of effective cation exchange capacity and base saturation level using barium chloride solution. Beuth Verlag GmbH, Berlin, Deutsches Institut für Normung

DIN EN ISO 14688–1:2003–01 (2003) Geotechnical investigation and testing—identification and classification of soil—part 1: identification and description. Beuth Verlag GmbH, Berlin, Deutsches Institut für Normung

Felix-Henningsen P, Urushadze TF, Narimannidze EI, Wichmann L, Steffens D, Kalandadze B (2007) Heavy metal pollution of soils and food crops due to mining wastes in an irrigation district south of Tbilisi, eastern Georgia. Annal Agrarian Sci 5(3):11–27

Ford RG (2007) Structural dynamics of metal partitioning to mineral surfaces. In: Hamon R, Mc Laughlin M, Lombi E (eds) Natural attenuation of trace element availability. Taylor and Francis, New York, pp 73–88

Friesl W, Horak O (2006) Immobilisierung von Schwermetallen (Metalloide) oberflächennaher, großflächiger Kontaminationen—Technischer Leitfaden. ARC Seibersdorf Research GmbH, Seibersdorf

Friesl W, Friedl J, Platzer K, Horak O, Gerzabek MH (2006) Remediation of contaminated agricultural soils near a former Pb/Zn smelter in Austria: batch pot and field experiments. Environ Pollut 144:40–50

Guo G, Zhou Q, Ma LQ (2006) Availability and assessment of fixing additives for the in situ remediation of heavy metal contaminated soils: a review. Environ Monit Assess 116:513–528

Hanauer T, Shnell S, Steffens D, Kalandadze B, Navrozashvili L, Felix-Henningsen P (2011) In situ remediation of Cd, Cu and Zn contaminated topsoils by different amendments, EGU General Assembly, vol 13. XY488 EGU2011, pp 155–159

Hanauer T, Felix-Henningsen P, Shteffens D, Kalandadze B, Navrozashvili L, Urushadze T (2011a) In situ stabilization of metals (Cu, Cd und Zn) in contaminatedsoils in the region of Bolnisi, Georgia. Plant Soil 341:193–208

Hartley W, Lepp NW (2008) Remediation of arsenic contaminated soils by iron-oxide application, evaluated in terms of plant productivity, arsenic and phytotoxic metal uptake. Sci Total Environ 390:35–44

Janssen RPT, Peijnenburg WJGM, Posthuma L, Magt VDH (1997) Equilibrium partitioning of heavy metals in Dutch fields. I. Relationship between metal partition coefficients and soil characteristics. Env Tox Chem 16:2470–2488

Kabata-Pendyas A, Pendyas H (1989) Microelements in soils and plants. Mir, Moscow

Kalandadze B, Felix-Henningsen P (2014) Pollution of irrigation soils and cultural plants with heavy metals in the basin of the river Kvirila due to mining industry (West Georgia). Eurasian Soil Congress Abs, Istambul, Turkey

Kalandadze B, Matchavariani L (2011) Impact of heavy metals on soils and plants in Mashavera river lowland, Georgia. In: Proceedings of international conference "soil, plant and food interactions. Mendel University, Brno, Chech Rep., 6–8 Sept, pp 587–598

Kalandadze B, Trapaidze V (2015) Quantitative evaluation of the impact of heavy metals on soil productivity on the example of ore-dressing and processing production in east Georgia, SGEM, vol 1I, soils, forest ecosystems, marine and ocean ecosystems, pp 271–278

Kalandadze B, Hanauer T, Felix-Henningsen P, Urushadze T, Narimanidze E, Steffens D (2009) Mining and agriculture in the Mashavera valley (South-East Georgia)—a land use conflict with severte consequences. Biol J Armenia LXI 2:22–29. National Academy of Sciences of the Republic of Armenia

Kandeler E (2010) Bodenorganismen und ihr Lebensraum. In: Schachtschabel et al (eds) Lehrbuch der Bodenkunde, 16th edn. Spektrum, Heidelberg/Berlin, pp 83–119

Khatisashvili G, Matchavariani L, Gakhokidze R (2015) Improving phytoremediation of soil polluted with oil hydrocarbons in Georgia. In: Chapter 19 in book "soil remediation and plants: prospects and challenges". Elsevier—Amsterdam, Boston, Heidelberg, London, New York, Oxford, Paris, San Diego, San Francisco, Singapore, Sydney, Tokyo, pp 547–569

Knox AS, Seaman J, Adriano DC (2000) Chemostabilization of metals in contaminated soils. In: Wise DL, Trantolo DJ, Cichon EJ, Inyang HI, Stottermeier U (eds) Bioremediation of contaminted soils. Marcel Dekker, New York, Basel, pp 811–836

Kovda V, Rozanov B (1988). Soils science. High school, Moscow (Pochvovedenie. Vishaya shkola, Moskva)

Kretzschmar R (2010) Chemische Eigenschaften. In: Schachtschabel et al (eds) Lehrbuch der Bodenkunde, 16th edn. Spektrum, Heidelberg/Berlin, pp 122–170

Kumpiene J, Ore S, Renella G, Mench M, Lagerkvist A, Maurice C (2006) Assessment of zerovalent iron for stabilization of chromium, copper and arsenic in soil. Environ Pollut 144:62–69

Lezhava VV, Matchavariani LG (1983) Preventing the adverse effects of pesticides on the environment. Georgian scientific-research institute for scientific-technical information, overview information (GSRISTI, OI), vol 3, issue 3, pp 44 (Predotvrashenie otritsatelnogo deistvia estitsidov na okruzhaiushuyu sredu. GruzNIINTI, OI)

Lombi E, Zhao F-J, Wieshammer G, Zhang G, McGrath SP (2002) In situ fixation of metals in soils using bauxite residue: biological effects. Environ Pollut 118:445–452

Marschner B, Müller I, Stolz R, Stempelmann I (2010) Immobilisierung von Schwermetallen in Gartenböden. Bodenschutz 02:34–41

Matchavariani L, Kalandadze B (2012) Pollution of soils by heavy metals from irrigation near mining region of Georgia. vol. Forum Geogr XI(2):127–137

Matchavariani L, Kalandadze B, Lagidze L, Gokhelishvili N, Sulkhanishvili N, Paichadze N, Dvalashvili G (2015) Soil quality changes in response to their pollution of heavy metals in Georgia. J Environ Biol Spec Issue 36(1):85–91

McBride MB (1989) Reactions controlling heavy metals solubility in soils. Adv Soil Sci 10:1–56

Mench M, Manceau A, Vangronsveld J, Clijsters H, Mocquot B (2000) Capacity of soil amendments in lowering the phytoavailability of sludge-born zinc. Agronomie 20:383–397

Müller I (2000) Einfluss eisenoxidhaltiger Bodenzusätze auf die Mobilität von Schwermetallen in kontaminierten Böden. Dissertation, Boden und Landschaft 27. Justus Liebig University

Nurzhanova A, Kalugin S, Zhambakin K (2013) Obsolete pesticides and application of colonizing plant species for remediation of contaminated soil in Kazakhstan. Environ Sci Pollut Res 20:2054–2063

Saet Yu, Basharkevich I, Revich B (1982) Methodical recommendations about a geochemical estimation of sources of environmental pollution. IMGRE

Sastre J, Hernadez E, Rodriguez R, Alcobe X, Vidal M, Rauret G (2004) Use of sorption and extraction tests to predict the dynamics of the interaction of trace elements in agricultural soils contaminated by a mine tailing accident. Sci Total Environ 329:261–281

Sauerbeck D (1982) Welche Schwermetallgehalte in Pflanzen dürfen nicht überschritten werden, um Wachstumsbeeinträchtigungen zu vermeiden? Landwirtsch Forsch Special Edition 39:108–129

Sayed MAHA (2006) Dynamik von Schwermetallen in belasteten schwarzerdeartigen Böden unter Bewässerung in Südost–Georgien. Dissertation, Boden und Landschaft 48, Justus Liebig University

Schatz M, Babaev E, Kalandadze B, Beuhm L, Deuring A (2015) Persistent organic pollutants in transcaucasian soils and sediments —the invisible heritage of former Soviet Union. In: 15th EuCheMS international conference on chemistry and the environment, Leipzig, Germany

VDLUFA (1976) Die chemische Untersuchung von Futtermitteln, 3. print. Verband Deutscher Landwirtschaftlicher Untersuchungs- und Forschungsanstalten, Speyer

Stahr K (2010) Anorganische Komponenten der Böden-Minerale und Gesteine. In: Schachtschabel et al (eds) Lehrbuch der Bodenkunde, 16th edn. Spektrum, Heidelberg/Berlin, pp 7–49

Usman ARAU (2004) Soil reclamation and conservation: evaluation of different additives for remediation and quality improvement of soil. Hohenheimer Bodenkundliche Hefte 73:Stuttgart

Vazhenin I (1987) Methods of definition the microelements in soils and plants. Kolos, Moscow

VDLUFA (1991) Methodenhandbuch I. Verband Deutscher Landwirtschaftlicher Untersuchungs- und Forschungsanstalten, Speyer

VDLUFA (1997) Phosphordüngung nach Bodenuntersuchung und Pflanzenbedarf. Verband Deutscher Landwirtschaftlicher Untersuchungs- und Forschungsanstalten, Darmstadt

Vinogradov AP (1957) Geochemistry of rare and absent-minded chemical elements in soils. ASUSSR, Moscow

Wilke B-M (2010) Gefährdung der Bodenfunktionen. In: Schachtschabel et al (eds) Lehrbuch der Bodenkunde, 16th edn. Spektrum, Heidelberg/Berlin, pp 449–520

Zeien H (1995) Chemische Extraktion zur Bestimmung der Bindungsformen von Schwermetallen in Böden Bonner. Dissertation, Bodenkundliche Abhandlungen, vol 17. Rheinische Friedrichs Wilhelm University

Appendix

See Figs. A.1, A.2, A.3 and A.4.

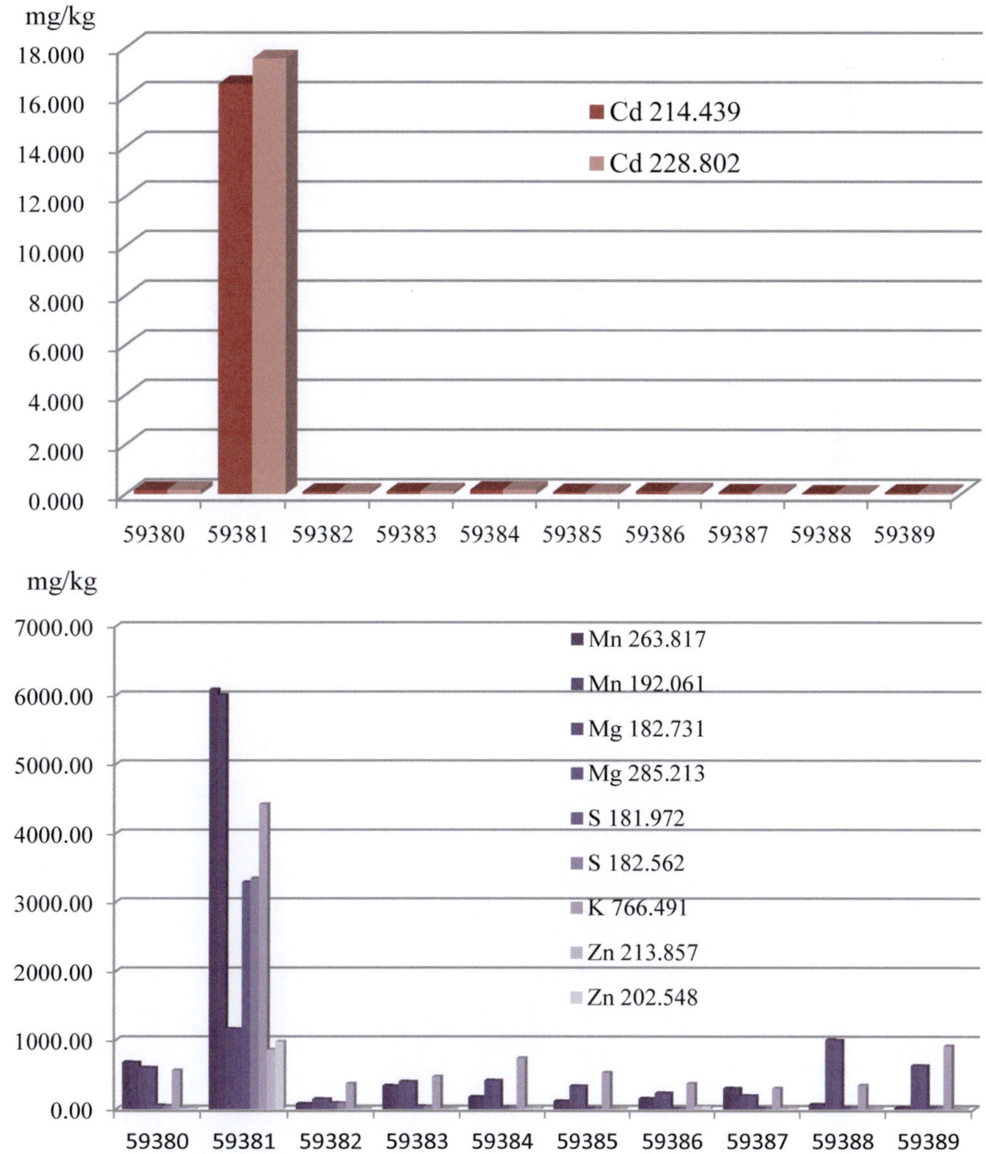

Fig. A.1 The content of heavy metals in soils of Zestaponi (EDTA), mg/kg (B. Kalandadze). Created by L. Matchavariani based on the data of B. Kalandadze

L. Matchavariani (ed.), *The Soils of Georgia*, World Soils Book Series,
https://doi.org/10.1007/978-3-030-18509-1

Fig. A.1 (continued)

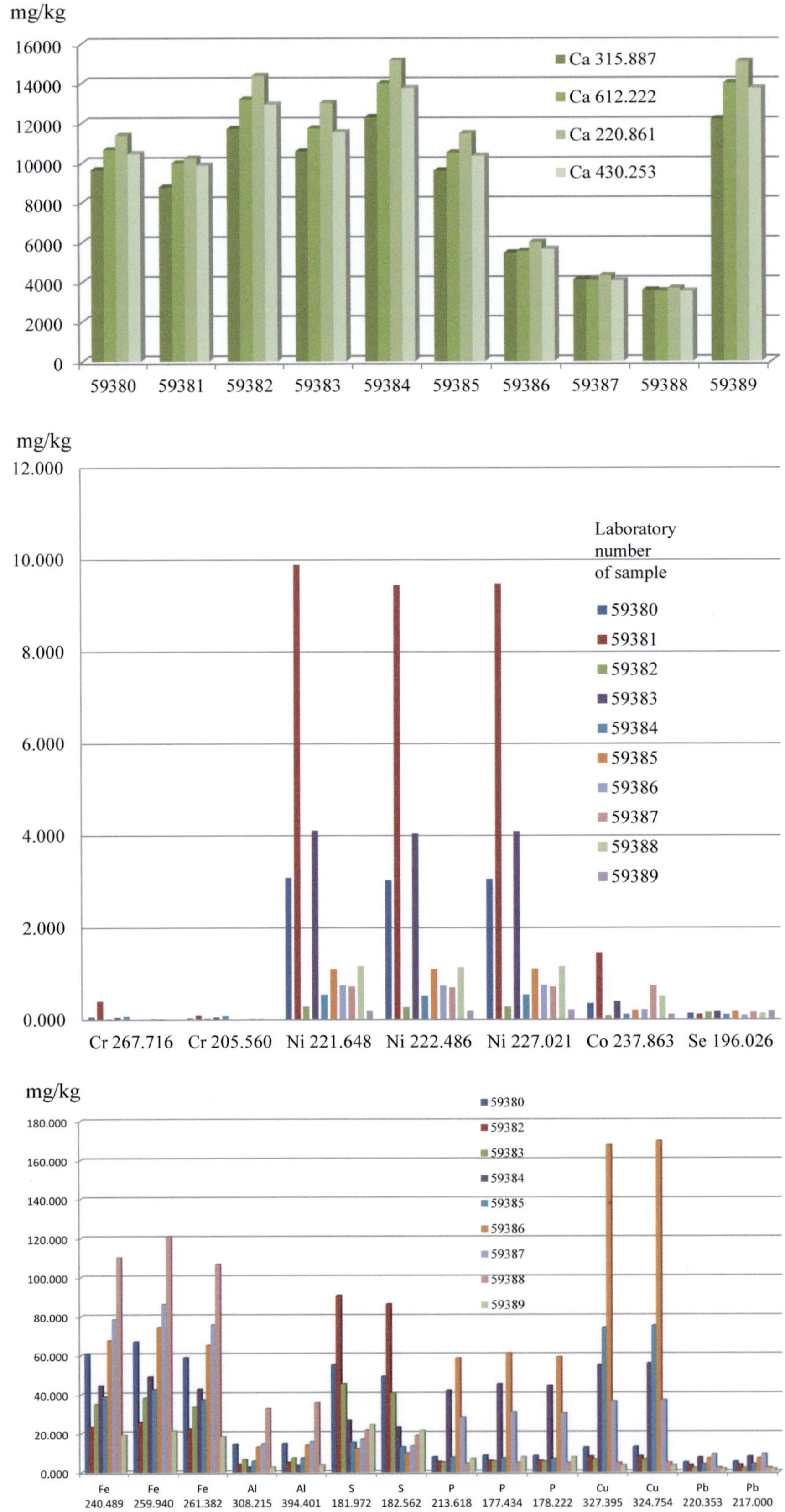

Fig. A.2 The content of heavy metals in soils of Zestaponi (KW), mg/kg (B. Kalandadze). Created by L. Matchavariani based on the data of B. Kalandadze

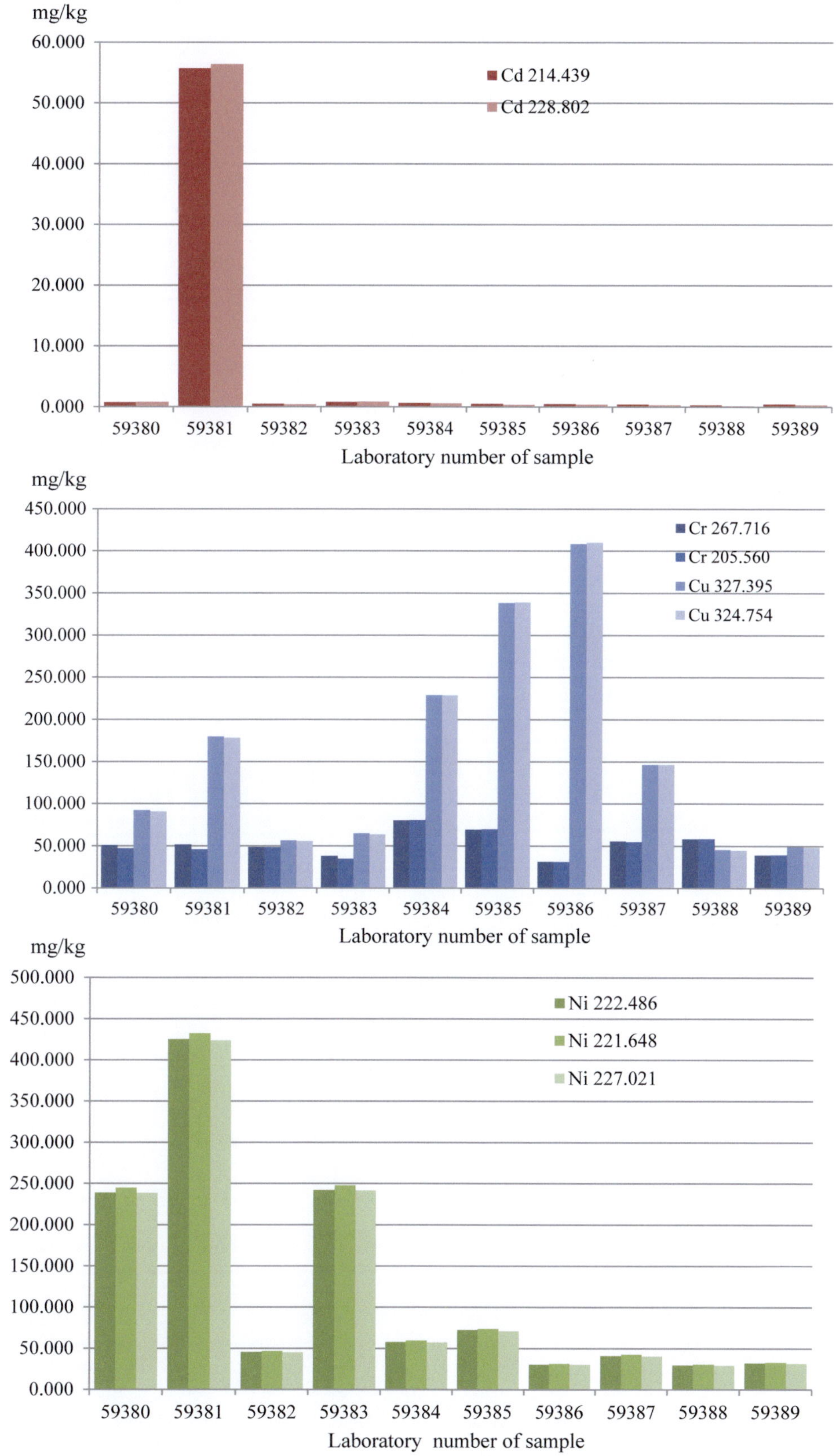

Fig. A.2 (continued)

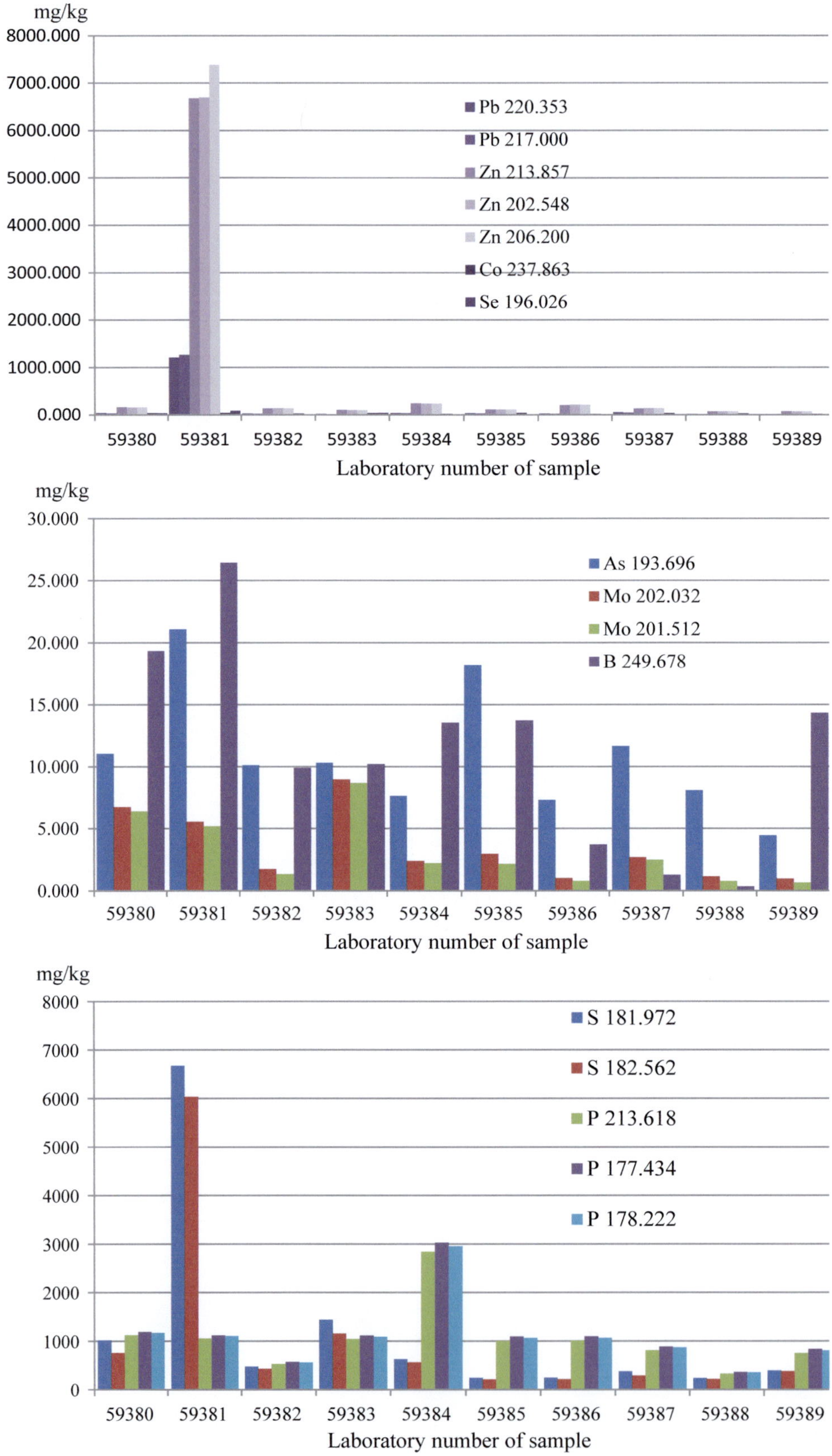

Fig. A.2 (continued)

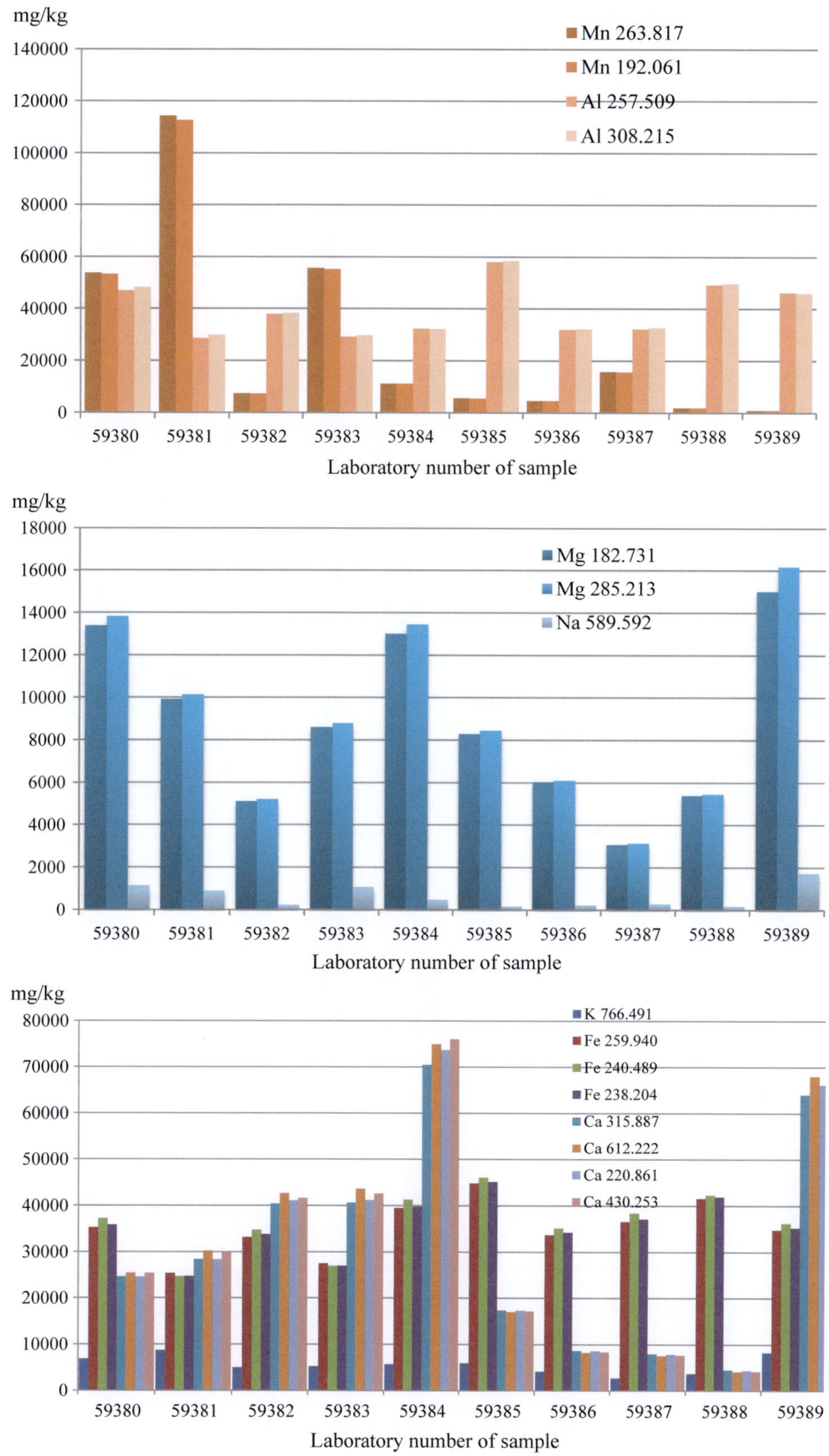

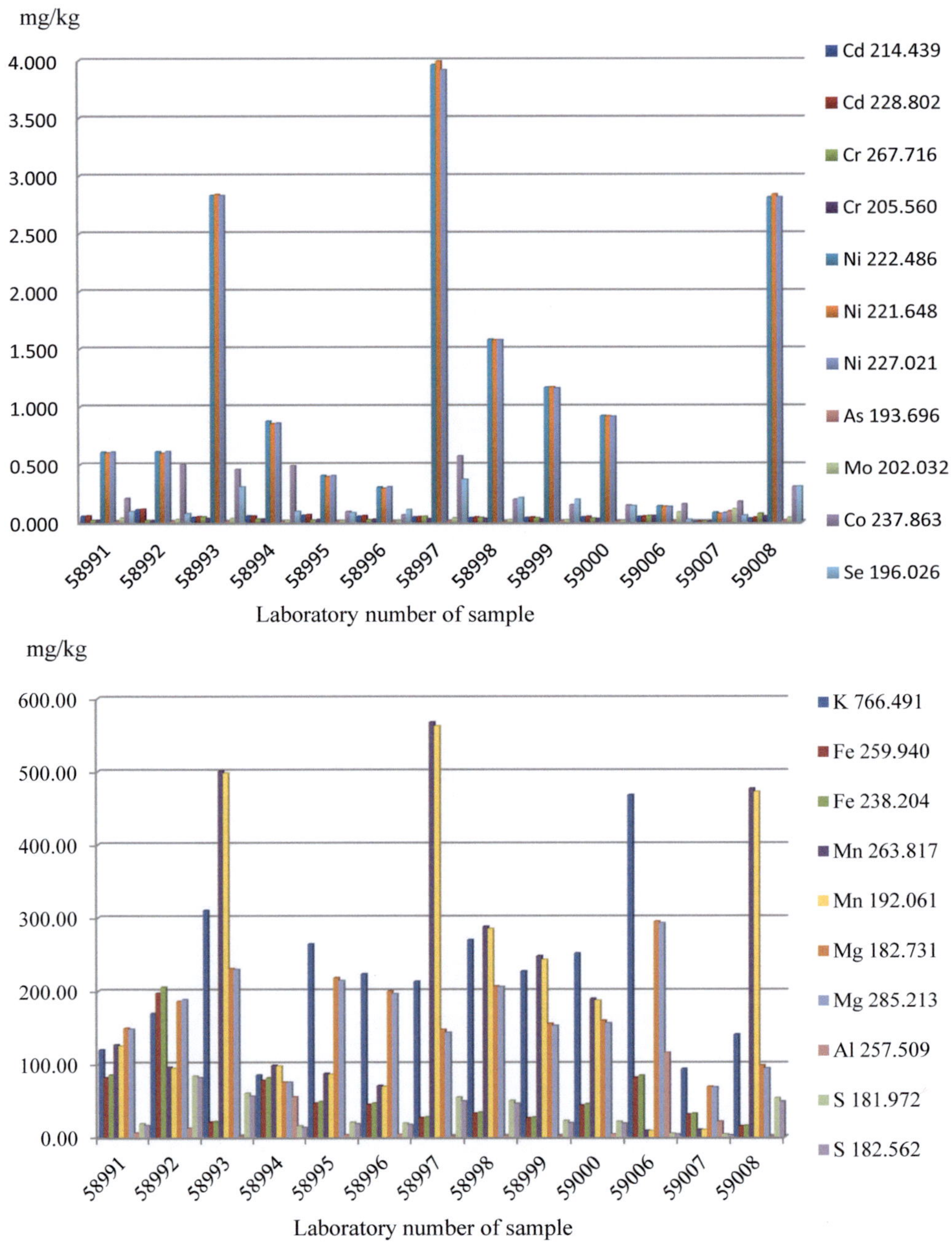

Fig. A.3 The content of heavy metals in soils of Chiatura (EDTA), mg/kg (B. Kalandadze). Created by L. Matchavariani based on the data of B. Kalandadze

Fig. A.3 (continued)

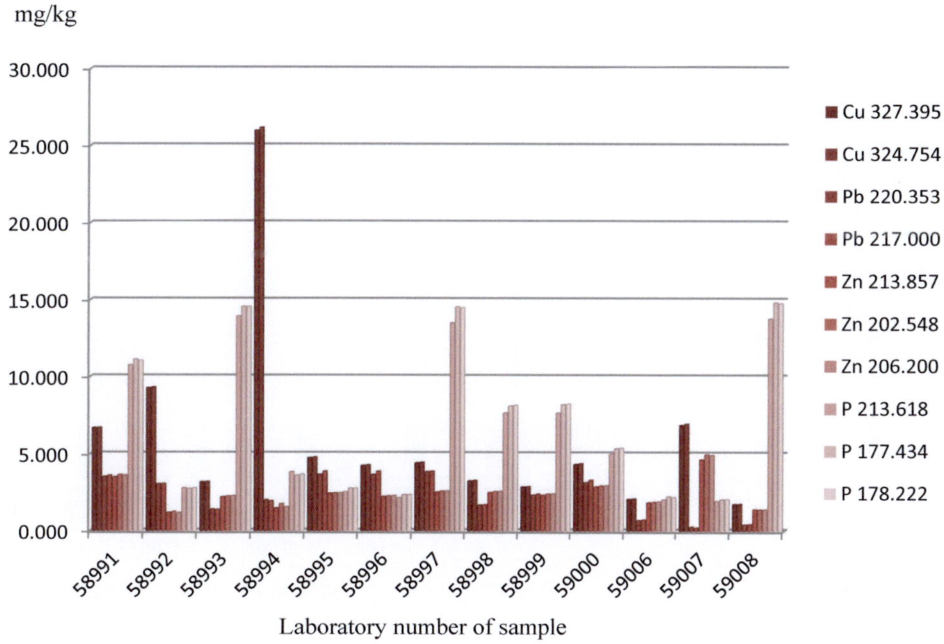

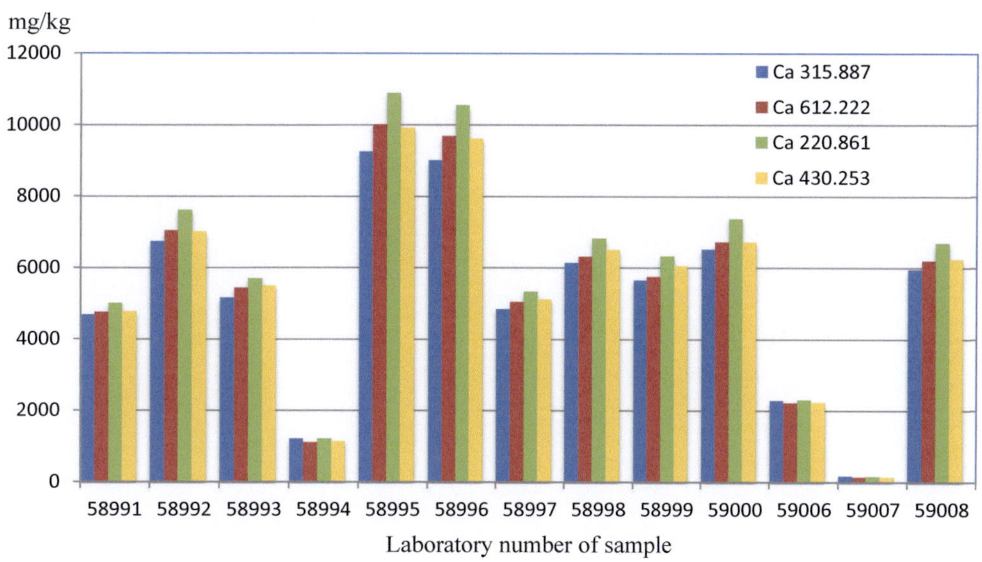

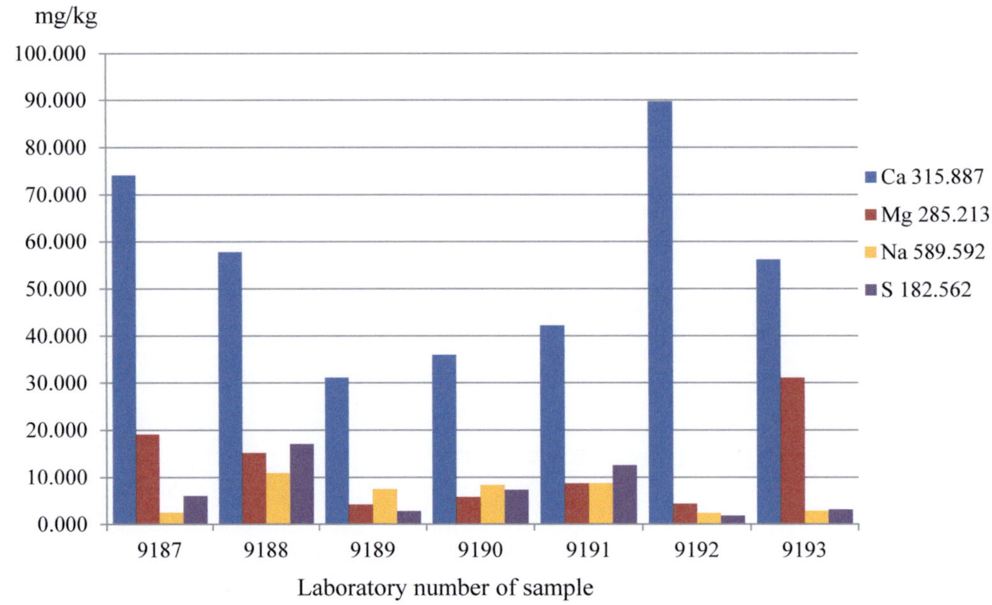

Fig. A.4 The content of heavy metals in water of Chiatura, mg/kg (B. Kalandadze). Created by L. Matchavariani based on the data of B. Kalandadze

Index

L. Matchavariani (ed.), *The Soils of Georgia*, World Soils Book Series,
https://doi.org/10.1007/978-3-030-18509-1

MIX
Papier aus verantwortungsvollen Quellen
Paper from responsible sources
FSC® C105338

If you have any concerns about our products,
you can contact us on
ProductSafety@springernature.com

In case Publisher is established outside the EU,
the EU authorized representative is:
Springer Nature Customer Service Center GmbH
Europaplatz 3, 69115 Heidelberg, Germany

Printed by Libri Plureos GmbH
in Hamburg, Germany